T

Crystallography and Crystal Defects

Crystallography and Crystal Defects

Revised Edition

A. Kelly
Churchill College, Cambridge, UK

G. W. Groves
Exeter College, Oxford, UK

and

P. Kidd
Queen Mary and Westfield College, London, UK

JOHN WILEY & SONS, LTD
Chichester • New York • Weinheim • Brisbane • Singapore • Toronto

Other Wiley Editorial Offices

John Wiley & Sons, Inc., 605 Third Avenue,
New York, NY 10158-0012, USA

WILEY-VCH Verlag GmbH, Pappelallee 3,
D-69469 Weinheim, Germany

Jacaranda Wiley Ltd, 33 Park Road, Milton,
Queensland 4064, Australia

John Wiley & Sons (Asia) Pte Ltd, Clementi Loop #02-01,
Jin Xing Distripark, Singapore 129809

John Wiley & Sons (Canada) Ltd, 22 Worcester Road,
Rexdale, Ontario M9W 1L1, Canada

Library of Congress Cataloguing-in-Publication Data

Kelly, A. (Anthony)
    Crystallography and crystal defects. — 2nd ed. / A. Kelly, G. W. Groves, and P. Kidd.
        p.   cm.
    Includes bibliographical references and index.
    ISBN 0-471-72043-7 (cloth : alk. paper). — ISBN 0-471-72044-5 (paper : alk. paper)
        1. Crystallography.   2. Crystals — Defects.   I. Groves, G. W.
    II. Kidd, P.   III. Title.
QD931.K4   2000
548'.8 — dc21   99-32444
                                                                            CIP

British Library Cataloguing in Publication Data

A catalogue record for this book is available from the British Library

ISBN 0 471 72043 7 (hardback)
ISBN 0 471 72044 5 (paperback)

Typeset in 10/12pt Times by Laser Words, Madras, India
Printed and bound in Great Britain by Biddles Ltd, Guildford, Surrey
This book is printed on acid-free paper responsibly manufactured from sustainable forestry, in which at least two trees are planted for each one used for paper production

Contents

Preface

This book contains an account of basic crystallography and of structural imperfections in crystals. One reason for treating these two subjects together is that the study of the first is necessary for a proper appreciation of the second. Because classical crystallography was developed nearly one hundred years ago, whilst imperfections in crystals have been intensively studied only in the last two decades, books on imperfections and on the solid state generally often give a brief summary of crystallographic theorems in early chapters and then attempt to deal in detail with specific properties of imperfections. In teaching classes we have found this approach unsatisfactory, firstly, because many properties of imperfections are unintelligible to the undergraduate if he or she lacks a *thorough* understanding of the idea of a lattice and of the crystal classes and, secondly, because a detailed study of the imperfections in a particular crystal, which the graduate student often undertakes, requires a knowledge of crystal structure and a facility in the calculation of such things as interplanar spacings, transformation of indices and manipulations involving the stereographic projection.

In this book we have attempted to give a more than superficial account of the basic theorems of classical crystallography in a way that makes their relevance to the study of dislocations, point defects, twinning and transformations clear whilst avoiding the wealth of nomenclature associated with the study of the symmetry of external forms. For this reason we start with the idea of a lattice. The basic theorems of crystallography and the description of simpler crystal structures occupy the first three chapters. We then describe the elementary tensor representation of physical properties. The second part of the book deals with specific imperfections and principally with the properties of these, which follow directly from their geometry. After an introductory chapter on elasticity, dislocations, point defects, twins, martensitic transformations and interfaces are dealt with in turn.

Only a knowledge of elementary mathematics is assumed to start with: some properties of vector, tensor and matrix methods are developed in the text and in appendices.

The book can be read by first year undergraduates and the chapters on elementary crystallography are based directly on courses taught to first and third year undergraduates. It is also sufficiently succinct on basic matter, whilst detailed in its description of individual imperfections, to be of use to final year undergraduates and to graduate students. Material from the later chapters has been used in final year courses for undergraduates at Cambridge University and the more advanced parts for graduate courses at Northwestern University and at the Carnegie Institute of Technology. We hope, therefore, that this text will be of use of students throughout their undergraduate and early postgraduate years.

A. Kelly   *G. W. Groves*
Churchill College   *Exeter College*
Cambridge   *Oxford*

Preface to Revised Edition

Over the years since the appearance of this book there have been many requests, received from many parts of the world, for a reprinted edition. This second edition has required very little modification of essential subject matter — the principal one being that the traditional view of the nature of a crystal must take account of the discovery of so-called quasicrystals.

We have also taken the opportunity to alter the units used and the description of angles so as to facilitate the use of computers and calculators, and have added a small number of sections on crystalline textures, polymer crystals and epitaxy.

A. Kelly   *P. Kidd*
Churchill College   *Queen Mary and Westfield College*
Cambridge   *London*

Acknowledgements for First Edition

We are grateful to many people for their interest in this book. In particular, helpful comments on drafts of chapters were received from M. F. Ashby, M. L. Brown, J. W. Christian, J. D. Embury, J. D. Eshelby, E. J. Freise, J. P. Hirth, C. L. Magee, T. E. Mitchell, H. Mykura, F. R. N. Nabarro, W. H. Taylor and M. J. Whelan, and from M. Cohen who also commented helpfully upon the arrangement of the book.

The problems at the end of Chapters 1, 2, 3 and 4 are based upon exercises set in the Departments of Metallurgy and of Mineralogy of the University of Cambridge as part of a joint course on the crystalline state. We are grateful to our colleagues for the use of these, and particularly to N. F. M. Henry and his colleagues of the Department of Mineralogy who developed most of those at the end of Chapter 2.

We are grateful to Mrs M. E. Harper, Miss M. Penfold, Mrs R. A. Ramsay and Mrs P. Smith for painstaking typing of the manuscript.

A special debt of gratitude must be recorded to the Carnegie Institute of Technology (now the Carnegie-Mellon University) and to its Departments of Metallurgy and Materials Science where most of the book was written, by both authors. A.K. also thanks the National Science Foundation of the United States of America for the award of a Senior Foreign Scientist Fellowship which made possible the close collaboration of the authors during the early months of 1967.

This book is the result of courses taught at Northwestern University, Cambridge University and the Carnegie Institute of Technology. We are grateful to successive departmental heads at these institutions, Professors Fine, Cottrell, Honeycombe, Mullins and Paxton, for the encouragement they have given us in this approach.

A. Kelly
National Physical Laboratory, Teddington
G. W. Groves
University of Oxford

PART I
Perfect Crystals

1

Lattice Geometry

1.1 THE UNIT CELL

Crystals are solids in which the atoms are regularly arranged with respect to one another. This regularity of arrangement can be described in terms of symmetry elements; these elements determine the symmetry of the physical properties of a crystal. For example, the symmetry elements show in which directions the electrical resistance of a crystal will be the same. Many naturally occurring crystals such as halite (sodium chloride) and calcite (calcium carbonate) have very well developed external faces. These faces show regular arrangements which indicate the regular arrangements of the atoms. Historically, such crystals are of great importance because the laws of crystal symmetry were deduced from measurements of the interfacial angles in such crystals; measurements were first carried out in the seventeenth century. The study of such crystals still possesses some heuristic advantages in learning about symmetry.

Nowadays the atomic pattern within a crystal can be studied directly. This is the fundamental pattern which is described by the symmetry elements and we shall begin with it.

In a crystal of graphite the carbon atoms are joined together in sheets. These sheets are only loosely bound to one another. A single sheet of such atoms provides an example of a two-dimensional crystal. The arrangement of the atoms within the sheets is shown in Fig. 1.1a. All of the atoms are identical. Each atom possesses three nearest neighbours. We describe this by saying that the coordination number is 3. In this case the coordination number is the same for all the atoms. It is the same for the two atoms marked A and B. However, note that the orientation of the neighbours is different at A and B. Atoms in a similar situation to those at A are found at N and Q; there is a similar situation to B at M and at P.

It is obvious that we can describe the whole arrangement of atoms and inter-atomic bonds shown in Fig. 1.1a by choosing a small unit such as OXAY, describing the arrangement of the atoms and bonds within it, then moving the unit so that it occupies the position NQXO and repeating the description and then moving it to ROYS and so on until we have filled all space with identical units

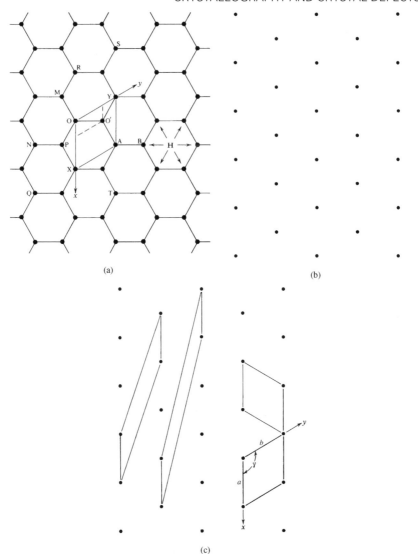

Figure 1.1

and described the whole pattern. If the repetition of the unit is understood to occur automatically, then to describe the crystal we need only describe the arrangement of the atoms and interatomic bonds within one unit. The unit chosen we would call the unit parallelogram in two dimensions (in three dimensions the unit cell). In choosing the unit we always choose a parallelogram in two dimensions or a parallelepiped in three dimensions. The reason for this will become clear later.

Having chosen the unit we describe the positions of the atoms inside it by choosing an origin O and taking axes Ox and Oy parallel to the sides. We state the lengths of the sides a and b, taking a equal to the distance OX, b equal to the distance OY (Fig. 1.1a), and we give the angle γ between Ox and Oy. In this case $a = b = 2.45$ Å (at $25\,^\circ$C) and $\gamma = 120^\circ$. To describe the positions of the atoms within the unit parallelogram, we note that there is one at each corner and one wholly inside the cell. The atoms at O, X, A, Y all have identical surroundings.[†] In describing the positions of the atoms we take the sides of the parallelogram, a and b, as units of length. Then the coordinates of the atom at O are (0, 0) that at X (1, 0), that at Y (0, 1) and that at A (1, 1). The coordinates of the atom at O' are obtained by drawing lines through O' parallel to the axes. The coordinates of O' are then $(\frac{1}{3}, \frac{2}{3})$. To fully describe the contents (i.e. the positions of the atoms) inside the unit parallelogram we need only give the coordinates of the atom at the origin, i.e. (0, 0), and those of that at O'. The reason is that the atoms at X, A and Y have identical surroundings to those at O and an atom such as O, X, A or Y is each shared between the four cells meeting at these points. The number of atoms contained within the area OXAY is two. O' is within the area, giving one atom. O, Y, A, X provide four atoms each shared between four unit cells, giving an additional $4 \times \frac{1}{4} = 1$. We note that the minimum number of atoms which a unit parallelogram could contain in this crystal structure is 2 since the atoms are situated in two differently surrounded positions.

To describe the atomic positions in Fig. 1.1a we chose OXAY as one unit parallelogram. We could equally well have chosen OXTA. The choice of a particular unit parallelogram or unit cell is arbitrary. NQPM is not a permissible choice — because the unit cannot be repeated to produce the graphite pattern; note that Q and P do not have identical surroundings.

The corners of the unit parallelogram OYAX in Fig. 1.1a all possess identical surroundings. We could choose and mark on the diagram all points with surroundings identical to those at O, Y and A, etc. Such points are N, Q, R, S, etc. The array of all such points with surroundings identical with those of a given point we call the mesh or net in two dimensions (a lattice in three dimensions). Each of the points is called a lattice point.

A formal definition of the lattice is as follows: *A lattice is a set of points in space such that the surroundings of one point are identical with those of all the others.* The type of symmetry described by the lattice is referred to as *translational* symmetry. One plane of the lattice of the graphite crystal structure drawn in Fig. 1.1a is shown in Fig. 1.1b. It consists of a set of points with identical surroundings: just a set of points — no atoms are involved. In Fig. 1.1c

---

[†] In choosing a unit parallelogram or a unit cell the crystal is always considered to be infinitely large. The pattern in Fig. 1.1a must then be thought of as extending to infinity. The fact that O, X, A and Y in a finite crystal are at slightly different positions with respect to the boundary of the pattern can then be neglected.

various primitive unit parallelograms are marked. A primitive unit parallelo-gram is defined as a unit parallelogram which *contains just one lattice point*. The conventional unit parallelogram for graphite corresponding to OXAY in Fig. 1.1a is outlined in Fig. 1.1c in heavy lines and the corresponding axes marked.

The conventional primitive unit parallelogram of a net which shows no obvious symmetry is taken with its sides as short and as nearly equal as possible, and γ, the angle between the x and y axes, is taken to be obtuse, if it is not equal to 90°. However, the symmetry of the pattern must always be taken into account. The net shown in Fig. 1.1b is very symmetric and in this case we can take the sides to be equal, so $a = b$, and we take $\gamma = 120°$.

Comparisons of Figs. 1.1a and 1.1b emphasize that the choice of the origin for the lattice is arbitrary. If we had chosen O' in Fig. 1.1a as the origin instead of O and then marked all corresponding points in Fig. 1.1a, we would have obtained the identical lattice with only a change of origin. The lattice then represents an essential element of the translational symmetry of the crystal however we choose the origin.

In three dimensions the definition of the lattice is the same as in two dimen-sions. The *unit cell* is now the parallelepiped containing just one lattice point. The origin is taken at a corner of the unit cell. The sides of the unit parallelepiped are taken as the axes of the crystal, x, y, z, using a right-handed notation. The angles α, β, γ between the axes are called the axial angles, (see Fig. 1.2). The smallest separations of the lattice points along the x, y, z axes are denoted by a, b, c respectively and called the lattice parameters.

Inspection of the drawing of the arrangement of the atoms in a crystal of caesium chloride in Fig. 3.1i shows that the lattice is an array of points such that $a = b = c, \alpha = \beta = \gamma = 90°$ so that the unit cell is a cube. There is one caesium atom and one chlorine atom associated with each lattice point. If we take the origin at the centre of a caesium atom, then there is one caesium atom in the unit cell with coordinates $(0, 0, 0)$ and one chlorine atom with coordinates $(\frac{1}{2}, \frac{1}{2}, \frac{1}{2})$. A projection along the z axis is shown in Fig. 1.3. It must be remembered that the caesium atoms at the corners of the unit cell project on top of one another; thus atoms at elevations 0 and 1 are superimposed as indicated in Fig. 1.3. When one is familiar with such drawings the atom at elevation 1 along the z axis can be omitted. The coordination number for caesium chloride is eight, each atom having eight of the other kind of atom as neighbours. The separation of these nearest neighbours, d, is easily seen to be given by

$$d = \left[\left(\frac{a}{2}\right)^2 + \left(\frac{b}{2}\right)^2 + \left(\frac{c}{2}\right)^2\right]^{1/2} = \frac{\sqrt{3}a}{2} \qquad (1.1)$$

since $a = b = c$.

The number of units of the formula CsCl per unit cell is clearly 1 in this case.

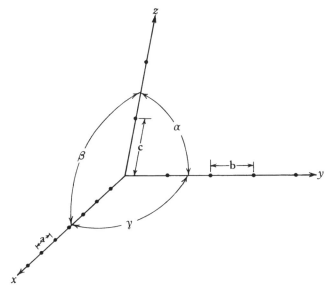

Figure 1.2

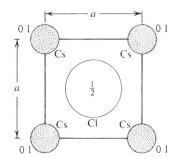

Figure 1.3 The numbers give the elevations of the centres of the atoms, along the z axis, taking the lattice parameter c as the unit of length

1.2 LATTICE PLANES AND DIRECTIONS

Figure 1.4 shows the rectangular mesh of a hypothetical two-dimensional crystal with mesh parameters a and b, of very different magnitude. Note that the mesh lines OB, O'B', O"B" all form part of a set and that the spacing of all lines in the set is quite regular; this is similar for the set of lines parallel to AB, i.e. A'B', A"B", etc. The spacing of each of these sets is determined only by a and b (and the angle between a and b). Also the angle between these two sets depends only on the *ratio* of a to b. If external faces of the crystal formed parallel to O"B and to AB, the angle between these faces would be uniquely related to the

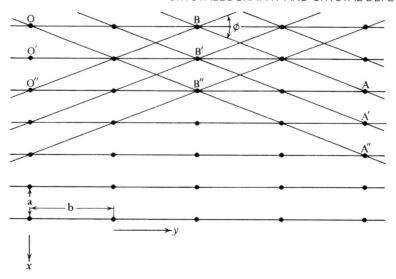

Figure 1.4

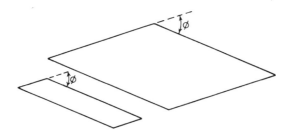

Figure 1.5

ratio $a:b$. Also, this angle would be independent of how large these faces were (see Fig. 1.5). This was recognized by early crystallographers who deduced the existence of the lattice structure of crystals from the observation of the constancy of angles between corresponding faces. This law of constancy of angle states: *In all crystals of the same substance the angle between corresponding faces has a constant value.*

The analogy between lines in a mesh and planes in a crystal lattice is very close. Crystal faces form parallel to lattice planes and important lattice planes contain a high density of lattice points. Lattice planes form an infinite regularly spaced set which collectively pass through all points of the lattice. The spacing of the members of the set is determined only by the lattice parameters and axial angles, and the angles between various lattice planes are determined only by the axial angles and the ratios of the lattice parameters to one another.

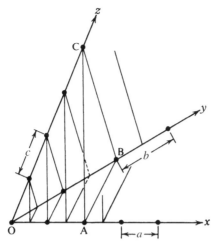

Figure 1.6

We designate a set of lattice planes as follows (see Fig. 1.6). Let one member of the set meet the chosen axes, x, y, z, at distances from the origin of A, B and C respectively. Express these intercepts in terms of the lattice parameters so that, for example, $A = ma$, $B = nb$ and $C = pc$. The reciprocals of the numbers m, n and p when cleared of fractions and without a common factor are called the *Miller indices* of the set of lattice planes. For example, the plane marked Y in Fig. 1.7 makes intercepts on the axes of infinity, $2b$, infinity, respectively, so $m = \infty$, $n = 2$ and $p = \infty$. Taking reciprocals we have 0, $\frac{1}{2}$, 0 and clearing of fractions gives $0\,1\,0$. The set of lattice planes parallel to Y is designated $(0\,1\,0)$. The triplet of numbers describing the Miller index is always enclosed in round brackets. Similarly, the plane marked P in Fig. 1.7 has intercepts $1a$, $2b$, $\frac{1}{3}c$, so $m = 1$, $n = 2$ and $p = \frac{1}{3}$. The reciprocals of these give 1, $\frac{1}{2}$, 3, and clearing of fractions we have $(2\,1\,6)$ as the Miller indices. The indices of some other planes are shown in Fig. 1.7.

The reason for using Miller indices is that they greatly simplify certain crystal calculations. The equation of the plane in Fig. 1.6 which meets the axes at A, B, C respectively is

$$\frac{x}{A} + \frac{y}{B} + \frac{z}{C} = 1 \qquad (1.2)$$

(see Appendix 1). Define $h'k'l'$ so that $A = a/h'$, $B = b/k'$ and $C = c/l'$. Substituting in Eqn (1.2), the equation to the plane is

$$\frac{x}{a/h'} + \frac{y}{b/k'} + \frac{z}{c/l'} = 1$$

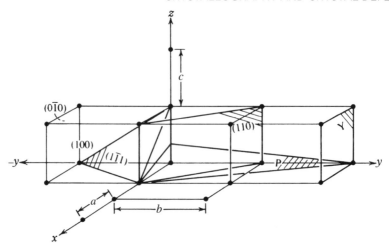

Figure 1.7

or

$$\frac{h'x}{a} + \frac{k'y}{b} + \frac{l'z}{c} = 1 \qquad (1.3)$$

We note that the parallel plane through the origin is

$$\frac{h'x}{a} + \frac{k'y}{b} + \frac{l'z}{c} = 0 \qquad (1.4)$$

which follows immediately from the derivation of Eqn (A1.2) in Appendix 1. If we now clear any fractions in $h'k'l'$ in Eqn (1.4) we obtain

$$\frac{hx}{a} + \frac{ky}{b} + \frac{lz}{c} = 0 \qquad (1.5)$$

where (hkl) are the Miller indices of the set of lattice planes, to which the particular plane we are considering belongs. The whole set of lattice planes is given by

$$\frac{hx}{a} + \frac{ky}{b} + \frac{lz}{c} = m \qquad (1.6)$$

where (hkl) are the Miller indices and m takes all integral values, both positive and negative. In the Miller index notation, negative values of the intercepts are indicated by a bar over the appropriate index (see the examples in Fig. 1.7). With a reasonable choice of unit cell, small values of the indices (hkl) belong to widely spaced planes containing a large areal density of lattice points. Well-developed crystals are usually bounded by such planes, so that it is found experimentally that prominent crystal faces have intercepts on the axes which

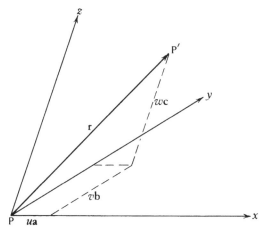

Figure 1.8

when expressed as multiples of a, b and c have ratios to one another that are small rational numbers.[†]

To specify a direction in a crystal we proceed as follows. A direction is a line in the crystal. Select any two points on the line, say P and P'. Choose one as the origin, say P (Fig. 1.8). Write the vector **r** between the two points in terms of translations along the x, y, z axes so that

$$\mathbf{r} = u\mathbf{a} + v\mathbf{b} + w\mathbf{c} \qquad (1.7)$$

where **a**, **b** and **c** are vectors along the x, y, z axes respectively and with magnitudes equal to the lattice parameters (Fig. 1.8). The direction is then denoted as $[u\,v\,w]$ — always cleared of fractions and reduced to its lowest terms. The triplet of numbers indicating a direction is always enclosed in square brackets. Some examples are given in Fig. 1.9. The equation of the line through the origin parallel to the direction $[u\,v\,w]$ is

$$\frac{x}{ua} = \frac{y}{vb} = \frac{z}{wc} \qquad (1.8)$$

If u, v and w are integers and the origin P is chosen at a lattice point, then P' is also a lattice point and the line PP' produced is a row of lattice points. Such a line is called a *rational* line, just as a plane of lattice points is called a rational plane.

[†] Formally the *law of rational indices* states that all planes which can occur as faces of crystals have intercepts on the axes which, when expressed as multiples of certain unit lengths along the axes (themselves proportional to a, b, c), have ratios which are small rational numbers. A rational number can always be written as p/q, where p and q are integers.

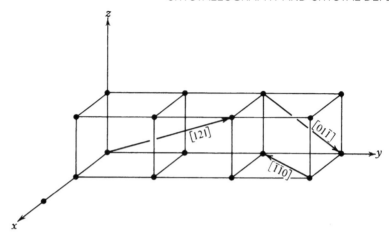

Figure 1.9

1.3 ZONES AND THE ZONE RULE

Any two lattice planes intersect in a line. This line, which lies in both, is said to be the zone axis of the zone in which the two planes are situated. Frequently a number of important crystal lattice planes all lie in the same zone, i.e. intersect one another in parallel lines. For instance, in Fig. 1.7 the planes $(1\,0\,0)$, $(0\,\bar{1}\,0)$ and $(1\,1\,0)$ are all parallel to the direction $[0\,0\,1]$. They would be said to lie in the zone $[0\,0\,1]$ since $[0\,0\,1]$ is a common direction lying in all of them. The normals to all of these planes are perpendicular to $[0\,0\,1]$.

Given the indices of any two planes, say $(h_1\,k_1\,l_1)$ and $(h_2\,k_2\,l_2)$, the indices of the zone in which they lie are given by

$$u = k_1 l_2 - l_1 k_2$$

$$v = l_1 h_2 - h_1 l_2 \tag{1.9}$$

$$w = h_1 k_2 - k_1 h_2$$

This is easily proved from Eqns (1.5) and (1.8). The equations of the planes through the origin parallel to the planes of Miller indices $(h_1\,k_1\,l_1)$, $(h_2\,k_2\,l_2)$ are

$$\frac{h_1 x}{a} + \frac{k_1 y}{b} + \frac{l_1 z}{c} = 0$$

and

$$\frac{h_2 x}{a} + \frac{k_2 y}{b} + \frac{l_2 z}{c} = 0$$

respectively. The equation of the line at the intersection of these two planes is obtained by eliminating first, say, x and then z between these two equations to give

$$\frac{x}{a(k_1 l_2 - l_1 k_2)} = \frac{y}{b(l_1 h_2 - h_1 l_2)} = \frac{z}{c(h_1 k_2 - k_1 h_2)} \qquad (1.10)$$

This is the equation of a line which passes through the origin, since it is satisfied by $x = y = z = 0$. Now let it also pass through the point with coordinates ua, vb, wc. If we substitute these values for x, y, z we obtain Eqns (1.9). There is an easy rule for remembering Eqns (1.9). Write down the two sets of Miller indices twice as follows:

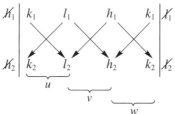

Draw vertical lines as shown and cross out the end four symbols. Multiplying according to the arrows and subtracting we obtain $k_1 l_2 - l_1 k_2, l_1 h_2 - h_1 l_2, h_1 k_2 - k_1 h_2$, which are the values of u, v and w respectively.

The condition that the plane of indices $(h k l)$ lies in the zone $[u v w]$ is that

$$hu + kv + lw = 0 \qquad (1.11)$$

This is called the *Weiss zone law*. It is the condition that the normal to $(h k l)$ be itself normal to the direction $[u v w]$. Equation (1.11) is easily derived as follows. The equation of the plane through the origin parallel to $(h k l)$ is given by Eqn (1.5). The line through the origin parallel to $(u v w)$ is given by Eqn (1.8). If this line is to lie in the plane $(h k l)$ we can eliminate x, y and z between Eqns (1.5) and (1.8) to obtain Eqn (1.11).

1.4 SYMMETRY ELEMENTS

The symmetrical arrangement of atoms in crystals is described formally in terms of elements of symmetry. The symmetry arises because an atom or group of atoms is repeated in a regular way to form a pattern. Any operation of repetition can be described in terms of one of the following three different types of pure symmetry elements or symmetry operators.

TRANSLATIONAL SYMMETRY

This describes the fact that similar atoms in identical surroundings are repeated at different points within the crystal and any one of these points can be brought into coincidence with another by an operation of translational symmetry. For

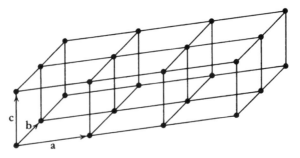

Figure 1.10

instance, in Fig. 1.1a the carbon atoms at O, Y, N and Q occupy completely similar positions. We use the idea of the lattice to describe this symmetry. The lattice is a set of points each with an identical environment which can be found by inspection of the crystal structure. We can define the arrangement of lattice points by saying that the vector **r** joining any two lattice points (or the operation of translational symmetry bringing one lattice point into coincidence with another) can always be written as

$$\mathbf{r} = u\mathbf{a} + v\mathbf{b} + w\mathbf{c} \tag{1.12}$$

where u, v, w are positive or negative integers or equal to zero. Inspection of Fig. 1.10 shows that to use this description we must be careful to choose **a, b** and **c** so as to include all lattice points. We do this by, say, making **a** the shortest vector between lattice points in the lattice, or one of several shortest ones. We then choose **b** as the shortest not parallel to **a**, and **c** the shortest not coplanar with **a** and **b**. Thus **a, b** and **c** define a primitive unit cell of the lattice in the same sense as in Section 1.1. Only one lattice point is included within the volume **a · [b × c]**,[†] which is the volume of the primitive unit cell; **a, b** and **c** are called the lattice translation vectors.

ROTATIONAL SYMMETRY

If one stood at the point marked H in Fig. 1.1a and regarded the surroundings in a particular direction, say that indicated by one of the arrows, then on turning through an angle of $60° = 360°/6$ the outlook would be identical. We say an axis of sixfold rotational symmetry passes normal to the paper through the point H. Similarly, at O' an axis of threefold symmetry passes normal to the paper since an identical outlook is found after rotation of $360°/3 = 120°$. A crystal possesses an n-fold axis of rotational symmetry if it coincides with itself upon rotation about the axis of $360°/n = 2\pi/n$ radians. In crystals, axes of rotational

---

[†] Volume **a · [b × c]** is equal to $abc \sin\alpha \cos\varphi$, where φ is the angle between **a** and the normal to the plane containing **b** and **c**. The notation of vector algebra is explained in Appendix 2, Section A2.1.

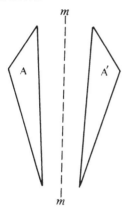

Figure 1.11

symmetry with values of n equal to one, two, three, four and six are the only ones found. These correspond to repetition every 360°, 180°, 120°, 90° and 60° and are called monad, diad, triad, tetrad and hexad axes respectively. The reasons for these limitations on the value of n are explained in Section 1.5.

A *centre of symmetry* is a point such that inversion through that point produces an identical arrangement. The operation of inversion moves the point with coordinates (x, y, z) to the position $(-x, -y, -z)$. For instance, in Fig. 1.3 (see also Fig. 3.1i), if we stood at the centre of the unit cell (coordinates $\frac{1}{2}, \frac{1}{2}, \frac{1}{2}$) and looked in any direction $[u\,v\,w]$ we would find an identical outlook if we looked in the direction $[\bar{u}\,\bar{v}\,\bar{w}]$. Clearly in a lattice all lattice points are centres of symmetry of the lattice. This follows from Eqn (1.12) since u, v, w can be positive or negative. Of course, since the origin of a set of lattice points can be arbitrarily chosen in a given crystal structure, we must not assume that in a given crystal with a centre of symmetry the centres of symmetry necessarily lie at the lattice points.

REFLECTION

A further type is reflection symmetry. The operation is that of reflection in a mirror. Mirror symmetry relates, for example, our left and right hands. The dotted plane running normal to the paper, marked m in Fig. 1.11, reflects object A to A' and vice versa. A' could not be moved about in the plane of the paper and made to superpose on A. The dotted plane is a mirror plane. It should be noted that the operation of inversion through a centre of symmetry also produces a right-handed object from a left-handed one.

1.5 RESTRICTIONS ON SYMMETRY ELEMENTS

All crystals show translational symmetry. A given crystal may or may not possess other symmetry elements. Axes of rotational symmetry must be consistent with

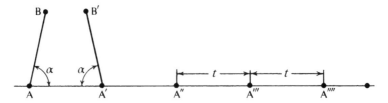

Figure 1.12

the translational symmetry of the lattice. A onefold rotation axis is obviously consistent. To prove that in addition only diads, triads, tetrads and hexads can occur in a crystal we consider just a two-dimensional lattice or net. In Fig. 1.12 let A, A', A", ... be lattice points of the mesh and let us choose the direction AA'A" so that the lattice translation vector **t** of the mesh in this direction is the shortest lattice translation vector of the net. Suppose an axis of n-fold rotational symmetry runs normal to the net at A. Then the point A' must be repeated at B by rotation through an angle $\alpha = $ A'AB $= 2\pi/n$. Also, since A' is a lattice point exactly similar to A there must also be an n-fold axis of rotational symmetry passing normal to the paper through A'. This repeats A at B', as shown in Fig. 1.12. Now B and B' define a lattice row parallel to AA'. Therefore the separation of B and B' by Eqn (1.12) must be an integral number times **t**. Call this integer N. From Fig. 1.12 the separation of B and B' is $(t - 2t \cos \alpha)$. Therefore the possible values of α are restricted to those satisfying the equation

$$t - 2t \cos \alpha = Nt$$

or

$$\cos \alpha = \frac{1 - N}{2} \tag{1.13}$$

where N is an integer. Since $-1 \leq \cos \alpha \leq 1$ the only possible solutions are shown in Table 1.1 and correspond to onefold, sixfold, fourfold, threefold and twofold axes of rotational symmetry. No other axis of rotational symmetry is consistent with the translational symmetry of a lattice and hence other axes do not occur in crystals.

Corresponding to the various allowed values of α derived from Eqn (1.13) the nets or two-dimensional lattices are defined. These are shown as the first three diagrams on the left-hand side of Fig. 1.15. It should be noted that the hexad

Table 1.1 Solutions of Eqn (1.13)

N	-1	0	1	2	3
$\cos \alpha$	1	$\frac{1}{2}$	0	$-\frac{1}{2}$	-1
α	$0°$	$60°$	$90°$	$120°$	$180°$

axis and the triad axis both require the same triequiangular mesh, the unit cell of which is a 120° rhombus (see Fig. 1.15c). In the same way that the possession of rotational symmetry axes perpendicular to the net places restriction on the net, so restrictions are placed upon the net by the possession of a mirror plane. There are just two types of net consistent with the possession of a mirror plane. To see this let A, A' be two lattice points of a net and let the vector **t** joining them be a lattice translation vector defining one edge of the unit cell. A mirror plane can be placed normal to the lattice row AA' as in Fig. 1.13a or as in Fig. 1.13b. It cannot be placed arbitrarily anywhere in between A and A'. It must either lie midway between A and A', as in Fig. 1.13a, or pass through a lattice point, as in Fig. 1.13b. Since AA' determines a row of lattice points one can build up a net consistent with mirror symmetry by placing a row identical to AA' parallel with AA' but displaced from it. There are just two possible arrangements, which are both shown in Fig. 1.14 with the original lattice vector **t** indicated and *all of the mirror planes* consistent with the arrangement of the lattice points marked in.

Figures 1.15a, b, c, d and e show the five essentially different two-dimensional lattices or nets (sometimes called meshes). The left-hand diagram of Fig. 1.15a shows the net of points consistent with a twofold axis of symmetry normal to the net and with no axis of higher symmetry. This net corresponds with the solutions $N = -1$ or 3, $\alpha = 0$ or 180° in Table 1.1 and is based on a *parallelogram*. The lengths of two adjacent sides of the parallelogram (a, b in Fig. 1.15a) and the value of the included angle γ can be chosen arbitrarily without removing the consistency with twofold symmetry. If a motif showing twofold symmetry normal to the net were associated with each lattice point then twofold symmetry axes would be present at all the points shown in the right-hand diagram of Fig. 1.15a.

All regular nets of points are consistent with twofold symmetry axes normal to the net because such a net of *points* is necessarily centrosymmetric and *in two*

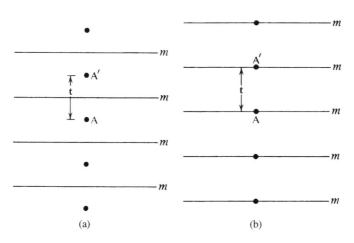

(a)   (b)

Figure 1.13

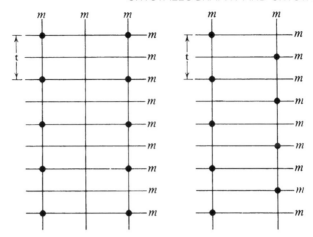

Figure 1.14

dimensions there is no difference between a centre of symmetry and a twofold or diad axis. The net corresponding to $N = 1$, $\alpha = 90°$ in Table 1.1 is based upon a square, shown in the left-hand diagram of Fig. 1.15b. If a two-dimensional crystal possesses fourfold symmetry it must necessarily possess this net. In addition, the atomic motif associated with each of the lattice or net points must also possess fourfold rotational symmetry. Provided the motif fulfils this condition there must be a fourfold axis at the centre of each of the basic squares of the net and twofold axes at the midpoints of the sides, as shown in the right-hand diagram of Fig. 1.15b. A two-dimensional crystal possessing fourfold rotational symmetry cannot possess fewer symmetry elements than those shown in the right-hand diagram of Fig. 1.15b.[†]

The net consistent with both $\alpha = 60°$ and $\alpha = 120°$, corresponding to the possession of hexad symmetry and triad symmetry respectively, is the triequiangular net shown in Fig. 1.15c. The primitive unit cell of this net has both sides equal and the included angle is necessarily 120°. It must be noted clearly that such a mesh of *points* is always consistent with sixfold symmetry. If the atomic motif associated with each lattice point is consistent with sixfold symmetry, then diad and triad axes are automatically present, as shown in the central diagram of Fig. 1.15c. A two-dimensional crystal will possess threefold symmetry provided

---

[†] Depending on how the motif showing fourfold symmetry is arranged with respect to the two axes of the crystal — sides of the square in Fig. 1.15b — additional symmetry elements may arise. A full discussion of this point for any of the arrangements of symmetry elements in Fig. 1.15, i.e. a discussion of consistent arrangements of symmetry elements in *space*, would take us immediately into the subject of *space groups*. We defer this until much later (see Section 2.14), but the discerning reader may wish to glance at Section 2.14 before proceeding further, both to gain reassurance that this subject matter is understood and to appreciate the mass of detail which is avoided by *not* following up this question now. Problem 1.14 illustrates how additional symmetry elements may arise.

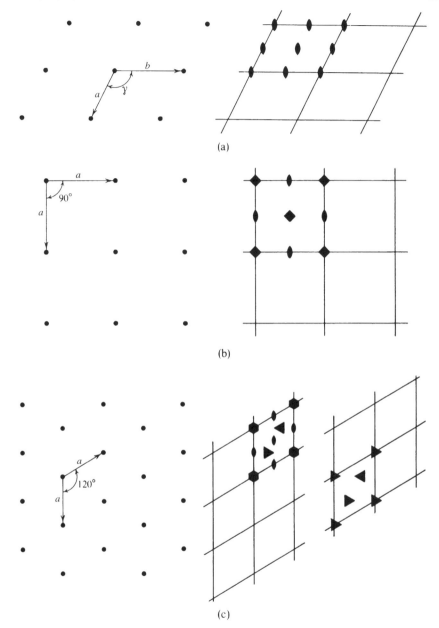

(a)

(b)

(c)

Figure 1.15 The five symmetrical plane lattices or nets. Rotational symmetry axes normal to the paper are indicated by the following symbols: ◖ = diad; ▲ = triad; ◆ = tetrad; ⬢ = hexad

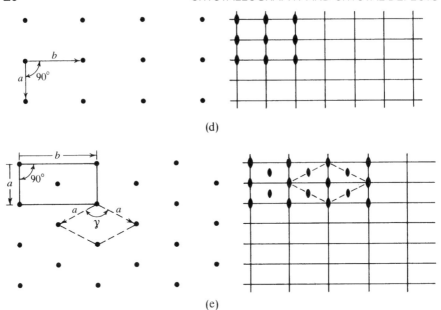

(d)

(e)

Figure 1.15 (*continued*)

the atomic motif placed at each lattice point of the lattice shown in Fig. 1.15c possesses threefold symmetry. The only symmetry elements *necessarily* present are then just the threefold axes arranged as indicated in the far right-hand diagram of Fig. 1.15c.

Two different types of net are both consistent with mirror symmetry. These are shown in Fig. 1.14. The left-hand one is a simple rectangular net and is repeated in Fig. 1.15d. The simplest primitive cell has a and b not necessarily equal and the angle between a and b is 90°. This *net* of points is consistent with the presence of diad axes at the intersection of the mirror planes (Fig. 1.15d), as is the mesh shown on the right-hand side of Fig. 1.14 reproduced in Fig. 1.15e. The simplest unit cell for the net in Fig. 1.15e is a rhombus indicated with the dotted lines. This has the two sides of the cell equal and the angle between them, γ, can take any value. When dealing with a net based on a rhombus it is, however, often *convenient* to choose as the unit cell a rectangle which contains an additional lattice point at its centre. This cell, outlined with full lines in Fig. 1.15e, has the angle between a and b necessarily equal to 90°. Hence it contains an additional lattice point inside it, which is called a *non-primitive* unit cell. The primitive unit cell is the dotted rhombus. The non-primitive cell clearly has twice the area of the primitive one and contains twice as many lattice points. It is chosen because it is naturally related to the symmetry, and is called the *centred rectangular cell*. This feature of choosing a non-primitive cell because it is more naturally related to the symmetry elements is one we shall meet often when dealing with the three-dimensional space lattices. The most symmetric arrangements of diad axes and mirror planes consistent with the rectangular net and with the centred rectangular net are shown in the right-hand diagrams of Figs. 1.15d and e respectively.

1.6 POSSIBLE COMBINATIONS OF ROTATIONAL SYMMETRIES

The axes of n-fold rotational symmetry which a crystal can possess are limited to values of n of 1, 2, 3, 4 or 6. These axes lie normal to a net. A crystal might conceivably be symmetric with respect to many intersecting n-fold axes. However, it turns out that the possible angular relationships between axes are severely limited. To discover these we need a method to combine the possible rotations; a convenient way to follow is that due to Euler and developed by Buerger.

Combinations of successive rotations about different axes are always inextricably related in groups of three. This arises because a rotation about an axis, say A, of amount α followed by a rotation about another axis, say B, of amount β can always be expressed as a single rotation about some third axis C of amount γ. Our strategy will be to find the angular relationships between A, B and C in terms of α, β and γ and then to take values of α, β and γ which correspond to those rotations possible in crystals, as given in Table 1.1.

Figure 1.16 represents a sphere of centre O. The axes of rotation are lines through the centre of the sphere, say OA and OB. We indicate the axis by a single letter denoting the point where the axis intersects the sphere. We study

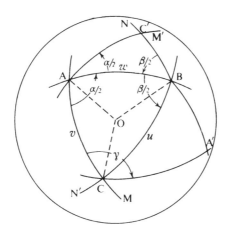

Figure 1.16

how points on the surface of the sphere are moved by successive rotations. A and B are thus axes, and we make successive rotations about them of amounts α and β. Join AB. Fix attention upon a line on the surface of the sphere such as AM which initially lies at an angle $\alpha/2$ to one side of AB. After rotation α about A it lies symmetrically on the other side of AB also at $\alpha/2$ from AB, in position AM'. Now consider BN, which is moved by rotation β about B from BN, initially at $\beta/2$ on one side of AB to BN' at $\beta/2$ on the other side of AB. There is a point C located at the intersection of AM and BN' unmoved by the successive rotations of α about A followed by β about B, because the first rotation moved it from C to C' and the second moved it back to C. During any rotation only one point is unmoved, namely the point where the axis of the rotation intersects the sphere. Since C is unmoved, therefore the successive rotations α about A followed by β about B must be equivalent to a rotation of amount γ (to be determined) about the axis OC. To find the value of γ we note that A is unmoved by the rotation about A and we let A' be the position to which A is moved by the rotation β about B. Join A' C. The angle γ is the angle between AC and A'C. The spherical triangles (triangles on the surface of a sphere) ABC and A'BC are congruent because AB = A'B, the angles $A\hat{B}C$ and $A'\hat{B}C$ are equal, both being of magnitude $\beta/2$, and the side BC is common to both triangles. Therefore the angles $A\hat{C}B$ and $A'\hat{C}B$ are both equal to $\gamma/2$.

Any spherical triangle can be solved if any three of the sides or the angles at the corners are known. If we assume values for α, β, γ corresponding to the allowed values for rotation axes in crystals, we can solve for the sides of the triangle ABC, thus obtaining the angles between these axes. To do this call the sides of the triangle ABC in Fig. 1.16 u, v and w respectively. Then we have (Appendix 1, Section A1.2)

$$\cos u = \frac{\cos \alpha/2 + \cos \beta/2 \cos \gamma/2}{\sin \beta/2 \sin \gamma/2} \tag{1.14a}$$

$$\cos v = \frac{\cos \beta/2 + \cos \alpha/2 \cos \gamma/2}{\sin \alpha/2 \sin \gamma/2} \tag{1.14b}$$

$$\cos w = \frac{\cos \gamma/2 + \cos \alpha/2 \cos \beta/2}{\sin \alpha/2 \sin \beta/2} \tag{1.14c}$$

To apply these results to crystals let us assume that A is a tetrad; then $\alpha = 90°$, $\alpha/2 = 45°$. Take B to be a diad; then $\beta = 180°$ and $\beta/2 = 90°$. From Eqn (1.14a),

$$\cos u = \frac{1/\sqrt{2}}{\sin \gamma/2} \tag{1.15a}$$

from (1.14b),

$$\cos v = \frac{1/\sqrt{2} \cos \gamma/2}{1/\sqrt{2} \sin \gamma/2} = \cot \gamma/2 \tag{1.15b}$$

and from (1.14c),

$$\cos w = \frac{\cos \gamma/2}{1/\sqrt{2}} = \sqrt{2}\cos \gamma/2 \qquad (1.15c)$$

We can now investigate what sort of rotational axis C could be by assuming values for γ and testing whether we get sensible results.

(a) Suppose C to be a diad; then $\gamma = 180°$, $\gamma/2 = 90°$. from (1.15a), $\cos u = 1/\sqrt{2}$ so $u = 45°$, from (1.15b), $\cos v = 0$, so $v = 90°$ and from (1.15c), $\cos w = 0$, so $w = 90°$. A possible combination of symmetry axes passing through a point in a crystal is then one tetrad and two diads, with the tetrad at 90° to both diads so that the diad axes lie in a plane normal to the tetrad and with an angle of 45° between the diads. This possibility is illustrated in Fig. 1.17a with axes A, B and C lettered according to these assumptions. The other diad axes marked in Fig. 1.17a must automatically be present since A is a tetrad.

(b) Suppose C to be a triad; then $\gamma = 120°$ and $\gamma/2 = 60°$. Solving Eqns (1.15) we find

$$\cos u = \sqrt{2/3}, \quad u = 35.27°$$
$$\cos v = 1/\sqrt{3}, \quad v = 54.73°$$
$$\cos w = 1/\sqrt{2}, \quad w = 45°$$

This arrangement is shown in Fig. 1.17b, again with the original axes marked. Again it should be noted that the presence of the tetrad at A automatically requires the presence of the other triad axes (and of other diads, not shown) since the fourfold symmetry about A must be satisfied. The triad axes lie at 70.53° to one another.

(c) Suppose C to be a hexad; then $\gamma = 60°$ and $\gamma/2 = 30°$. Equations (1.15a) and (1.15c) cannot then be solved. If one of them is insoluble the proposed combination of rotations is impossible. We have therefore shown that both a fourfold axis and a sixfold axis cannot exist together in a crystal.

Equations (1.14) can be studied to find the possible combinations of rotational axes in crystals. The resulting permissible combinations and the angles between the axes corresponding to these are listed in Table 1.2, following M. J. Buerger. In deriving these possibilities from Eqns (1.14) it is useful to note that $\cos^{-1}(1/\sqrt{3}) = 54.73°$, $\cos^{-1}\sqrt{(2/3)} = 35.26°$, and $\cos^{-1}(1/3) = 70.53°$. The sets of related rotations shown in Table 1.2 can always be designated by three numbers, e.g. 222, 233 or 234, each number indicating the appropriate rotational axis.

Table 1.2   Permissible combinations of rotation axes in crystals

Axes									
A	B	C	α	β	γ	u	v	w	System
2	2	2	180°	180°	180°	90°	90°	90°	Orthorhombic
2	2	3	180°	180°	120°	90°	90°	60°	Trigonal
2	2	4	180°	180°	90°	90°	90°	45°	Tetragonal
2	2	6	180°	180°	60°	90°	90°	30°	Hexagonal
2	3	3	180°	120°	120°	70.53°	54.73°	54.73°	Cubic
2	3	4	180°	120°	90°	54.73°	45°	35.27°	Cubic

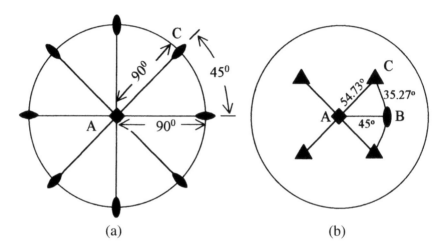

Figure 1.17

1.7 CRYSTAL SYSTEMS

The permissible combinations of rotation axes, listed in Table 1.2, are each identified with a crystal system in the far right-hand column of that table. A crystal
system contains all those crystals that possess certain axes of rotational symmetry.
In any crystal there is a necessary connection between the possession of an axis
of rotational symmetry and the geometry of the lattice of that crystal. We shall
explore this in the next section and we have seen some simple examples in
two dimensions in Section 1.5. Because of this connection between the rotational symmetry of the crystal and its lattice a certain convenient conventional
cell can always be chosen in each crystal system. These systems are listed in
Table 1.3, which gives the name of the system, the rotational symmetry element
or elements which define the system and the conventional unit cell which can
always be chosen. This cell is in many cases non-primitive, i.e. it contains more
than one lattice point. The symbol $\neq$ means 'not necessarily equal to'.

Table 1.3   The crystal systems

System	Symmetry	Conventional cell
Triclinic	No axes of symmetry	$a \neq b \neq c$; $\alpha \neq \beta \neq \gamma$
Monoclinic	A single diad	$a \neq b \neq c$; $\alpha = \gamma = 90° < \beta$
Orthorhombic	Three mutually perpendicular diads	$a \neq b \neq c$; $\alpha = \beta = \gamma = 90°$
Trigonal[a]	A single triad	$a = b = c$; $\alpha = \beta = \gamma < 120°$[b]
Tetragonal	A single tetrad	$a = b \neq c$; $\alpha = \beta = \gamma = 90°$
Hexagonal	One hexad	$a = b \neq c$; $\alpha = \beta = 90°, \gamma = 120°$
Cubic	Four triads	$a = b = c$; $\alpha = \beta = \gamma = 90°$

[a] Also sometimes called rhombohedral.
[b] The conventional cell of the hexagonal system is also frequently used for trigonal crystals.

1.8 SPACE LATTICES (BRAVAIS LATTICES)

All of the symmetry elements in a crystal must be mutually consistent. There are no fivefold axes of rotational symmetry because such axes are not consistent with the translational symmetry of the lattice. In Section 1.6 we derived the possible combinations of pure rotational symmetry elements that can pass through a point. These combinations are classified into different crystal systems and we will now investigate the types of space lattice (i.e. the regular arrangement of points in three dimensions as defined in Section 1.1) that are consistent with the various combinations of rotation axes. We shall find, as we did for the two-dimensional lattice (or net) consistent with mirror symmetry (Section 1.5), that more than one arrangement of points is consistent with a given set of rotational symmetry elements. However, the number of essentially different arrangements of points is limited to fourteen. These are the fourteen Bravais or space lattices. Our derivation is by no means a rigorous one for we do not show that our solutions are unique. This derivation of the Bravais lattices is introduced to provide a background for a clear understanding of the properties of imperfections studied in Part II of this book. The lattice is the most important symmetry element for the discussion of dislocations and martensitic transformations.

We start with the planar lattices or nets illustrated in Fig. 1.15. To build up a space lattice we stack these nets regularly above one another to form an infinite set of parallel sheets of spacing z. All of the sheets are in identical orientation with respect to an axis of rotation normal to their plane, so that corresponding lattice vectors t_1 and t_2 in the nets are always parallel. The stacking envisaged is shown in Fig. 1.18. The vector t_3 joining lattice points in adjacent nets is held constant from net to net. The triplet of vectors t_1, t_2, t_3 defines a unit cell of the Bravais lattice.

We start with the net based on a parallelogram, (Fig. 1.15a). If we stack nets of this form so that the points of intersection of twofold axes in successive nets do *not* lie vertically above one another, then we destroy the twofold symmetry axes normal to the nets. We have a lattice of points showing no rotational symmetry.

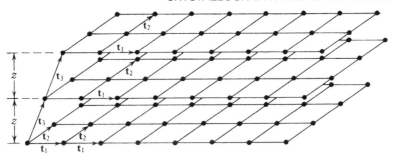

Figure 1.18

The unit cell is an arbitrary parallelepiped with edges a, b, c, no two of which are necessarily equal, and the angles of the unit cell α, β, γ can take any value; the cell is shown in Fig. 1.19a. By proper choice of a, b, c we can always ensure that the cell is primitive. Although this lattice contains no axis of rotational symmetry, the set of points is of course necessarily centrosymmetric.

To preserve twofold symmetry we can proceed in one of two different ways. We can arrange parallelogram nets vertically above one another so that t_3 is normal to the plane of the sheets, as in Fig. 1.20a, or we can produce the staggered arrangement shown in plan, viewed perpendicular to the nets, in Fig. 1.20b. In the first of these arrangements the twofold axes at the corners of the unit parallelogram of the nets all coincide and we produce a lattice of which one unit cell is shown in Fig. 1.19b. This has no two sides of the primitive cell necessarily equal but two of the axial angles are 90°. A frequently used convention is to take α and γ as 90° so that y is normal to x and to z; β is then taken as the obtuse angle between x and z. The staggered arrangement of the parallelogram nets in Fig. 1.20b is such that the twofold axes at the corners of the unit parallelograms of the second net coincide with those at the centres of the sides of the unit parallelogram of those of the first (or zero level) net. A lattice is then produced of which a possible unit cell is shown in Fig. 1.21. This is multiply-primitive, containing two lattice points per unit cell, and the vector t_4 is normal to t_1 and t_2. Such a cell with lattice points at the centres of a pair of opposite faces parallel to the diad axis is also consistent with twofold symmetry. The cell centred on opposite faces shown in Fig. 1.21 is chosen to denote the lattice produced from the staggered nets because it is more naturally related to the twofold symmetry than a primitive unit cell would be for this case. The staggered arrangement of nets shown in Figs. 1.20b and 1.21 could also have shown diad symmetry if we had arranged that the corners of the net at height z had lain not above the midpoints of the side containing t_1 in Fig. 1.20b but vertically above either the centre of the unit parallelogram of the first net or above the centre of the side containing t_2 in Fig. 1.20b. These two staggered arrangements are not essentially different from the first one since, as Fig. 1.22 shows, a new choice of axes in the plane of the nets is all that is needed to make them completely equivalent.

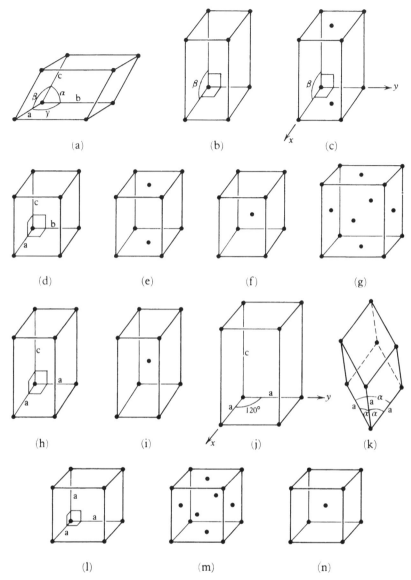

Figure 1.19 Unit cells of the fourteen Bravais space lattices. (a) Primitive triclinic. (b) Primitive monoclinic. (c) Side-centred monoclinic — conventionally the twofold axis is taken parallel to y and the $(0\,0\,1)$ face is centred (C centred). (d) Primitive orthorhombic. (e) Side-centred orthorhombic — conventionally centred on $(0\,0\,1)$, i.e. C centred. (f) Body-centred orthorhombic. (g) Face-centred orthorhombic. (h) Primitive tetragonal. (i) Body-centred tetragonal. (j) Primitive hexagonal. (k) Primitive rhombohedral (trigonal). (l) Primitive cubic. (m) Face-centred cubic. (n) Body-centred cubic

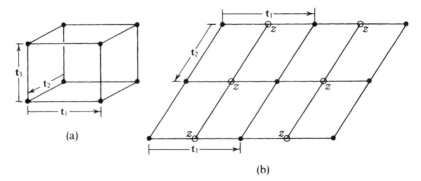

Figure 1.20 Lattice points in the net at height zero are marked as dots, those at height z with rings

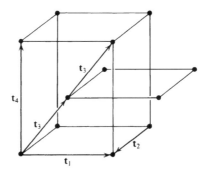

Figure 1.21

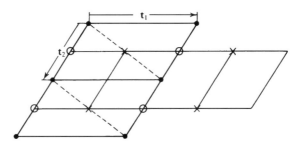

Figure 1.22 Lattice points in the net at height zero are marked with dots. The rings and crosses indicate alternative positions of the lattice points in staggered nets at height z arranged so as to preserve twofold symmetry. The dotted lines show an alternative choice of unit cell

There are then two lattices consistent with monoclinic symmetry: the primitive one with the unit cell shown in Fig. 1.19b and a lattice made up from staggered nets of which the conventional unit cell is centred on a pair of opposite faces. The centred faces are conventionally taken as the faces parallel to the x and y axes, i.e. $(0\,0\,1)$ with the diad parallel to y (see Fig. 1.19c). This lattice is called the monoclinic C lattice. The two lattices in the monoclinic system can be designated P and C respectively.

The two tetragonal lattices can be rapidly developed. The square net in Fig. 1.15b has fourfold symmetry axes arranged at the corners of the squares and also at the centres. This fourfold symmetry may be preserved by placing the second net with a corner of the square at $00z$ with respect to the first (t_3 normal to t_1 and t_2) or with a corner of the square at $(\frac{1}{2}, \frac{1}{2}, z)$ with respect to the first. The unit cells of the lattices produced by these two different arrangements are shown in Figs. 1.19h and i respectively. They can be designated P and I. The symbol I indicates a lattice with an additional lattice point at the centre of the unit cell (German: *innenzentrierte*). In the tetragonal system the tetrad axis is usually taken parallel to c, so a and b are necessarily equal and all of the axial angles are 90°.

The nets shown in Figs. 1.15d and e are each consistent with the symmetry of a diad axis lying at the intersection of two perpendicular mirror planes. In Section 2.4 it is shown that a mirror plane is completely equivalent to what is called an inverse diad axis, i.e. a diad axis involving the operation of rotation plus inversion. This inverse diad axis, given the symbol $\bar{2}$, lies normal to the mirror plane. The symmetry of a diad axis at the intersection of two perpendicular mirror planes could therefore be described as $2\,\bar{2}\,\bar{2}$, indicating the existence of three orthogonal axes: one diad and two inverse diads. The lattice consistent with this set of symmetry elements will also be consistent with the arrangement $2\,2\,2$ in the orthorhombic crystal system (Table 1.3).[†] To develop the lattices consistent with orthorhombic symmetry, therefore, the two relevant nets are the rectangular net (Fig. 1.15d) and the rhombus net (Fig. 1.15e). The positions of diad axes are shown on the right-hand sides of Figs. 1.15d and e. The rhombus net can also be described as the centred rectangular net.

If we stack rectangular nets vertically above one another so that a corner lattice point of the second net lies vertically above a similar lattice point in the net at zero level (t_3 normal to t_1 and t_2) then we produce the primitive lattice P. The unit cell is shown in Fig. 1.19d. It is a rectangular parallelopiped.

If we stack rhombus nets (centred rectangles) vertically above one another we obtain the lattice shown in Fig. 1.23. This is the orthorhombic lattice with centring on one pair of faces.

[†] The validity of this statement does not follow immediately at this point. Its truth is plausible if one notes that it will be shown later (Section 2.4) that an inverse diad axis plus a centre of symmetry is equivalent to a diad axis normal to a mirror plane, and that the lattice points of a lattice are centres of symmetry of the lattice.

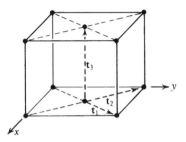

Figure 1.23

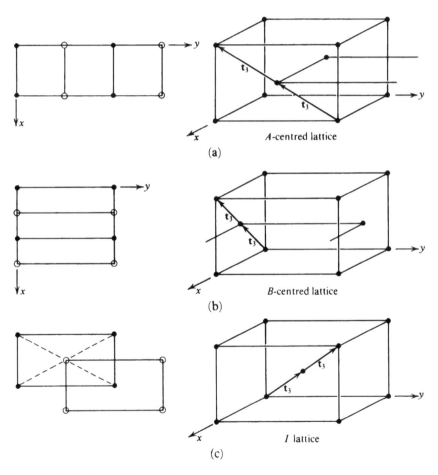

Figure 1.24 In the left-hand diagrams, lattice points in the net at zero level are denoted with dots and those in the net at level *z* with open circles

We can also preserve the symmetry of a diad axis at the intersection of two mirror planes by stacking the rectangular nets in three staggered sequences. These are shown in Figs. 1.24a, b and c. The lattices designated A centred and B centred are not essentially different since they can be transformed into one another by appropriate relabelling of the axes.[†] The staggered sequence shown in Fig. 1.24c is described by the unit cell shown in Fig. 1.19f. It is the orthorhombic body-centred lattice, symbol I. There is only one possibility for the staggered stacking of rhombus nets. Careful inspection of the right-hand side of Fig. 1.15e shows that the only places in the net where a diad axis lies at the intersection of two perpendicular mirror planes is at points with coordinates $(0, 0)$ and $(\frac{1}{2}, \frac{1}{2})$ of the *rhombus primitive cell*. We have already dealt with the vertical stacking of the rhombus nets. If we take the only staggered sequence possible, where the second net lies with a lattice point of the rhombus net vertically above the centre of the rhombus in the zero level net (so that the end of t_3 has coordinates $\frac{1}{2}, \frac{1}{2}, z$ in the rhombus net), then we produce the arrangement shown in Fig. 1.25. This is most conveniently described in terms of a unit cell shown in Fig. 1.19g, which is a rectangular parallelepiped with lattice points at the corners and also in the centres of all faces of the parallelepiped. This is the orthorhombic F cell. The symbol F stands for face-centred, indicating additional lattice points at the centres of all faces of the unit cell.

All of the lattices consistent with $2\,2\,2$, i.e. orthorhombic symmetry, are shown in Figs. 1.19d, e, f and g. The unit cells are all rectangular parallelepipeds so that the crystal axes can always be taken at right angles to one another, i.e. $\alpha = \beta = \gamma = 90°$, but the cell edges a, b and c may all be different. The primitive lattice P can then be described by a unit cell with lattice points only at the

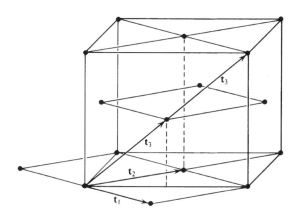

Figure 1.25

[†] A means a lattice point on the $(1\,0\,0)$ face, B a lattice point on $(0\,1\,0)$ and C a lattice point on $(0\,0\,1)$, in all crystal systems.

corners, the body-centred lattice I by a cell with an additional lattice point at its centre and the F lattice by a cell centred on all faces. The A-, B- and C-centred lattices, shown in Figs. 1.24a and b and Fig. 1.23 respectively, are all described by choosing the axes so as to give a cell centred on the $(0\,0\,1)$ face, i.e. a C-centred cell.

We have so far described nine of the Bravais space lattices. All further lattices are based upon the stacking of triequiangular nets of points. The triequiangular net is shown in Fig. 1.15c. There are sixfold axes only at the lattice points of the net. To preserve sixfold rotational symmetry in a three-dimensional lattice such nets must be stacked vertically above one another so that t_3 is normal to t_1 and t_2. The lattice produced has the unit cell shown in Fig. 1.19j. The unique hexagonal axis is taken to lie along the z axis so $a = b \neq c$, $\gamma = 120°$ and α and β are both $90°$. This lattice (i.e. the array of points) possesses sixfold rotational symmetry and is the only lattice to do so. However, it is also consistent with threefold rotational symmetry about an axis parallel to z. A crystal in which an atomic motif possessing threefold rotational symmetry was associated with each lattice point of this lattice would belong to the trigonal crystal system.

A lattice consistent with a single threefold rotational axis can be produced by stacking triequiangular nets in a staggered sequence. A unit cell of the triequiangular net of points is shown outlined in Fig. 1.26 by the vectors t_1, t_2 along the x and y axes. Axes of threefold symmetry pierce the net at the origin of the cell $(0, 0)$, i.e. points such as A, and also at two positions within the cell with coordinates $(\frac{1}{3}, \frac{2}{3})$ and $(\frac{2}{3}, \frac{1}{3})$ respectively, which are labelled B and C respectively in Fig. 1.26. We can preserve the threefold symmetry (while of course destroying

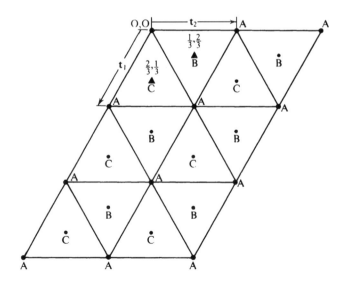

Figure 1.26

the sixfold one) by stacking nets so that the extremity of $\mathbf{t}_3$ has coordinates of either $(\frac{2}{3}, \frac{1}{3}, z)$ or $(\frac{1}{3}, \frac{2}{3}, z)$. The two positions B and C in Fig. 1.26 are equivalent to one another in the sense that the same lattice is produced whatever the order in which these two positions are used. A plan of the lattice produced, viewed along the triad axis, is shown in Fig. 1.27 and a sketch of the relationship between the triequiangular nets and the primitive cells of this lattice is shown in Fig. 1.28. In

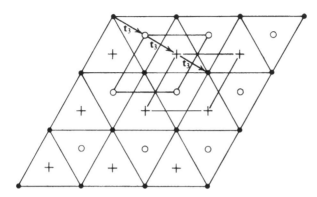

Figure 1.27 Lattice points in the net at level zero are marked with a dot, those in the net at height z by an open circle and those at $2z$ by a cross. The projection of $\mathbf{t}_3$ on to the plane of the nets is shown

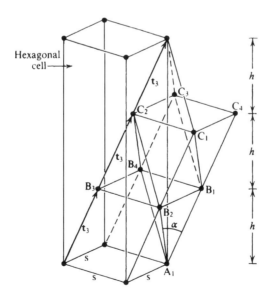

Figure 1.28 The relationship between a primitive cell of the trigonal lattice and the triply-primitive hexagonal cell

Figs. 1.27 and 1.28 the stacking sequence of the nets has been set as ABC ABC ABC .... Exactly the same lattice but in a different orientation (rotated 60° clockwise looking down upon the paper in Fig. 1.27) would have been produced if the sequence ACBACBACB ... had been followed. The primitive cell of the trigonal lattice in Fig. 1.28 is shown in Fig. 1.19k. It can be given the symbol R. It is a rhombohedron, the edges of the cell being of equal length, each equally inclined to the single threefold axis. To specify the cell we must state $a = b = c$ and the angle $\alpha = \beta = \gamma < 120°$.

An alternative cell is sometimes used to describe the trigonal lattice R because of the inconvenience in dealing with a lattice of axial angle α which may take any value between 0 and 120°. The alternative cell is shown in Fig. 1.28 and in plan viewed along the triad axis in Fig. 1.29. It is a triply-primitive cell, three mesh layers high, with internal lattice points at elevations of $\frac{1}{3}$ and $\frac{2}{3}$ of the repeat distance along the triad axis. The cell is of the same shape as the conventional unit cell of the hexagonal Bravais lattice and to specify it we must know $a = b \neq c$; $\alpha = \beta = 90°$ and $\gamma = 120°$.

Crystals belonging to the cubic system possess four threefold axes of rotational symmetry. Reference to Table 1.2 and Fig. 1.17b shows that four threefold axes cannot exist alone in a crystal. They must be accompanied by at least three twofold axes. The angles between the four threefold axes are such that these triad axes lie along the body diagonals of a cube (Fig. 1.30), with angles of 70.53° ($\cos^{-1}\frac{1}{3}$) between them. To indicate how the lattices consistent with this arrangement of threefold axes arise, we start with the R lattice shown in Fig. 1.28 and call the separation of nearest neighbour lattice points in the triequiangular net s and the vertical separation of the nets along the triad axis h. The positions of the lattice points in the successive layers when all are projected on to the plane perpendicular to the triad axis can be designated ABC ABC ... , as in Figs. 1.26 and 1.28. In a trigonal lattice the spacing of the nets h is unrelated to the separation of the lattice points within the nets s. If we make the spacing of the nets such that $h = \sqrt{2/3}s(=2s/\sqrt{6})$, the angle α in Fig. 1.28 becomes 60° and triangles $A_1B_1B_2$, $A_1B_2B_4$, $A_1B_1B_4$ all become equilateral. Planes such as $A_1B_1C_1B_2$, $A_1B_2C_2B_4$, $A_1B_1C_3B_4$ all contain triequiangular nets of points. Planes parallel to each of these three planes also contain triequiangular nets of points and they are also stacked so as to preserve triad symmetry along lines normal to them. When $\alpha = 60°$ then the original trigonal lattice becomes consistent with the possession of four threefold axes. The conventional unit cell of this lattice is shown in

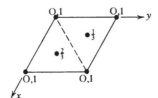

Figure 1.29

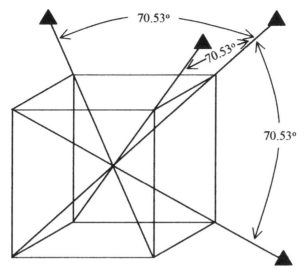

Figure 1.30

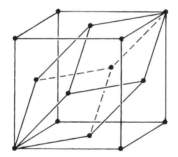

Figure 1.31   The relationship between the primitive unit cell and the conventional cell in the face-centred cubic lattice

Fig. 1.19m; it is a cube centred on all faces. The relationship between this cell and the primitive one with $\alpha = 60°$ is shown in Fig. 1.31.

The large non-primitive unit cell in Figs. 1.19m and 1.31 is the face-centred cubic lattice which can be designated F. It contains four lattice points. These are at the corners and centres of each of the faces.

When h in Fig. 1.28 becomes equal to $s/\sqrt{6}$ the primitive unit cell of the R lattice becomes a cube with $\alpha = 90°$. This is the cubic primitive lattice P shown in Fig. 1.19l, containing lattice points at the corners of the cubic unit cell.

Lastly, if in Fig. 1.28 h takes the value $\frac{1}{6}\sqrt{3/2}s = s/(2\sqrt{6})$, the angle α is equal to $109.47°(=180° - 70.53°) = \cos^{-1}(-\frac{1}{3})$. The lattice formed by such an array

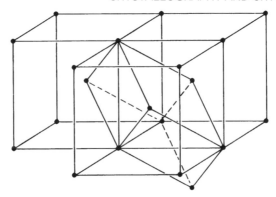

Figure 1.32 The relationship between the primitive unit cell and the conventional cell in the body-centred cubic lattice

of points also contains four threefold axes of symmetry. The conventional unit cell of this lattice is shown in Fig. 1.19n. It is a cube with lattice points at the cube corners and also one at the centre of the cube. It can be designated I, the cubic body-centred lattice. The relationship between the doubly-primitive unit cell shown in Fig. 1.19n and the primitive unit cell, which is a rhombohedron with axial angles of 109.47°, is shown in Fig. 1.32.

We have just described the three lattices consistent with the possession of four threefold axes of rotational symmetry. They are shown together in Figs. 1.19l, m and n. The unit cell of each can be taken as a cube with $a = b = c$; $\alpha = \beta = \gamma = 90°$. The primitive cell contains one lattice point, the face-centred cell four and the body-centred cell two.

The unit cells of the fourteen Bravais space lattices are shown together in Fig. 1.19. All crystals possess one or other of these lattices with an identical atomic motif associated with each lattice point. In some crystals a single spherically symmetric atom is associated with each lattice point. In this case the lattice itself possesses direct physical significance because the lattice and the crystal structure are identical. In other cases the lattice is a very convenient framework for describing the translational symmetry of the crystal. If the lattice is given and the arrangement of the atomic motif about a *single* lattice point is given, the crystal structure is fully described. The lattice is the most important symmetry element for describing the properties of imperfections in crystals.

PROBLEMS

(The material in Section A1.1 in Appendix 1 may assist in some of these exercises.)

1.1 (a) Select any convenient point on the plane pattern appearing on the following diagram and mark all the corresponding points, thus indicating the lattice.

(b) Outline the unit cell in several different ways, mark in the x and y axes in each case and measure the cell dimensions a, b and γ in each case.

(c) Draw a line parallel to MN through any one lattice point and add all the lines of this set. Determine the indices of this set of lines for each of your different choices of unit cell.

(d) Repeat (c) for the set of lines parallel to PQ.

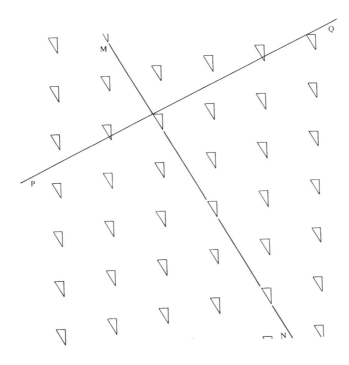

1.2 Rutile, TiO_2, has $a = b = 4.58$ Å, $c = 2.95$ Å and $\alpha = \beta = \gamma = 90°$. The atomic coordinates are:

Ti : $0, 0, 0; \frac{1}{2}, \frac{1}{2}, \frac{1}{2}$

O : $u, u, 0; -u, -u, 0; \frac{1}{2} + u, \frac{1}{2} - u, \frac{1}{2}; \frac{1}{2} - u, \frac{1}{2} + u, \frac{1}{2}$, where $u = 0.31$

(a) Draw an accurate projection of a unit cell on the plane containing the x and y axes.

(b) Determine the number of formula units per cell.

(c) Find the number of oxygen atoms surrounding each titanium atom. Calculate the interatomic distances Ti–O for the titanium atom at $\frac{1}{2}, \frac{1}{2}, \frac{1}{2}$.

1.3 The unit cells of several orthorhombic crystals are described below. What is the Bravais lattice of each?

(a) Two atoms of the same kind per unit cell located at

$$0, \tfrac{1}{2}, 0; \tfrac{1}{2}, 0, \tfrac{1}{2}.$$

(b) Four atoms of the same kind per unit cell located at

$$xyz; \bar{x}\,\bar{y}z; (\tfrac{1}{2}+x)(\tfrac{1}{2}-y)\bar{z}; (\tfrac{1}{2}-x)(\tfrac{1}{2}+y)\bar{z}.$$

(c) Two atoms of one kind per unit cell located at $\tfrac{1}{2}$, 0, 0; 0, $\tfrac{1}{2}$, $\tfrac{1}{2}$ and two of another kind located at 0, 0, $\tfrac{1}{2}$; $\tfrac{1}{2}$, $\tfrac{1}{2}$, 0.

1.4 Show with a sketch that a face-centred tetragonal lattice is equivalent to a body tetragonal lattice in a different orientation.

1.5 The spinel $MgAl_2O_4$ has $a = b = c = 8.11$ Å and $\alpha = \beta = \gamma = 90°$. Crystals show the faces (1 1 1), ($\bar{1}$ 1 1), (1 $\bar{1}$ 1) and (1 1 $\bar{1}$) and the faces parallel to these, namely ($\bar{1}$ $\bar{1}$ $\bar{1}$), (1 $\bar{1}$ $\bar{1}$), ($\bar{1}$ 1 $\bar{1}$) and ($\bar{1}$ $\bar{1}$ 1). Such crystals look like regular or distorted octahedra.

(a) Find the zone axis symbol (indices) of the zone containing (1 1 1) and ($\bar{1}$ $\bar{1}$ 1). Which of the other faces listed above also lie in this zone?

(b) Show that the zone containing (1 1 1) and ($\bar{1}$ $\bar{1}$ 1) also contains (0 0 1) and (1 1 0).

(c) Sketch the projection of the lattice on (0 0 1). Draw in the trace of the (1 1 0) lattice plane and the normal to the plane through the origin. Find the perpendicular distance of the first plane of this set from the origin in terms of the cell side.

(d) Draw the Section of the lattice through the origin which contains the normal to (1 1 0) and the normal to (0 0 1). Mark in on this section the traces of the (1 1 1), ($\bar{1}$ $\bar{1}$ 1), ($\bar{1}$ $\bar{1}$ $\bar{1}$) and (1 1 $\bar{1}$) planes. Measure the angles between adjacent planes.

1.6 Find the angle between [1 1 1] and the normal to (1 1 1) in (a) a cubic crystal and (b) a tetragonal one with $a = 5.67$ Å, $c = 12.70$ Å.

1.7 Show that the faces (1 1 1), (2 3 1) and (1 $\bar{2}$ 4) all lie in a zone. (a) Find the zone symbol (indices) of this zone and (b) calculate the angle between this zone and the zone [1 0 0] in the cubic system.

1.8 In a tetragonal crystal $CuFeS_2$ the angle between (1 1 1) and ($\bar{1}$ $\bar{1}$ 1) is 108.67°. Calculate (a) the ratio of the axial lengths and (b) the interzone angle [2 3 6]^[0 0 1].

1.9 Following the method of Section 1.6 determine the angles between the rotational axes in the axial combinations of (a) two triads and a diad and (b) a tetrad, a triad and a diad. Draw a sketch to show the arrangement of the symmetry elements in each case.

1.10 Investigate whether the axial combinations of (a) two tetrad axes and one triad axis and (b) a diad, a triad and a hexad axis can occur in crystals.

1.11 The unit cell of a two-dimensional lattice has $a = b$, $\gamma = 120°$. Dispose three atoms about each lattice point in different ways so that the two-dimensional crystal may show the following symmetries in turn: (a) a sixfold axis with mirror planes parallel to it, (b) a threefold axis, (c) a diad axis at the intersection of two mirror planes, (d) a twofold axis and (e) a onefold axis.

1.12 CdI_2 and $CdCl_2$ both belong to the trigonal crystal system. The former has a primitive hexagonal lattice and the latter a rhombohedral lattice. The coordinates of the atoms associated with each *lattice point* are given below, using a hexagonal unit cell for both crystals.

$$CdI_2 \qquad\qquad CdCl_2$$
$$Cd; 0, 0, 0 \qquad\quad Cd; 0, 0, 0$$
$$I; \pm \left(\tfrac{2}{3}, \tfrac{1}{3}, \tfrac{1}{4}\right) \qquad Cl; \pm \left(\tfrac{2}{3}, \tfrac{1}{3}, \tfrac{1}{12}\right)$$

Draw plans of the two structures on $(0\,0\,0\,1)$ and outline on your diagram for $CdCl_2$ the projections of the cell edges of the true rhombohedral primitive unit cell.

1.13 Show that there are three cubic lattices by following the procedure used in Section 1.8 and finding the conditions that in Fig. 1.28 triad axes lie normal to $(0\,1\,0)$, $(1\,0\,0)$ and $(0\,0\,1)$ of the rhombohedral primitive unit cell. *Hint.* Find the condition that successive planes project along their normal so that the lattice points in one lie at the centroids of triangles of lattice points in the plane below.

1.14 A two-dimensional crystal possesses fourfold rotational symmetry. Sketch the net. Position four atoms of an *element* at the net points so that the arrangement is consistent with (a) just fourfold symmetry and (b) fourfold symmetry with mirror planes parallel to the 4. Are there two different arrangements of mirror planes possible in (b)?

SUGGESTIONS FOR FURTHER READING

Borchardt-Ott, W., *Crystallography*, Springer, Berlin, London (1995).
Buerger, M. J., *Elementary Crystallography*, Wiley, New York (1963).
Giacovazzo, C. (ed.), *Fundamentals of Crystallography* (International Union of Crystallography Texts on Crystallography — 2), Oxford University Press, Oxford (1992).
Hilton, H., *Mathematical Crystallography*, Dover Publications (1963).
Jaswon, M. A., *An Introduction to Mathematical Crystallography*, American Elsevier (1965).

2

The Stereographic Projection and Point Groups

2.1 PRINCIPLES

In the study of crystallography it is often useful to be able to represent crystal planes and crystal directions on a diagram in two dimensions so that angular relationships and the symmetrical arrangements of crystal faces can be *measured* and discussed upon a flat piece of paper. Clearly the most useful type of diagram will be one in which the angular relationships in three dimensions in the crystal are faithfully reproduced in a plane.

Imagine the crystal to be positioned with its centre at the centre of a sphere which we call the *sphere of projection* (Fig. 2.1a) and draw normals to crystal planes through the centre of the sphere to intersect the surface of the sphere, say at P. P is called the *pole* of the plane of which OP is the normal. A direction is similarly represented by a point on the surface of the sphere defined as the point where the line parallel to the given direction, passing through the centre of the sphere, strikes the surface of the sphere. A crystal plane can also be represented by drawing the parallel plane through the centre of the sphere and extending it until it strikes the sphere (Fig. 2.1a). Since the plane passes through the centre of the sphere it is a *diametral plane* and the line of intersection of the sphere with such a plane is called a *great circle*. A great circle is a circle on the surface of a sphere with the radius equal to the radius of the sphere.

At this stage we have represented directions in the crystal, i.e. normals to lattice planes or lattice directions, by points (poles) on the surface of the sphere. We have a *spherical projection* of the crystal. The angle between two planes of which the normals are OP and OQ (Fig. 2.1b) is equal to the angle between these normals, which is the angle subtended at the centre of the sphere of projection by the arc of the great circle drawn through the poles P and Q. To make a drawing in two dimensions we now project the poles on to a flat piece of paper.

The spherical projection is like a terrestrial globe. Let us define north and south poles, N and S in Fig. 2.1a in analogy with the north and south poles of a globe. The equatorial plane passes through the centre of the sphere normal to

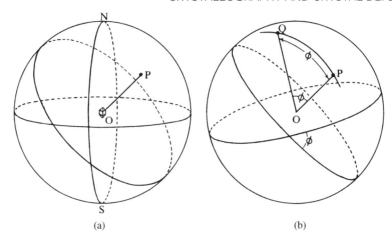

(a)   (b)

Figure 2.1 (a) Sphere of projection. (b) Angle between two planes is equal to the angle ϕ between the two poles

the line NS and cuts the sphere in a great circle called the equator. There are various ways of projecting points on the sphere on to a flat piece of paper. These are shown in Fig. 2.2. In the *orthographic* projection a pole P is projected from a point at infinity on to a plane parallel to the equatorial plane to form P'_o. In the *gnomonic* projection the point of projection is the centre of the sphere giving the projected pole at P'_G. In the *stereographic* projection the pole P is projected from a point on the surface of the sphere, say S, called the pole of projection on to a plane normal to OS. This plane can pass through any point on NS. If it passes through N the point P projects to P'_S. The most convenient plane for our purpose is the equatorial plane normal to SO and if we project the point P from S on to this plane we define the point P' so produced as the *stereographic projection* of P. In what follows we shall only consider the stereographic projection and we shall always take the plane of projection as the equatorial plane. The line of intersection of the plane of projection with the sphere of projection is a great circle called the primitive circle, or, for brevity, the primitive. Figure 2.3a shows the method of projection we shall adopt. A pole P_1 in the northern hemisphere projects to P'_1, inside the primitive, and is marked with a dot on the paper. All poles in the northern hemisphere project inside the primitive. Poles in the southern hemisphere, say P_2, give a projection P'_2 outside the primitive. The point P'_2 is the *true* projection of P_2. It is often inconvenient to work with projected poles outside the primitive and to avoid this a pole P_2, in the southern hemisphere, may be projected from the north pole N (diametrically opposite S) to give the projected pole at P''_2. The projected pole P''_2 is then distinguished from the *true projection* of P_2 (at P'_2) by marking the point P''_2 with a ring instead of with a dot.

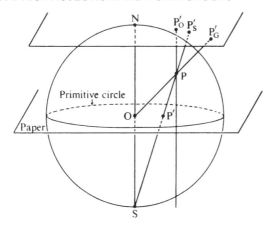

Figure 2.2 Projections of poles on the surface of a sphere on to a flat piece of paper

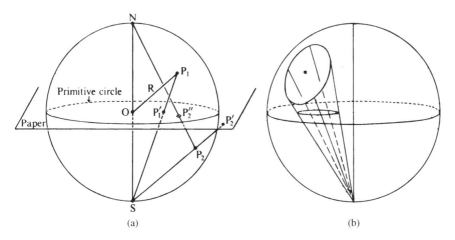

(a)   (b)

Figure 2.3 (a) Stereographic projection. (b) A small circle projects as a circle

The importance of the stereographic projection in crystallography arises because the projection is angle true, i.e. angles on the sphere of projection project as equal angles. Also, all circles (great or small) on the surface of the sphere of projection project as circles. This is illustrated for a small circle in Fig. 2.3b. A proof of these properties is given in Appendix 1, Section A-1.5.

We can now proceed to draw the stereographic representation or stereogram of the poles of crystal planes in a cubic crystal. The crystal axes are positioned with respect to the pole and plane of projection as in Fig. 2.4a. The three axes are orthogonal and of equal length (Table 1.3). In the *standard* projection shown in

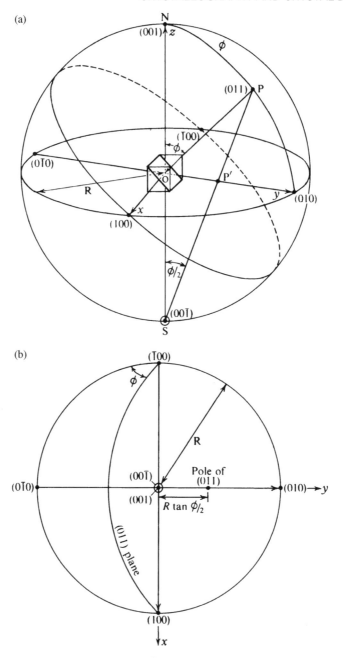

Figure 2.4 (a) Poles of a cubic crystal. (b) Stereogram of a cubic crystal

Fig. 2.4b the z axis of the crystal is taken normal to the plane of projection and, since the axes are orthogonal, the x and y axes then lie in the plane of projection at 90° to one another. The pole of the $(0\,0\,1)$ planes coincides with N and projects to the centre of the primitive (Fig. 2.4b). The poles of $(1\,0\,0)$, $(0\,1\,0)$, $(\bar{1}\,0\,0)$ and $(0\,\bar{1}\,0)$ lie on the primitive equally spaced at angles of 90°. The $(0\,0\,\bar{1})$ planes would lie at infinity if projected from S so we project it from N and denote it by the ring. The $(0\,1\,1)$ planes are represented by the pole P; $(0\,1\,1)$ lies in the zone of which the x axis is the zone axis. The poles of all planes in the zone with axis Ox lie on the great circle defined by the locus of all points 90° from the $(1\,0\,0)$ pole. This great circle projects as the *line* on the stereogram joining $(0\,\bar{1}\,0)$, $(0\,0\,1)$ and $(0\,1\,0)$. Therefore, P projects somewhere between $(0\,0\,1)$ and $(0\,1\,0)$. The angle ϕ in Fig. 2.4a is the angle between $(0\,0\,1)$ and $(0\,1\,1)$; for the cubic crystal $\phi = 45°$. From Fig. 2.4a the distance OP' is given by

$$OP' = R \tan \phi/2$$

where R is the radius of the sphere of projection. This follows since SONP and P' all lie in the same plane and the angle OSP is equal to $\phi/2$ since OSP is the angle at the circumference standing on the same arc NP as the angle NOP at the centre. We can therefore insert the $(0\,1\,1)$ pole on the stereogram at a distance $R \tan \phi/2$ (in this case $R \tan 45°/2 = R \tan 22.5°$ from the $(0\,0\,1)$ pole along the radius of the primitive joining $(0\,0\,1)$ and $(0\,1\,0)$.

The plane $(0\,1\,1)$ itself can be drawn upon the stereogram instead of just the pole of $(0\,1\,1)$ by drawing the projection of the great circle which is the locus of points 90° from the pole $(0\,1\,1)$. This is drawn in Fig. 2.4b. Drawing a great circle of which the pole is given can be accomplished either by construction or by using graphical aids. We now deal with some constructions on the stereogram and then with the use of the graphical aid called the Wulff net.

2.2 CONSTRUCTIONS

To obtain a thorough understanding of the stereogram it is wise for the beginner to carry out a number of constructions accurately and without any graphical aid. We will now describe some of these. However, since all of these constructions can be accomplished with graphical aids, this section can be omitted without prejudice to the rest of the book.

TO CONSTRUCT A SMALL CIRCLE

About the Centre of the Primitive (Fig. 2.5)

The stereographic projection of the required angular radius ϕ of the small circle is plotted on either side of N, at X'Y', so that $NY' = NX' = R \tan \phi/2$, where R is the radius of the primitive. A circle is then described with N as centre, and

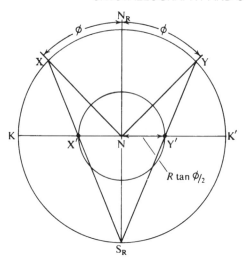

Figure 2.5 Construction of a small circle about the centre of the primitive

NX' (or NY') as the radius. This is the *only* case where the centre of the small circle in projection coincides with the stereographic projection of the centre of the small circle. Alternatively, we could locate the points X' and Y' in projection solely by construction as follows. Draw the diameter of the primitive upon which we wish to locate X' and Y' and then draw the diameter of the primitive normal to this, $N_R NS_R$ (Fig. 2.5). Find the point X, on the primitive, such that the arc XNN_R subtends an angle ϕ at the centre of the primitive. If we join XS_R, then X' is located where XS_R cuts the first diameter of the primitive. The justification for this construction is shown in detail in Fig. 2.6, where it is seen that if we imagine keeping the plane of projection fixed but rotate the sphere of projection through 90° about KK' (a line lying in the plane of projection), then the pole of projection S comes to lie on the primitive at S_R. Similarly, N lies at N_R and the validity of the construction follows. This useful trick of imagining the whole sphere of projection rotated through 90° is often used to establish constructions on the stereogram.

About a Pole within the Primitive — say about P (Fig. 2.7)

Draw the diameter through P' and the diameter normal to this to locate B and A. Project P' from A to the primitive to give P, say. Measure off ϕ (the required radius of the small circle) along the primitive on either side of P to obtain the points X, Y on the primitive (Fig. 2.7), noting that ϕ is the angle subtended at the centre of the primitive by the arc XP. Reproject X and Y from A to obtain the points X', Y', which are opposite ends of the diameter of the required small circle. It should be noted that ϕ is the angle subtended at the centre of the primitive

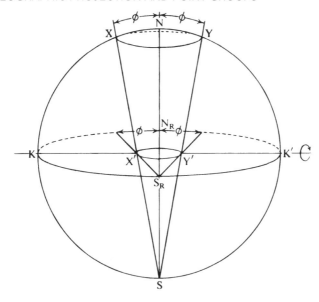

Figure 2.6

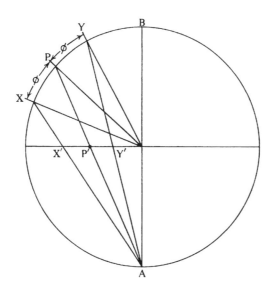

Figure 2.7

and that P' is not half-way between X' and Y'. It is a useful exercise to draw a diagram similar to Fig. 2.6 to justify this construction.

About a Pole on the Primitive — say about P (Fig. 2.8)

Draw the radius NP and from P measure off the angle ϕ (equal to the angular radius of the required circle) subtended at the centre of the primitive to locate the point M. At M draw the tangent to the primitive to meet NP produced in S. S is the *centre* of the required small circle and SM its radius. This construction is a little quicker than that described previously for the case of a small circle about a pole upon the primitive.

TO FIND THE OPPOSITE OF A POLE

The opposite of a given pole is the pole 180° away from the given pole, i.e. the other end of a diameter of the sphere of projection passing through a given

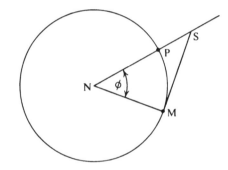

Figure 2.8

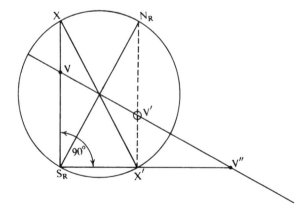

Figure 2.9   To find the opposite of a given pole

pole. Suppose that we wish to find the stereographic projection of the opposite of the pole V in Fig. 2.9. Draw the diameter of the primitive through V. Then the opposite of V, say V", clearly lies on this diameter and 180° from V. Draw the diameter of the primitive normal to the diameter through V and mark opposite ends of this, S_R and N_R. Project V from S_R to the primitive to find the point X. Draw the diameter of the primitive through X and mark the other end of this diameter at X'. X' is projected from S_R on to the diameter through V to give the required opposite of V at V" (Fig. 2.9). The justification for this construction is easily seen from Fig. 2.10, where we use the same device as in Fig. 2.6 and imagine the sphere of projection rotated 90° about the diameter of the primitive containing the projection of V, so that S comes to lie on the primitive at S_R but V and V", lying on the axis of rotation, do not move. The angle XS_RX' in Fig. 2.9 is clearly 90°, so in practice V" is easily located by putting a 90° set-square at S_R with one edge running through V and using the other edge to locate V".

If the given pole V lies inside the primitive in projection then the true opposite V" lies outside the primitive circle. In some cases, therefore, one works with the opposite V' obtained by projecting not from S but from N on the sphere of projection (Fig. 2.3a). Clearly V' is always found on the diameter of the primitive through V at an equal distance from the centre on the opposite side (Fig. 2.9).

To draw a great circle through two poles, find the true opposite of one of them, using the last construction, and then construct the circle passing through the two given points and this opposite. This is the required great circle.

To find the pole of a great circle — say through ACB in Fig. 2.11 — draw the diameter AB of the primitive which is a chord of the great circle and draw the diameter of the primitive normal to AB. Let this diameter intersect the great circle at T. Project T from B to the primitive at X and measure 90° from X

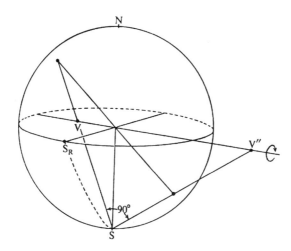

Figure 2.10

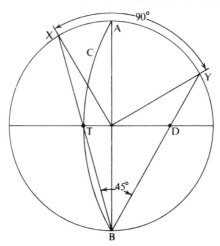

Figure 2.11 To find the pole of a great circle

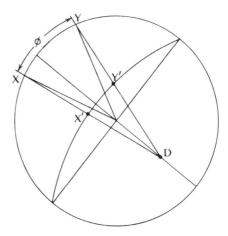

Figure 2.12 To find the angle between two poles

over the pole A to find the pole Y. Project Y from B onto the diameter of the primitive through T to find the pole D which is the required pole. Since the arc XY subtends an angle of 90° at the centre of the primitive, clearly the angle TBD is 45° and so D may be rapidly located once T is found by placing the 45° angle of a set-square as shown in Fig. 2.11. This construction is easily justified by drawing a diagram similar to Fig. 2.10.

To measure the angle between two poles on an inclined great circle, let X'Y' be the two projected poles (in Fig. 2.12). Locate the pole D of the great circle upon which they lie by the previous construction. From D project X' and Y' to the primitive to locate X and Y respectively. The angle ϕ subtended by the arc XY *at the centre of the primitive* is the angle between the two poles. It is a useful exercise to draw a diagram similar to Fig. 2.10 to prove this.

2.3 CONSTRUCTIONS WITH THE WULFF NET

A graphical aid for the construction of stereograms which is also very useful for taking angular measurements upon them is called a Wulff net. A net is shown in Fig. 2.13b. The net is the projection of one half of the terrestrial globe with lines of latitude and longitude marked upon its surface and with the north and south poles lying in the plane of projection. The relationship between the lines of latitude (which are all small circles except the equator) and the lines of longitude (all great circles) on the surface of the sphere of projection and their representation on the net is shown in Figs. 2.13a and 2.13b. The radius of the net and of the sphere of the projection are of course equal and equal to the radius of the primitive circle of the projection. The net is used by placing it under the stereogram, which is drawn on transparent tracing paper, and the centres of the two are located by a pin. The stereogram can be rotated above the net. The angle between poles within the primitive is measured by rotating the stereogram until the two poles lie on the same great circle and the angle is measured by counting the small circles between the poles (Figs. 2.14a and b). Angles between poles on the primitive are measured directly.

The great circle corresponding to the locus of points 90° from a given pole is found by placing the pole on the equator of the net (the line marked 0°0° in Figs. 2.13a and 2.15) and tracing out the great circle 90° from the given pole as indicated in Fig. 2.15. This great circle is the trace of the plane of which P'$_2$ is the pole.

It is often useful to be able to rotate a stereogram about a given axis. To rotate any pole, say A$_1$ in Fig. 2.16, about the pole B, lying on the primitive, the net is rotated until the axis NS of the net lies along the diameter of the primitive through B. A$_1$ is then rotated the required number of degrees about B by moving A$_1$ to A$_2$ along the small circle as shown in Fig. 2.16. A pole which will pass outside the primitive upon rotation is also shown in Fig. 2.16. The true projection of C$_2$ as distinct from its opposite (shown) would be on the same small circle as C_1. If B does not lie on the primitive the procedure shown in Fig. 2.17 can be followed. The net is rotated until NS lies perpendicular to the radius of the primitive through B. B and A are then both rotated about the axis NS until B lies at the centre of the primitive (B' in Fig. 2.17) and A moves to A'. If it is required to rotate A, say, 40° clockwise about B we now rotate A' 40° clockwise about B' to A". This is easily done as shown in Fig. 2.17. We now rotate B' back to B by rotation about NS and rotate A" about NS by the same amount in the same sense to obtain A''', the required rotated position of A.

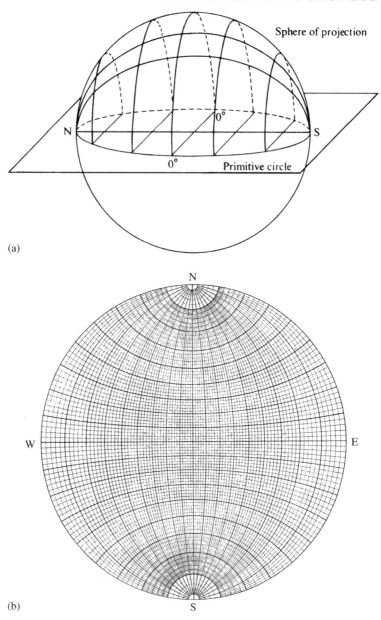

Figure 2.13   (a) Projection of lines of latitude and longitude to make the Wulff net. (b) Wulff net drawn to 2° intervals

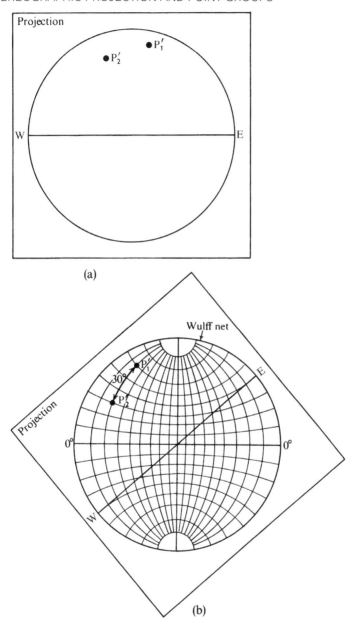

Figure 2.14 (a) Stereographic projection of two poles P'_1 and P'_2. (b) Rotation of the projection to put both poles on the same great circle of the Wulff net. Angle between poles = 30°. (Based on Cullity [1])

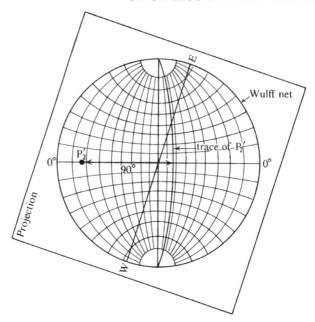

Figure 2.15 To find the trace of a pole using the Wulff net. (Based on Cullity [1])

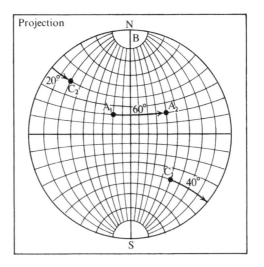

Figure 2.16 Rotation of poles about an axis in the plane of projection. (Based on Cullity [1])

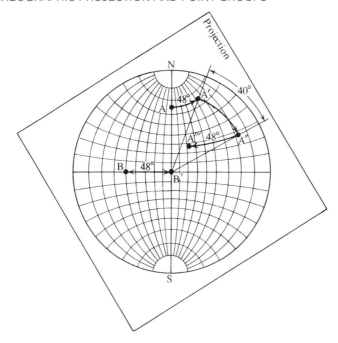

Figure 2.17   Rotation of poles about an inclined axis. (Based on Cullity [1])

TWO-SURFACE ANALYSIS

In the identification and subsequent study of planar imperfections in crystals it is often necessary to identify a crystal plane from the linear traces which it makes in two (or more) other non-parallel planes. The procedure is illustrated in Fig. 2.18. Suppose the planes A and B (which are flat surfaces of a crystal) intersect along the line PQ. Let us draw a stereogram with the plane of projection parallel to B so that the pole of B lies at the centre of the primitive. The angle between A and B is ϕ (the angle between the outward normals to A and B) and so the pole of A lies on the stereogram as shown in Fig. 2.18b. The planes A and B intersect along PQ, and so this intersection can be marked upon the stereogram. The plane we are interested in is MNT, which makes an angle θ_A with PQ in the face A and θ_B with PQ in the face B. Consider the trace TT' in the face B. The direction TT' lies in plane B and at angle θ_B from PQ measured counterclockwise from P, as shown in Fig. 2.18a. We can therefore plot the direction TT' on the stereogram in Fig. 2.18b at an angle θ_B to PQ. The plane MNT when drawn on the stereogram must project as a great circle which passes through the points T and T'. An infinite number of great circles pass through TT', corresponding to all the possible planes that intersect the plane B in a direction parallel to TT'. To locate another point on the stereogram through which the projection of the plane

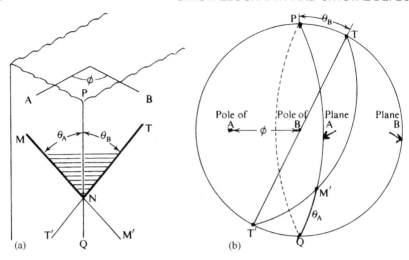

Figure 2.18 Two-surface analysis

MNT must pass we consider the trace MM' of the plane MNT in the face A. The trace MM' makes a planar angle in the face A of θ_A with PQ. We have already plotted the direction of PQ in the stereogram so θ_A must be measured in the appropriate sense from PQ and *in the plane* A. The point M' giving the direction of the trace of MM' in A is at an angle θ_A measured counterclockwise from Q along the great circle representing the plane A in Fig. 2.18b. The angle θ_A can be measured on the stereogram either by using the method of construction to measure the angle between two poles on an inclined great circle (see Section 2.2) or more simply using the Wulff net so that NS in Fig. 2.13b lies along the line PQ in Fig. 2.18b and θ_A is measured off by counting the lines of latitude along the great circle coinciding with the trace of the plane A. We have now located T, M' and T' as poles on the stereogram through which the projection of the plane MNT must pass. We can, therefore, draw in MNT as the great circle passing through these poles, and use the Wulff net or a construction to find the pole of MNT and hence the indices of MNT if those of A and B are known.

2.4 MACROSCOPIC SYMMETRY ELEMENTS

The macroscopically measured properties of a crystal, e.g. electrical resistance, thermal expansion, magnetic susceptibility or the elastic constants, show a symmetry which can be defined and understood without reference to the translational symmetry elements defined by the lattice. If the translational symmetry of the crystal is disregarded the remaining symmetry elements (such as axes of rotational symmetry, mirror planes and the centre of inversion), themselves consistent with the translational symmetry of the lattice, can be arranged into 32 consistent groups. These are the 32 crystallographic point groups. They are so-called because all of the symmetry elements in a group pass through a single point and the operation of these elements leaves just one point unmoved — the point through which they pass.

The axes of rotational symmetry, the mirror plane and the centre of inversion are called macroscopic symmetry elements because their presence or absence in a given crystal can be decided in principle by macroscopic tests such as etching of the crystal, the arrangement of the external faces or the symmetry of the physical properties, without any reference to the atomic structure of the crystal. The macroscopic symmetry elements are of two kinds. The *first kind*, e.g. a pure rotation axis, when operating on a right-handed object (say) produces a right-handed object from it and all subsequent repetitions of this object are also right-handed. A symmetry operation of the *second kind* repeats an *enantiomorphous* object from an original object. The left and right hands of the human body are enantiomorphously related. The operation of reflection illustrated in Fig. 1.11 is an example of a symmetry operation of the second kind since a left-handed object is repeated from an original right-handed object. Subsequent operation of the same symmetry element would produce a right-handed object again and then a left-handed object and so forth. A symmetry operation of the second kind thus involves a reversal of sense in the operation of repetition. Inversion through a centre is also an operation of the second kind.

In developing the 32 crystallographic point groups it is convenient to have all the macroscopic symmetry elements represented by axes and to do this we define what are called *improper rotations*. These produce repetition by a combination of a rotation and an operation of inversion. We shall use *rotoinversion* axes. These involve rotation coupled with inversion through a centre.[†] A pure rotation axis is said to produce a proper rotation.

The basic operation of repetition by a rotation axis is shown in Fig. 2.19. An n-fold axis repeats an object by successive rotations through an angle of $2\pi/n$. In the example shown in Fig. 2.19 the axis is one of fourfold symmetry. The operation of monad, diad, triad, tetrad and hexad pure rotation axes on a single initial pole is shown in Fig. 2.20. These diagrams are stereograms with the pole of the axis at the centre of the primitive circle. The numbers below the stereograms give the shorthand labels for the axes 1, 2, 3, 4 and 6, indicating a onefold, twofold, threefold, fourfold and sixfold axis respectively. Figure 2.21 shows the repetition of an object by a mirror plane, symbol m, and by a centre of symmetry (or centre of inversion). In Fig. 2.21a the mirror plane lies normal to the primitive circle. It is denoted by a strong line I, coinciding with the mirror in

---

[†] It is also possible to use rotoreflection axes in developing the point groups. These repeat an object by rotation coupled with reflection in a plane normal to the axis. Onefold, twofold, threefold, fourfold and sixfold rotoreflection axes are possible, usually denoted $\tilde{1}, \tilde{2}, \tilde{3}, \tilde{4}$ and $\tilde{6}$ respectively. $\tilde{1}$ is clearly equivalent to a mirror plane.

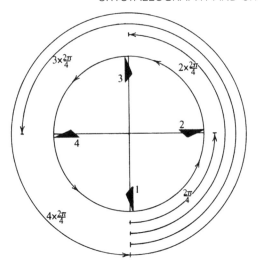

Figure 2.19

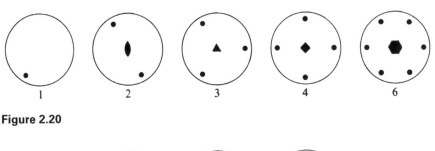

Figure 2.20

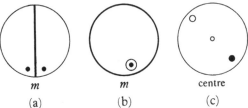

Figure 2.21

the stereographic projection. In Fig. 2.21b the mirror coincides with the primitive. In Fig. 2.21c the centre of inversion is at the centre of the sphere of projection.

The onefold inversion axis is a centre of symmetry. The operation of the other rotoinversion axes is explained in Figs. 2.22 and 2.23. The twofold rotation–inversion axis $\bar{2}$ shown in Fig. 2.22 repeats an object by rotation through 180° (360°/2) to give the dotted circle, followed by inversion to give

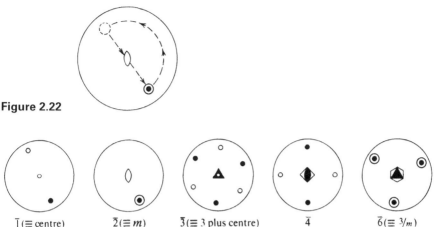

Figure 2.22

| $\bar{1}(\equiv \text{centre})$ | $\bar{2}(\equiv m)$ | $\bar{3}(\equiv 3 \text{ plus centre})$ | $\bar{4}$ | $\bar{6}(\equiv {}^3/m)$ |

Figure 2.23

the full circle. Similarly, the threefold inversion axis $\bar{3}$ involves rotation through $360°/3 = 120°$ coupled with an inversion. In general, an n-fold rotoinversion axis $\bar{n}$ involves rotation through an angle of $2\pi/n$ coupled with inversion through a centre. The rotation and inversion are both part of the operation of repetition and must not be considered as separate operations. The operation of the various rotoinversion axes, which can occur in crystals, on a single initial pole is shown in Fig. 2.23, with the pole of the rotoinversion axis at the centre of the primitive circle. The following symbols are used: inversion monad, $\bar{1}$, symbol O; inversion diad, $\bar{2}$, symbol 0; inversion triad, $\bar{3}$, symbol $\blacktriangle$; inversion tetrad, $\bar{4}$, symbol $\blacklozenge$; inversion hexad, $\bar{6}$, symbol $\spadesuit$. Inspection of Figs. 2.20, 2.21 and 2.23 shows that $\bar{1}$ is identical to a centre of symmetry, $\bar{2}$ is identical to a mirror plane normal to the inversion diad, $\bar{3}$ is identical to a triad axis plus a centre of symmetry and $\bar{6}$ is identical to a triad axis normal to a mirror plane (symbol $3/m$, the sign '$/m$' indicating a mirror plane normal to an axis of symmetry).[†] Only $\bar{4}$ is unique. The operation of repetition described by $\bar{4}$ cannot be reproduced by any combination of a proper rotation axis and a mirror plane or a centre of symmetry.

The various different combinations of 1, 2, 3, 4 and 6 pure rotation axes and $\bar{1}$, $\bar{2}$, $\bar{3}$, $\bar{4}$ and $\bar{6}$ rotoinversion axes constitute the 32 point groups or crystal classes. These 32 classes are grouped into systems according to the presence of defining symmetry elements (Table 1.3). Stereograms of each of the 32 point groups or crystal classes are given in Fig. 2.24, following the conventions of the *International Tables for X-Ray Crystallography*. Each point group except the

---

[†] Our symbols are identical to those used in the *International Tables for X-ray Crystallography* [2], with the exception of '0', which in the *International Tables* is always indicated by the symbol for a mirror plane.

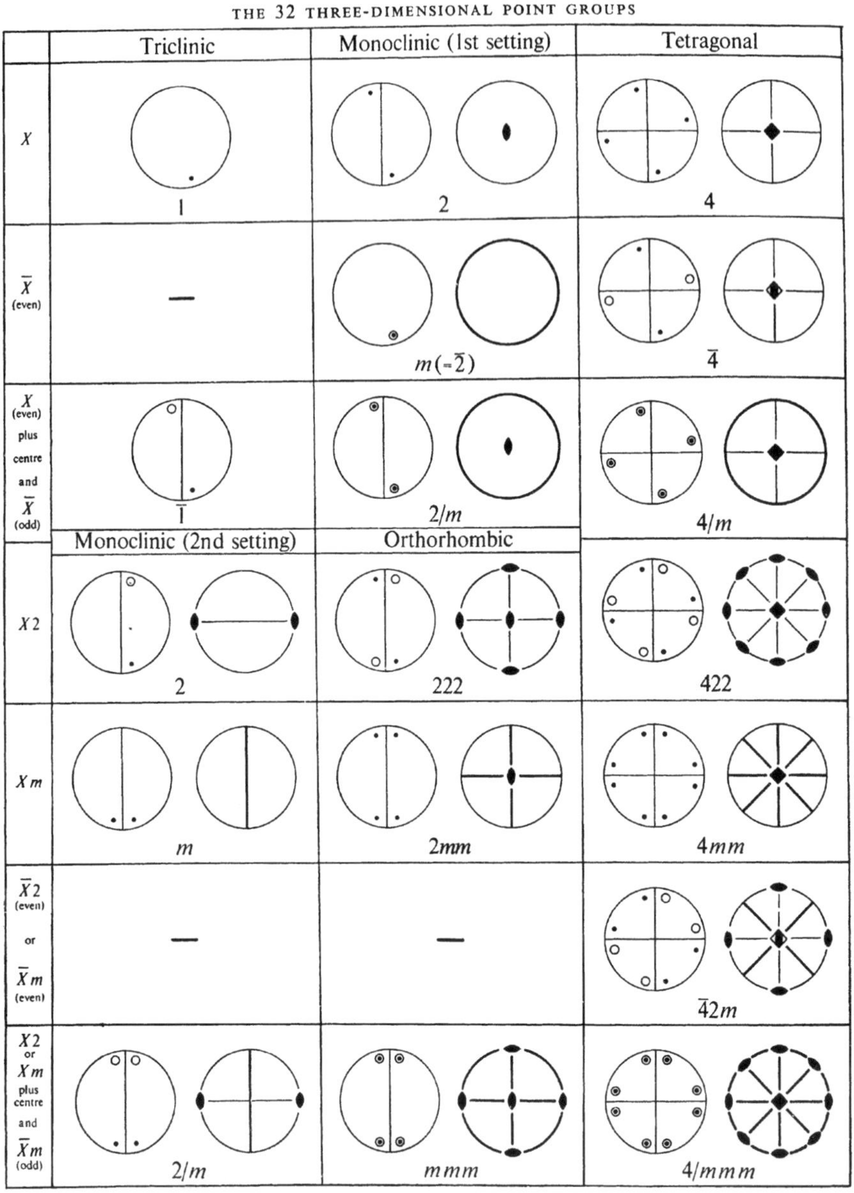

THE 32 THREE-DIMENSIONAL POINT GROUPS

Figure 2.24 Stereograms of the poles of equivalent general directions and of the symmetry elements of each of the 32 point groups. The z-axis is normal to the paper. (Taken from the *International Tables for X-ray Crystallography*, Vol. 1 [2])

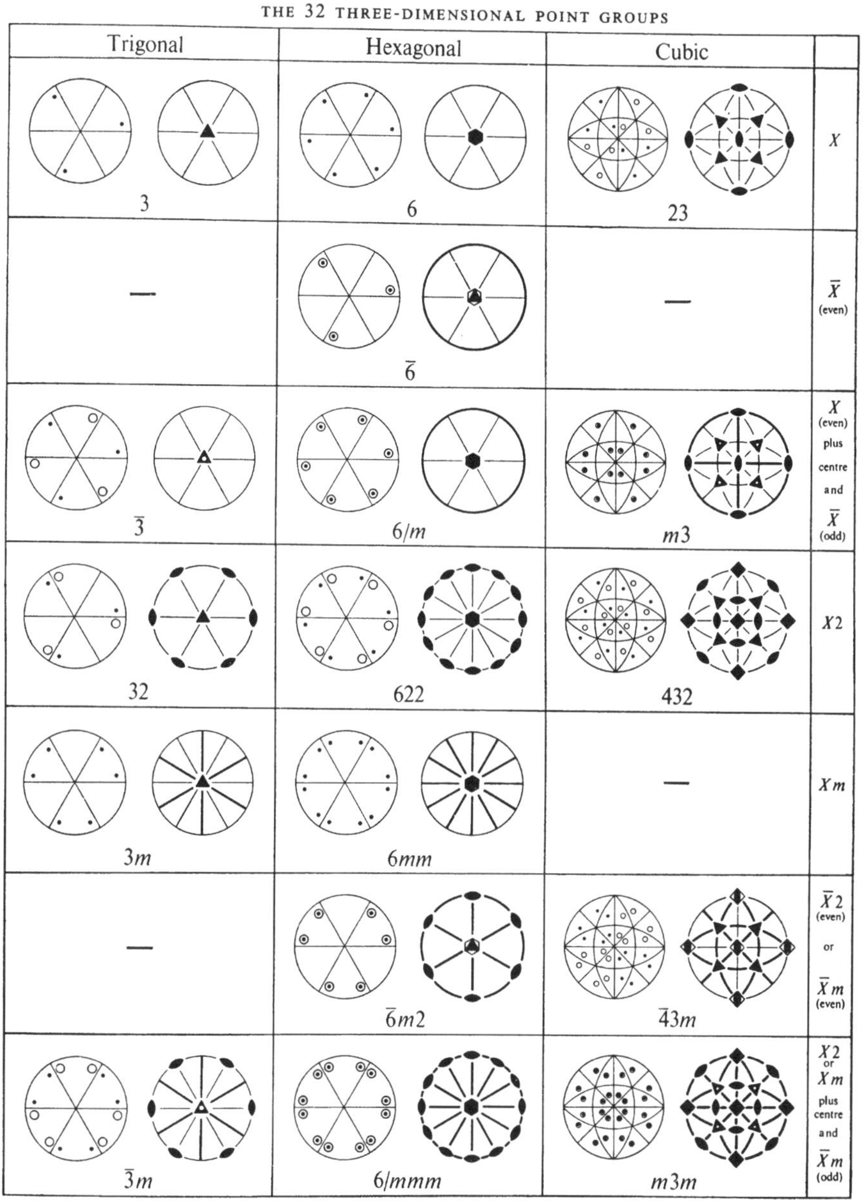

THE 32 THREE-DIMENSIONAL POINT GROUPS

Trigonal	Hexagonal	Cubic	
3	6	23	X
—	$\bar{6}$	—	$\bar{X}$ (even)
$\bar{3}$	$6/m$	$m3$	X (even) plus centre and $\bar{X}$ (odd)
32	622	432	$X2$
$3m$	$6mm$	—	Xm
—	$\bar{6}m2$	$\bar{4}3m$	$\bar{X}2$ (even) or $\bar{X}m$ (even)
$\bar{3}m$	$6/mmm$	$m3m$	$X2$ or Xm plus centre and $\bar{X}m$ (odd)

Figure 2.24 (continued)

triclinic ones is depicted by two stereograms. The first shows how a single initial pole is repeated by the operations of the point group and the second stereogram shows all of the symmetry elements present. The nomenclature for describing the crystal classes is as follows. X indicates a rotation axis and $\overline{X}$ an inversion axis. X/m is a rotation axis normal to a mirror plane, Xm a rotation axis with a mirror plane parallel to it and $X2$ a rotation axis with a diad normal to it. X/mm indicates a rotation axis with a mirror plane normal to it and another parallel to it. $\overline{X}m$ is an inversion axis with a parallel plane of symmetry. A plane of symmetry is an alternative description of a mirror plane.

We shall describe each of the classes in Sections 2.5 to 2.11. A derivation of the 32 classes follows the lines of noting from Sections 1.5 and 1.6 that the rotation axes consistent with translational symmetry are 1, 2, 3, 4 and 6. Each of these existing alone gives in total five crystal classes. The consistent combinations of these axes gives another six classes (see Table 1.2), viz. 2 2 2, 3 2 2, 4 2 2, 6 2 2, 3 3 2 and 4 3 2, thus totalling eleven. All of these eleven involve only operations of the first kind. A lattice is inherently centrosymmetric (Section 1.4) and so each of the rotation axes could be replaced by the corresponding rotoinversion axis, thus giving another five classes, viz. $\overline{1}$, $\overline{2}$, $\overline{3}$, $\overline{4}$ and $\overline{6}$. The remaining sixteen can be described as combinations of the proper and improper rotation axes.

2.5 ORTHORHOMBIC SYSTEM

A crystal in this system contains three diads, which must be at right angles to one another. In Fig. 2.24 they are the crystal axes. From Table 1.3 and Figs. 1.19d,e,f,g, the lattice parameters may be all unequal.

The point group containing just three diad axes at right angles is shown in Fig. 2.24 and is designated 2 2 2. In general a pole is repeated four times. If the indices of the initial pole are $(h\,k\,l)$, where there is no special relationship between h, k and l, then the operation of all of the symmetry elements of the point group on this one initial pole produces other poles. These are $(\overline{h}\,\overline{k}\,l)$, $(h\,\overline{k}\,\overline{l})$ and $(\overline{h}\,k\,\overline{l})$ (Fig. 2.25). The assemblage of crystal faces produced by repetition of an initial crystal face with indices $(h\,k\,l)$ is called the *form hkl* and is given the symbol $\{h\,k\,l\}$.[†] If the assemblage of faces encloses space the form is said to be closed; otherwise it is open. In this case $\{h\,k\,l\}$ is closed. The symbol $\{h\,k\,l\}$ with curly brackets means all faces of the form *hkl*, i.e. in this case, for the point group 2 2 2, $(h\,k\,l)$, $(\overline{h}\,\overline{k}\,l)$, $(h\,\overline{k}\,\overline{l})$ and $(\overline{h}\,k\,\overline{l})$. The form $\{h\,k\,l\}$ is said to show a *multiplicity* of four. Then $\{h\,k\,l\}$ would be said to be a *general form*, i.e. a form that bears no special relationship to the symmetry elements of the point group.

[†] All the individual faces of the form $\{h\,k\,l\}$ are crystallographically identical. In a similar way, if a single direction is given, say $[u\,v\,w]$, then all the directions produced if the repetition operations of the point group are carried out on the initial direction are called the family of directions of the type $u\,v\,w$ and are given the symbol $\langle u\,v\,w\rangle$.

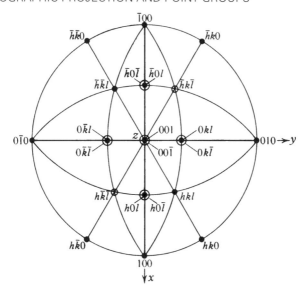

Figure 2.25 A stereogram of an orthorhombic crystal of point group 222. The diad axes are parallel to the x, y and z axes

Special forms in this crystal class would be $\{1\,0\,0\}$, $\{0\,1\,0\}$ and $\{0\,0\,1\}$; each are giving just two faces, e.g. for $\{1\,0\,0\}$, the two faces $(1\,0\,0)$ and $(\bar{1}\,0\,0)$ (Fig. 2.25). These forms are all open ones. They are easily recognized as special forms since their multiplicity is less than that of the general form. Forms such as $\{h\,k\,0\}$ $\{h\,0\,l\}$ and $\{0\,k\,l\}$ having one index zero and no special relationship between the other two would also be spoken of as special even though, as Fig. 2.25 shows, the multiplicity of each is four. The reason for this is that these poles lie normal to a diad axis. This is a special position with respect to this axis and has the result that if a crystal grew with faces parallel only to the planes of indices $\{h\,k\,0\}$, $\{h\,0\,l\}$ and $\{0\,k\,l\}$ it would appear to show mirror symmetry as well as the three diad axes. Special forms usually correspond to poles lying normal to or on an axis of symmetry, and normal to or on mirror planes and sometimes to poles lying midway between two axes of symmetry. However, the best definition of a special form is as follows. A form is special if the development of the complete form shows a symmetry of arrangement of the poles which is higher than the crystal actually possesses. Special forms in all of the crystal classes are listed later in Table 2.3.

The orthorhombic system also contains the classes 2*mm* and *mmm* (Fig. 2.24). The former contains two mirror planes which must be at right angles. Two mirror planes at right angles automatically show diad symmetry along the line of intersection (Fig. 2.24). Since $m \equiv \bar{2}$ this group could be designated $2\,\bar{2}\,\bar{2}$ and this is why it appears in the orthorhombic system, which is defined as possessing three diad axes. This crystal class could simply be designated *mm* since the diad is automatically present. However, *mm2* is usually used for the later development of space groups (see Section 2.11). A crystal containing three diad axes can also contain mirrors normal to all of these without an axis of higher symmetry. Such a point group is designated *mmm*, or could be designated 2/*mm*. As Fig. 2.24 shows, the multiplicity of the general form is now eight. The special forms {$hk0$}, {$h0l$} and {$0kl$} now show a lower multiplicity than the general one. The point group *mmm* shows the highest symmetry in the orthorhombic system. The point group showing the highest symmetry in a system is said to be the *holosymmetric* class.

To plot a stereogram of an orthorhombic crystal if given the lattice parameters *a*, *b* and *c*, we proceed as shown in Fig. 2.26. In Fig. 2.26a the poles of the (001), (010) and (100) planes are immediately inserted at the centre of the

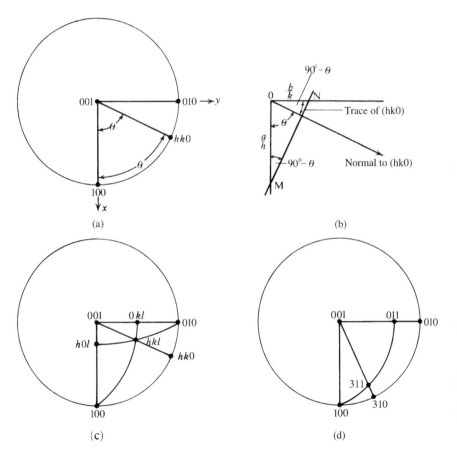

Figure 2.26

primitive and where the y and x axes cut the primitive respectively. A pole $(h\,k\,0)$ can be inserted at an angle θ along the primitive to $(1\,0\,0)$, as shown in Fig. 2.26a, by noting from Fig. 2.26b (which is a section of the crystal parallel to $(0\,0\,1)$ showing the intersection of the plane $(h\,k\,0)$ with the crystal axes) that the pole of $(1\,0\,0)$ lies along OM, that $(h\,k\,0)$ makes intercepts on the crystal axes of a/h along the x axis and b/k along the y axis and that θ is the angle between the normal to $(h\,k\,0)$ and the normal to $(1\,0\,0)$ angle θ is indicated in Figs. 2.26a and b. From Fig. 2.26b we have

$$\tan\theta = \cot(90 - \theta) = \frac{a/h}{b/k} = \frac{a}{b}\frac{k}{h}$$

i.e.

$$\tan(1\,0\,0)\widehat{}(h\,k\,0) = (a/b)(k/h) \tag{2.1a}$$

Similarly, if the lattice parameters are given we can locate $(0\,k\,l)$ and $(h\,0\,l)$ since

$$\tan(0\,0\,1)\widehat{}(0\,k\,l) = (c/b)(k/l) \tag{2.1b}$$

and

$$\tan(0\,0\,1)\widehat{}(h\,0\,l) = (c/a)(h/l) \tag{2.1c}$$

$(0\,0\,1)\widehat{}(0kl)$ means the angle between the pole of $(0\,0\,1)$ and the pole of $(0\,k\,l)$, etc. These relations can be seen immediately by drawing diagrams similar to Fig. 2.26b but looking along the x and y axes respectively. When poles such as $(h\,0\,l)$, $(0\,k\,l)$, $(h\,k\,0)$ are inserted we can insert a pole such as $(h\,k\,l)$ immediately by use of the zone rule, given in Eqn (1.11). From Eqn (1.11) if two poles $(h_1\,k_1\,l_1)$, $(h_2\,k_2\,l_2)$ both lie in the same zone with the indices $[u\,v\,w]$, then

$$h_1 u + k_1 v + l_1 w = 0 \tag{2.2}$$

and

$$h_2 u + k_2 v + l_2 w = 0 \tag{2.3}$$

If we multiply Eqn (2.2) by a number m and Eqn (2.3) by a number n and add them we have

$$(mh_1 + nh_2)u + (mk_1 + nk_2)v + (ml_1 + nl_2)w = 0 \tag{2.4}$$

Therefore the pole of the plane $(mh_1 + nh_2, mk_1 + nk_2, ml_1 + nl_2)$ also lies in $[u\,v\,w]$. In other words, the indices formed by taking linear combinations of the indices of two planes in a given zone provide the indices of a further plane in that same zone. In general m and n can be positive or negative. In Fig. 2.26c, after $(0\,0\,1)$, $(0\,1\,0)$, $(1\,0\,0)$ and $(h\,k\,0)$, $(h\,0\,l)$, $(0\,k\,l)$ are plotted, then to plot, say, $(h\,k\,l)$ we note that $(h\,k\,l)$ must lie in the zone containing $(0\,0\,1)$ and $(h\,k\,0)$, since if we multiply $(0\,0\,1)$ by the number l and add the indices $(0\,0\,l)$ and $(h\,k\,0)$ we obtain $(h\,k\,l)$. Similarly, $(h\,k\,l)$ lies in the zone containing $(0\,k\,l)$ and $(1\,0\,0)$, since h times $(1\,0\,0)$ gives $(h\,0\,0)$ and this added to $(0\,k\,l)$ gives $(h\,k\,l)$. We then draw the zone containing $(0\,0\,1)$ and $(h\,k\,0)$ and that containing $(0\,0\,1)$ and $(0\,k\,l)$ and we know that $(h\,k\,l)$ is situated where these intersect.

A particular example may make the procedure clear. Suppose we wish to locate $(3\,1\,1)$ after plotting $(0\,0\,1)$, $(0\,1\,0)$ and $(1\,0\,0)$ (Fig. 2.26d). One way to proceed would be to locate $(0\,1\,1)$ using Eqn (2.1b), setting $k = 1$ and $l = 1$ and using the known lattice parameters. We then find $(3\,1\,0)$, on the primitive, by finding the angle between $(1\,0\,0)$ and $(3\,1\,0)$ from Eqn (2.1a), setting $h = 3$ and $k = 1$. Finally, we note that $(3\,1\,1)$ lies in the zone containing $(0\,0\,1)$ and $(3\,1\,0)$ since $(0\,0\,1)$ plus $(3\,1\,0)$ yields $(3\,1\,1)$. Also $(3\,1\,1)$ lies in the zone containing $(1\,0\,0)$ and $(0\,1\,1)$ since 3 times $(1\,0\,0)$ plus $(0\,1\,1)$ yields $(3\,1\,1)$. The pole of $(3\,1\,1)$ is then immediately located by drawing the great circle through $(0\,0\,1)$ and $(3\,1\,0)$ and that through $(0\,1\,1)$ to $(1\,0\,0)$; $(3\,1\,1)$ is located where these great circles meet.

Using the above procedure for locating poles is usually the quickest way to draw an accurate stereogram when key poles have been located by calculation. It must be strongly emphasized that, although we have chosen the orthorhombic system as an example, the use of Eqn (2.4) to locate poles applies to any crystal system and does not depend on the crystal axes being at any particular angle to one another. The utility of Eqn (2.4) is one of the great advantages of the Miller index for denoting crystal planes, and arises naturally from the properties of a space lattice.

Equations like (2.1) can of course be used to find the ratio of the lattice parameters, i.e. the axial ratios, from measurements of the angles between poles.

2.6 TETRAGONAL SYSTEM

The tetrad axis is always taken parallel to the z axis in this system. The lattice parameters a and b are equal.

The holosymmetric point group is $4/mmm$, showing three mutually perpendicular mirror planes with a tetrad normal to one of them (Fig. 2.24). If a single pole is repeated according to the presence of these symmetry elements it will be found that diad axes are necessarily present normal to the mirrors and that, in addition, a second pair of diad axes also normal to mirror planes automatically arises. One of the pairs of mutually perpendicular diads is chosen as defining the directions of the x and y axes. The general form is $\{h\,k\,l\}$ with a multiplicity of 16. Special forms are $\{0\,0\,1\}$, $\{1\,0\,0\}$, $\{1\,1\,0\}$, $\{h\,k\,0\}$, $\{h\,0\,l\}$ and $\{h\,h\,l\}$. The last of these, viz. $\{h\,h\,l\}$, indicates a face making equal intercepts on the x and y axes.

The point group 422 could be specified simply as 42 since if 4 and 2 are present at right angles a second pair of diad axes arises and one of these pairs is chosen to define the x and y axes. The group $\bar{4}2m$ can be developed as $\bar{4}m$. It is then found that diad axes automatically arise at $45°$ to the two, mutually perpendicular, mirror planes. The pair of diad axes are taken to define the x and y axes. The other tetragonal point groups are straightforward and no diad axes arise.

The angles between poles on a stereogram and the ratio of the lattice parameters (i.e. the ratio a/c in this crystal system) are easily related by using equations like (2.1) with $a = b$.

2.7 CUBIC SYSTEM

Cubic crystals possess four triad axes. These are arranged as in Fig. 1.30 and are always taken to lie along the $\langle 1\,1\,1\rangle$ type directions of the unit cell, which is a cube, so $a = b = c$. This is the *only* crystal system in which the direction $[u\,v\,w]$ necessarily coincides with the normal to the plane $(u\,v\,w)$ for all u, v, w.

If we put four triad axes to coincide with the $\langle 1\,1\,1\rangle$ directions on a stereogram and allow these to operate on a single pole as in Fig. 2.27, we find that diad axes automatically arise along the crystal axes. The presence of the diads also follows from Table 1.2, with the row starting 233. The point group symbol used to describe the combination of a diad and two triads, shown in Table 1.2, is just 23. This is the cubic point group of lowest symmetry. The multiplicity of the general

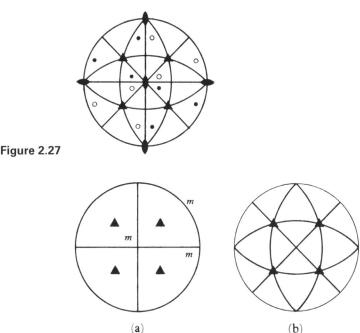

Figure 2.27

       (a)   (b)

Figure 2.28

form is 12. A feature of the symbols for describing point groups in the cubic system is that the symbol 3, even though it indicates the defining axis for the system, is never placed first because the triads are always at $54.73°(\cos^{-1}(1/\sqrt{3}))$ to the crystal axes. (In all other systems the defining axis comes first in the symbol — to distinguish between monoclinic and orthorhombic point groups the diad or inverse diads parallel to at least two axes must be stated.) The figure 3 always occurs second in the symbol for cubic point groups and this enables a point group in the cubic system to be distinguished from those in all other crystal systems. In a cubic crystal mirror planes can run either parallel to $\{1\,0\,0\}$ planes as in Fig. 2.28a or else parallel to $\{1\,1\,0\}$ planes as in Fig. 2.28b. The first alternative is described by putting the symbol m before the 3 (to give $m3$) and the second by placing m after the 3 to give $X3m$, where X is an axis other than 3.

If we add mirror planes parallel to $\{1\,0\,0\}$ to the class 23 we obtain $2/m3$, conventionally denoted $m3$. As Fig. 2.24 shows, this contains a centre of symmetry in addition to the three diads at the intersection of three mutually perpendicular mirror planes and the four triads. The triads therefore become inversion triads — $\bar{3}$. The multiplicity of the general form $\{h\,k\,l\}$ is 24. It is worth noting from Table 2.3 that when there are no diads along $\langle 110 \rangle$ nor mirrors parallel to $\{1\,1\,0\}$, $\{h\,k\,0\}$ and $\{k\,h\,0\}$ are separate special forms. This occurs in the classes 23 and $m3$.

Replacement of the diads in 23 by tetrads gives 43. Here we notice that diads automatically arise along the $\langle 1\,1\,0 \rangle$ directions. This class is denoted 432 to indicate the diads because of later development of space groups (see Section 2.14). However, 43 is sufficient to identify it.

Replacement of 2 by $\bar{4}$ in 23 will be found to automatically produce mirror planes parallel to the $\{1\,1\,0\}$ planes and hence passing through the triads. Correspondingly, if mirrors parallel to $\{1\,1\,0\}$ are added to 23 then the diad axes become $\bar{4}$ axes. If we have mirror planes passing through the triad axes, then parallel to the crystal axes we can have either $\bar{4}$ or 4. The first of these classes is $\bar{4}3m$ and the second $m3m$. In $\bar{4}3m$ there is no centre of symmetry and no additional symmetry elements, other than those indicated in the symbol. The multiplicity of the general form is 24. In $\bar{4}3m$ (as in 23) $\{1\,1\,1\}$ and $\{1\,\bar{1}\,1\}$ are separate special forms each shows four planes parallel to the surfaces of a regular tetrahedron.

Class $m3m$ has mirror planes parallel to $\{1\,0\,0\}$ and to $\{1\,1\,0\}$; thus nine are present in all. There are six diads, three tetrads, a centre and of course the four triads. All of these can be produced by putting mirrors parallel to both $\{1\,1\,0\}$ and $\{1\,0\,0\}$ coupled with the four triads. Hence the symbol $m3m$ is used to describe this point group which is the cubic holosymmetric class. The general form has 48 faces.

Figure 2.29 shows a stereogram of a cubic crystal with the poles of a number of faces indicated and the zones in which they lie. Table A3.1 in Appendix 3, gives values of the angles between a number of poles of different indices in the

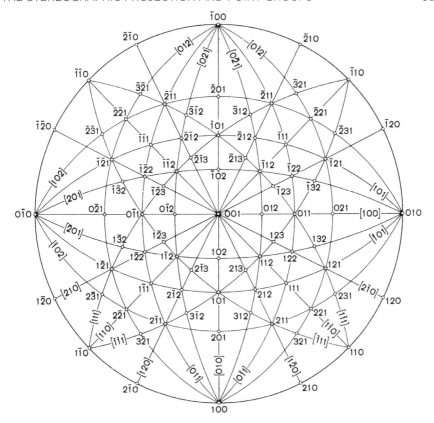

Figure 2.29 Stereogram of a cubic crystal. (From de Jong [3]. Copyright © 1959.)

cubic system. Additional poles would be easily located on a stereogram, such as that in Fig. 2.29, by use of the zone addition rule explained in Section 2.5.

2.8 HEXAGONAL SYSTEM

Crystals possessing a hexad axis have a Bravais lattice illustrated in Fig. 1.19j in Section 1.8. The x and y axes are at 120° to one another and perpendicular to the hexagonal axis along z. The holosymmetric class of this system, 6/*mmm*, possesses a hexad at the intersection of two sets of three vertical planes of symmetry, two sets of three diad axes normal to these, a plane of symmetry normal to the hexad axis and a centre of symmetry (Fig. 2.24). These symmetry elements are shown in Fig. 2.30a. Diad axes at 120° to one another are taken as the crystal axes. If this is done the indices of a number of faces are as

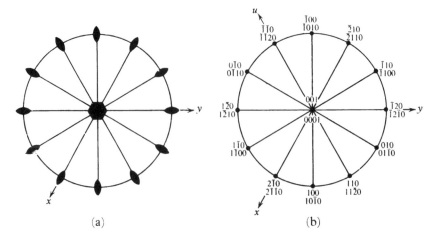

Figure 2.30

marked in Fig. 2.30b. The plane of index $(1\,0\,0)$ is repeated by the hexad axis to give $(0\,1\,0)$, $(\bar{1}\,1\,0)$, $(\bar{1}\,0\,0)$, $(0\,\bar{1}\,0)$, $(1\,\bar{1}\,0)$. All of these are identical crystallographic planes and yet their Miller indices appear quite dissimilar: note $(0\,1\,0)$ and $(\bar{1}\,1\,0)$. To avoid the possibility of confusion because planes of the same form have quite different indices, it is customary in the hexagonal system to employ Miller–Bravais indices. To do this we choose a third crystal axis u normal to the hexagonal axis and at $120°$ to both the x and y axes. The lattice repeat distance u along $\mathbf{u}$ is equal to $a(= b)$ from Fig. 1.19j in Section 1.8. To state the Miller–Bravais indices of a plane we then take the intercepts along all three axes, x, y and u, express these in terms of the lattice parameters and proceed exactly as in Section 1.2. The result is that a plane always has four indices $(hkil)$. It is obvious that there is a necessary relationship between h and k and i. This can be deduced from Fig. 2.31 and is

$$i = -(h + k)$$

i.e. the third index is always the negative of the sum of the first two. The indices of a number of poles are given in both Miller–Bravais and Miller indices in Fig. 2.30b. The hexagonal symmetry is then apparent in the former from the indices of planes of the same form.

A Miller–Bravais system is also used to specify directions in hexagonal crystals so that the index appears as $[u\,v\,t\,w]$ instead of $[u\,v\,w]$. When this is done a little care is needed in relating the results to those obtained using a three-index notation. So that directions of a given family should always have indices of similar appearance, directions are specified by taking steps along all three axes and arranging that the step along the u axis is of such length that the number of unit repeat vectors moved along this direction, i.e. t, is equal to the negative of

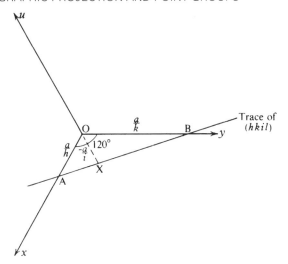

Figure 2.31

$(u + v)$. The direction corresponding to the x axis in Fig. 2.32 is then $[2\,\overline{1}\,\overline{1}\,0]$ in the four-index notation and $[1\,0\,0]$ in the three-index system. It should be noted that the Miller–Bravais index cannot be immediately written down from the first two indices of the three-index triplet, as is the case for a plane. Some indices of direction specified in both ways are given in Fig. 2.32. In relating planes and zone axes using Eqn (1.11) it is usually best to work entirely in the three-index notation for both planes and directions and to translate the three-index notation for a direction into the four-index system at the end of the calculation. The four-index notation both for planes and for directions is very widely used in dealing with both hexagonal and trigonal crystals.

Other point groups besides $6/mmm$ in the hexagonal system are shown in Fig. 2.24. We note that $\overline{6}\,(\equiv 3/m)$ is placed in this system because of the use of rotoinversion axes to describe symmetry operations of the second sort; 6, $\overline{6}$, $6/m$ and $6mm$ show no diad axes. The crystal axes for $6mm$ are usually chosen to be perpendicular to one set of mirrors (they then lie in the other set) and $\overline{6}m2$ could be developed as $\overline{6}m\,(\equiv 3/mm)$. The diads automatically arise and are chosen as crystallographic axes. Of course, 622 contains diads. It could be developed as 62 since the second set of diads arises automatically. The axes are chosen parallel to one set of diads. Only $6/m$ and $6/mmm$ are centrosymmetric in this system.

In the hexagonal system stereograms are most easily plotted and the angles between poles of planes are related to the ratio of the lattice parameters (a/c) by finding the angle between faces such as $(0\,0\,0\,1)$ and $(h\,h\,2\overline{h}\,l)$, e.g. $(1\,1\,\overline{2}\,1)$. $(h\,h\,2\overline{h}\,l)$ is chosen because it is equally inclined to the x and y axes. For example, from Fig. 2.33 the angle θ between the $(0\,0\,0\,1)$ pole and the $(h\,h\,2\overline{h}\,l)$ pole is

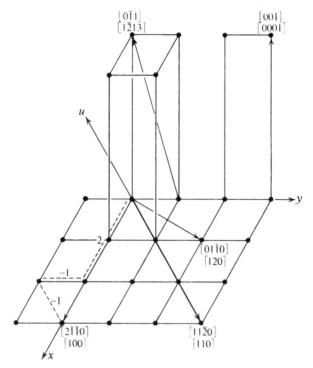

Figure 2.32

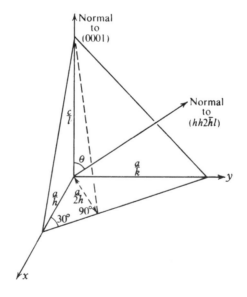

Figure 2.33

seen to be given by

$$\tan\theta = \frac{c}{a}\frac{2h}{l}$$

Sometimes $(h\,0\,\bar{h}\,l)$ is used, e.g. $(1\,0\,\bar{1}\,1)$. The angle between $(0\,0\,0\,1)$ and $(h\,0\,\bar{h}\,l)$ is $\tan^{-1}[(c/a)(2/\sqrt{3})(h/l)]$. The special forms in the various classes are listed later in Table 2.3.

2.9 TRIGONAL SYSTEM (RHOMBOHEDRAL SYSTEM)

This crystal system is defined by the possession of a single triad axis. It is closely related to the hexagonal system. The possession of a single triad axis by a crystal does not, by itself, indicate whether the *lattice* considered as a set of points is truly hexagonal, or whether it is based on the staggered stacking of triequiangular nets. In either case one can use a cell of the shape of Fig. 1.19k (if the lattice is hexagonal this will then of course not be primitive). The cell in Fig. 1.19k is a rhombohedron and the angle α ($<120°$) is characteristic of the substance. The symmetry elements in the holosymmetric class $\bar{3}m$ are shown in Fig. 2.24 and the repetition of a single pole in accordance with this symmetry is also demonstrated. Three diad axes arise automatically from the presence of $\bar{3}$ and the three mirrors lying parallel to $\bar{3}$. These diad axes, which intersect in the inverse triad axis, do not lie in the mirror planes. If the rhombohedral cell is used for such a crystal then the axes cannot be chosen parallel to prominent axes of symmetry. A stereogram of a trigonal crystal indexed according to a rhombohedral unit cell is shown in Fig. 2.34. The value of α is 98°. The x, y and z axes are taken to lie in the mirror planes and the inverse triad is a body diagonal of the cell therefore lying along the direction $[1\,1\,1]$. It is clear that the x, y and z axes, i.e. directions $[1\,0\,0]$, $[0\,1\,0]$ and $[0\,0\,1]$, do not lie normal to the $(1\,0\,0)$, $(0\,1\,0)$ and $(0\,0\,1)$ planes respectively. However, these directions are easily located since the z axis $[0\,0\,1]$ is the pole of the zone containing $(1\,0\,0)$ and $(0\,1\,0)$, the y axis is the pole of the zone containing $(1\,0\,0)$ and $(0\,0\,1)$ and the x axis is the pole of the zone containing $(0\,0\,1)$ and $(0\,1\,0)$. In plotting a stereogram of a trigonal crystal from a given value of the angle α it is worth noting that the angle γ between any of the crystal axes and the unique triad axis is given by

$$\sin\gamma = \frac{2}{\sqrt{3}}\sin\alpha/2 \tag{2.5}$$

(see problem 2.7).

For many purposes it is more convenient to use a hexagonal cell when dealing with trigonal crystals. The shape of the cell chosen is the same as for hexagonal crystals (Fig. 1.19j) and, since chosen without reference to the lattice, may or may not be primitive. The value a/c is characteristic of the substance. A stereogram of a crystal of the point group $\bar{3}m$ with planes indexed according to the

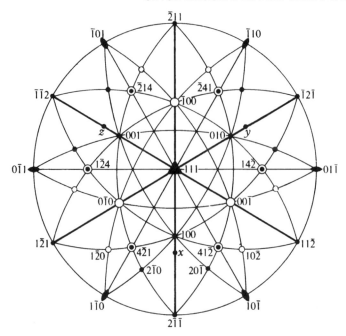

Figure 2.34 A stereogram of a trigonal crystal of class $\bar{3}m$. When poles in the upper and lower hemisphere coincide in projection, the indices refer to the poles in the upper hemisphere

Miller–Bravais scheme is shown in Fig. 2.35. This is the same crystal as that in Fig. 2.34 ($c/a = 1.02$). When using the hexagonal cell the x, y and u axes are chosen parallel to the diads. The relationship between the indices in the two stereograms can be easily worked out using the zone addition rule and Eqn (1.11) from the orientation relationship of the two cells. In Figs. 2.34 and 2.35 this is $(0\,0\,0\,1)\|(1\,1\,1)^\dagger$ and $(1\,0\,\bar{1}\,1)\|(1\,0\,0)$. The plane $(1\,0\,\bar{1}\,0)$ in Fig. 2.35 then has indices $(2\,\bar{1}\,\bar{1})$ in Fig. 2.34. The indices $(2\,\bar{1}\,\bar{1})$ would be deduced by noting that $(2\,\bar{1}\,\bar{1})$ is parallel to the $[1\,1\,1]$ direction, is equally inclined to the y and z axes and lies in the zone containing $(1\,1\,1)$ and $(1\,0\,0)$. The plotting of a stereogram and the determination of axial ratios for a trigonal crystal referred to hexagonal axes proceeds as for the hexagonal system.[‡]

In the class $\bar{3}m$ special forms lie normal to the triad $\{0\,0\,0\,1\}$, parallel to the triad $\{h\,k\,i\,0\}$, normal to mirror planes $\{h\,0\,\bar{h}\,l\}$ and equally inclined to two diads

---

[†] The symbol $\|$ means 'parallel to'.

[‡] The relationship between the two cells used in Figs. 2.34 and 2.35 is shown in Fig. A4.2a in Appendix 4, where general methods of transforming the indices of planes and directions, when different unit cells are chosen for the same crystal, are dealt with.

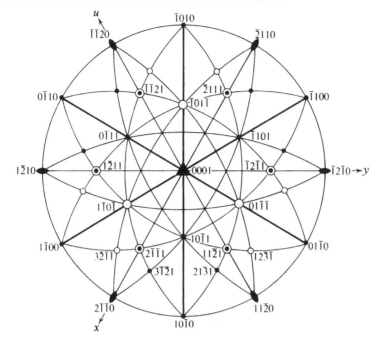

Figure 2.35 The same crystal as in Fig. 2.34 indexed using a hexagonal cell. When poles in the upper and lower hemispheres superpose in projection, the indices refer to poles in the upper hemisphere

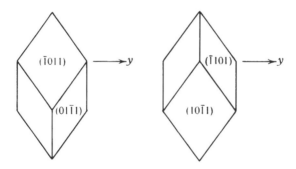

Figure 2.36

$\{h\,h\,2\,\overline{h}\,l\}$. The *six faces* in the form $\{h\,0\,\overline{h}\,l\}$ make a rhombohedron; $\{1\,0\,\overline{1}\,2\}$ would be an example. This form is similar in appearance to $\{0\,k\,\overline{k}\,l\}$ which is also a rhombohedron, rotated 60° with respect to the first one. Figure 2.36 shows the relationship between the two. They are quite separate forms and each is special. We therefore add $\{0\,k\,\overline{k}\,l\}$ to the list of special forms in addition to $\{h\,0\,\overline{h}\,l\}$.

Figure 2.34 shows that using the rhombohedral cell the two forms $\{1\,0\,\bar{1}\,1\}$ and $\{0\,1\,\bar{1}\,1\}$ have different indices since the face above $(0\,0\,\bar{1})$ in the projection in Fig. 2.34 would have indices $(2\,2\,\bar{1})$.

The other trigonal point groups are shown in Fig. 2.24. In 3 and in $\bar{3}$ there are neither mirrors nor diads. The class $3m$ has three mirrors intersecting in the triad axis; the x, y and u axes of the hexagonal cell are taken to lie perpendicular to the mirror planes. 32 contains diads normal to 3, which are taken as the crystallographic axes. If three mirror planes intersect in an inversion triad as in $\bar{3}m$ then diads automatically arise normal to the mirror planes; the diads are chosen as the u, x and y axes. The class $3/m$ is placed in the hexagonal system because it is equivalent to $\bar{6}$.

2.10 MONOCLINIC SYSTEM

Crystals in the monoclinic system possess a single twofold axis. Since a mirror plane is equivalent to an inverse diad the class $m(\equiv \bar{2})$ is put into this system. As Fig. 2.24 shows, the monoclinic point groups can be derived systematically in more than one way (e.g. we can take a diad normal to a monad or we can take

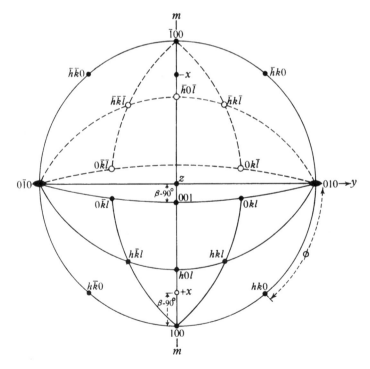

Figure 2.37

the diad as the defining axis x). We shall describe the point groups taking the diad axis along the y axis to correspond with Fig. 1.19b and c and the second setting in Fig. 2.24. The sides of the unit cell are in general all unequal to one another; $\alpha = \gamma = 90°$ and β is chosen to be obtuse. Figure 2.37 shows the symmetry elements in the holosymmetric point group $2/m$. The pole of $(0\,1\,0)$ and the y axis coincide on the stereogram. $[0\,0\,1]$, i.e. the z axis, is at the centre of the primitive and so $(1\,0\,0)$ lies on the primitive $90°$ from $(0\,1\,0)$. The x axis, i.e. the direction $[1\,0\,0]$, is in the lower hemisphere. β is the angle between $[0\,0\,1]$ and $[1\,0\,0]$ and the angle between $[0\,0\,1]$ and $(0\,0\,1)$ is $(\beta - 90°)$, which is of course equal to the angle between $(1\,0\,0)$ and $[1\,0\,0]$. Poles such as $\{h\,k\,0\}$ lie around the primitive. A pole such as $(h\,k\,0)$ lies at angle ϕ to $(0\,1\,0)$, given by

$$\cot\varphi = \frac{(a/h)\sin\beta}{(b/k)}$$

(see Fig. 2.38), the factor $\sin\beta$ arising because the normals to $(h\,k\,0)$ and $(0\,1\,0)$ and the x axis do not lie in the same plane. In the point group $2/m$ the only special

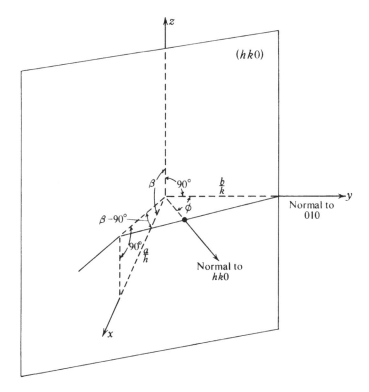

Figure 2.38 The angle ϕ in Fig. 2.37 can be derived from this figure and the given indices of the plane $(h\,k\,0)$

form besides $\{0\,1\,0\}$ is $\{h\,0\,l\}$, with a multiplicity of two. The two faces in this latter form are parallel, both lying normal to the mirror plane and parallel to the diad axis. The general form is $\{h\,k\,l\}$ with a multiplicity of four (see Fig. 2.37). $\{h\,k\,0\}$ and $\{0\,k\,l\}$ are also general forms.

The other point groups in the monoclinic system are 2 and m. Again, $\{h\,0\,l\}$ is a special form in both and so is $\{0\,1\,0\}$. However, 2 does not possess a centre and so $\{0\,1\,0\}$ and $\{0\,\bar{1}\,0\}$ must each be listed as separate special forms.

2.11 TRICLINIC SYSTEM

There are no special forms in either of the point groups in this system. The unit cell is a general parallelepiped. The drawing of a stereogram can be carried out with the aid of Figs. 2.39a and b. This diagram is completely general and can be specialized to apply to any crystal system more symmetric than the triclinic by setting one or more of the axial angles to particular values and by setting two or more of the lattice parameters to be equal. Figure 2.39b shows the plane of indices $(h\,k\,l)$ and the angles marked in Fig. 2.39b are easily related to those marked in Fig. 2.39a by noting that ϕ_1 is a plane angle of $(0\,0\,1)$ and is the angle between the zone containing $(h\,k\,l)$ and $(0\,0\,1)$ and the zone containing $(0\,0\,1)$ and $(1\,0\,0)$. We can therefore mark ϕ_1 on the stereogram. Similarly, ϕ_5 is a plane angle of $(1\,0\,0)$ and is the angle between the zone containing $(h\,k\,l)$ and $(1\,0\,0)$ and that containing $(1\,0\,0)$ and $(0\,1\,0)$; ϕ_5 can therefore be marked on the stereogram. Proceeding in this way we can mark in all of the angles ϕ_1 to ϕ_6. We have, from Fig. 2.39a,

$$\alpha = 180° - (\phi_5 + \phi_6)$$

$$\beta = 180° - (\phi_3 + \phi_4)$$

$$\gamma = 180° - (\phi_1 + \phi_2)$$

From the plane triangle of the $(0\,0\,1)$ face in Fig. 2.39b we have

$$\frac{a}{h}\sin\phi_2 = \frac{b}{k}\sin\phi_1$$

Therefore,

$$\frac{a/h}{b/k} = \frac{\sin\phi_1}{\sin\phi_2} \tag{2.6a}$$

and similarly

$$\frac{c/l}{b/k} = \frac{\sin\phi_6}{\sin\phi_5} \tag{2.6b}$$

and

$$\frac{a/h}{c/l} = \frac{\sin\phi_4}{\sin\phi_3} \tag{2.6c}$$

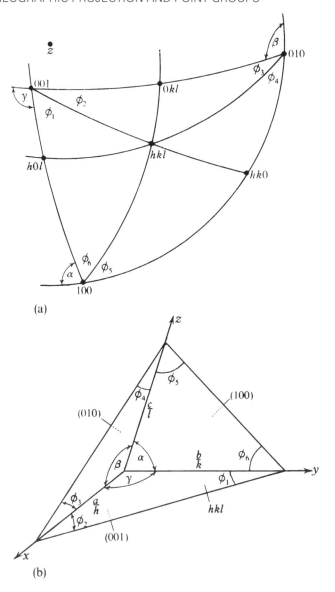

Figure 2.39

These relations plus the zone addition rule, explained in Section 2.5, allow additional faces to be inserted and hence a complete stereogram to be drawn. Equations (2.6) are also of use in finding axial ratios and axial angles from measured angles between planes.

2.12 ENANTIOMORPHOUS CRYSTAL CLASSES

In some crystal classes the possibility exists of crystals being found in either a right-handed or left-handed modification with the two not being superposable in the sense that the right hand of the body may not be superposed upon the left hand. This occurs if the crystal possesses no symmetry operation of the second kind, i.e. no mirror plane nor rotoinversion axis of any degree appears in the point group. There are eleven such classes, viz. 1, 2, 3, 4, 6, 23, 222, 32, 422, 622 and 432.

2.13 LAUE GROUPS

Some physical methods of examination of crystals cannot determine whether or not there is a centre of symmetry present. The Laue group is the crystal class to which a given crystal belongs if a centre of symmetry be added to the symmetry elements already present. The eleven Laue groups are shown in Table 2.1.

2.14 SPACE GROUPS

We noted in Section 1.4 that repetition of an object (e.g. an atom or group of atoms) in a crystal is carried out by operations of rotation, reflection (or inversion) and translation. We have described the consistent combinations of rotations and inversions that can occur in crystals and in Section 1.8 we described the possible translations that can occur in crystals. We have so far made no attempt to combine the operations of rotation (and inversion) with translation,

Table 2.1 Laue groups

System	Class	Laue group
Cubic	432, $\bar{4}3m$, $m3m$, 23, $m3$	$m3m$ $m3$
Hexagonal	62, $6mm$, $\bar{6}m2$, $6/mmm$ 6, $\bar{6}$, $6/m$	$6/mmm$ $6/m$
Tetragonal	422, $4mm$, $\bar{4}m2$, $4/mmm$ 4, $\bar{4}$, $4/m$	$4/mmm$ $4/m$
Trigonal	32, $3m$, $\bar{3}m$ 3, $\bar{3}$	$\bar{3}m$ $\bar{3}$
Orthorhombic	222, $2mm$, mmm	mmm
Monoclinic	2, m, $2/m$	$2/m$
Triclinic	1, $\bar{1}$	$\bar{1}$

except briefly in Section 1.5. A full description of the symmetry of a crystal involves a description of the way in which all of the symmetry elements are distributed in space. This is called the *space* group. There are 230 different crystallographic space groups and each one gives the fullest description of the symmetry elements present in a crystal possessing that space group.

The rotation axes and rotoinversion axes possible in crystals were discussed in Sections 2.4 and the possible translations in Section 1.8. An enumeration of the way these axes can be consistently combined with the translation is an enumeration of the possible space groups. Space groups are most important in the solution of crystal structures. For a discussion of imperfections in crystals they have not to date been of great importance, so we give here just a short description of some features of space group theory because it is useful in the description of crystal structures.

When we attempt to combine the operations of rotation and translation the possibility arises of what is called a *screw axis* which involves repetition by rotation about an axis together with translation parallel to that axis. Similarly, a repetition by reflection in a mirror plane may be combined with a translational component parallel to that plane to produce a *glide plane*. If a twofold rotational axis (say) occurs in a crystal, then this means some structural unit or motif is arranged about this direction so that it is repeated by a rotation of 180° about the axis. The repetition shown in Fig. 2.40a corresponds to a pure rotational diad axis. However, the rotation of 180° could be coupled with a translation of one-half of the lattice repeat distance in the direction of the axis to give the screw diad axis shown in Fig. 2.40b, denoted by the symbol 2_1. The translation $t/2$ will be of length of the order of the lattice parameters of the crystal and hence of the order of a few angstrom units (10^{-10} m), and so quite undetected by the naked eye. The operation of glide plane repetition is shown in Fig. 2.40c. Again the magnitude of the translation will be of the order of the lattice parameters. Macroscopically the crystal containing a 2_1 axis would then show diad symmetry about that axis in the symmetry of its external faces or of its physical properties. Similarly, the glide plane in Fig. 2.40c would show itself macroscopically as a mirror plane.

Although the presence of a screw axis, say a screw diad as in Fig. 2.40b, indicates the presence of identical atomic motifs arranged about it so that they are related by a rotation plus translation, the screw axis must not be thought of as translating the translation vectors of the lattice. A screw axis and a pure rotation axis of the same order n repeat a translation in the same way. It follows that the rotational components of screw axes can only be $2\pi/1$, $2\pi/2$, $2\pi/3$, $2\pi/4$, $2\pi/6$ and that an n-fold screw and an n-fold pure rotation axis must have similar locations with respect to a similar set of translations. The angles between screws and between screws and rotations must therefore be the same as the permissible combinations listed in Table 1.2.

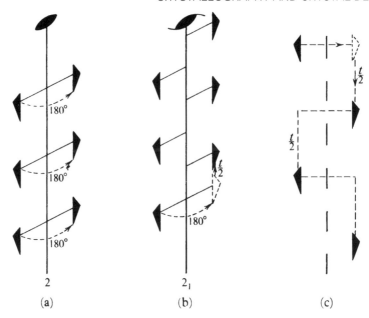

Figure 2.40

When we combine the operations of rotation with translation and look for the possible consistent combinations we can start with the point group and associate the symmetry elements of the point group with each lattice point of the lattices consistent with that point group. This will in fact give 72 space groups.[†] We then consider the possibility of introducing screw axes instead of pure rotational axes and of glide planes to replace mirror planes. The result of this is to produce in all 230 different space groups.

The various kinds of screw axis possible are shown in Table 2.2. An n_N-fold screw axis involves repetition by rotation through $360°/n$ with a translation of tN/n, where $\mathbf{t}$ is a lattice repeat vector parallel to the axis. There are five types of screw hexad; e.g. 6_1 involves rotation through $60°$ and translation of $\mathbf{t}/6$ whilst 6_5 involves rotation through $60°$ and translation of $5\mathbf{t}/6$. By drawing a diagram showing the repetition of objects by these axes it is easily seen that 6_1 is a screw of opposite hand to 6_5, 4_1 to 4_3 and 6_2 to 6_4. A diagram to show this for 3_1 and 3_2 is given in Fig. 2.41. A and A' are lattice points. The operation of 3_1 is straightforward, as shown in Fig. 2.41a. When repeating an object according

---

[†] From the number of Bravais lattices and number of point groups detailed in Fig. 1.19 and 2.24 one would expect to obtain 66 space groups in this way. Six additional ones arise because of cases in which the glide planes or screw axes *automatically* arising are different for different orientations of the point group with respect to the lattice; e.g. in the tetragonal system $I\bar{4}m2$ and $I\bar{4}2m$ are distinct space groups.

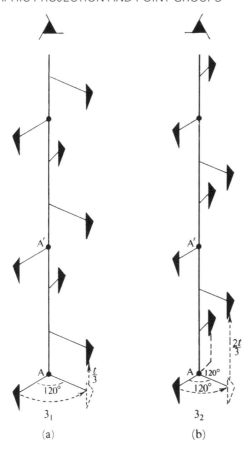

Figure 2.41

to 3_2 note that the object at height $\frac{4}{3}$ must occur also at height $\frac{1}{3}$, since A and A' are both lattice points and the pattern is infinitely long parallel to AA'. It should also be noted that the axes 4_2 and 6_3 include the pure rotation axes 2 and 3 respectively in the lattice. 6_2 and 6_4 also contain twofold rotation axes.

Glide planes are described as *axial glide planes* if the translation parallel to the mirror is parallel to a single axis of the unit cell and equal to one-half of the lattice parameter in that direction. Glide planes are given the symbols a, b or c corresponding to the directions of the glide translations. A *diagonal glide plane* involves a translation of one-half of a face diagonal or one-half of a body diagonal (the latter in the tetragonal and cubic systems), given the symbol n, and a *diamond glide plane* involves a translation of one-quarter of a face diagonal, given the symbol d. The diamond glide plane involves a translation of one-quarter of a body diagonal in the tetragonal and cubic systems.

Table 2.2 Screw axes in crystals

Name	Symbol	Graphical symbol	Right-handed screw translation along axis in units of the lattice parameter
Screw diad	2_1		$\frac{1}{2}$
Screw triads	3_1		$\frac{1}{3}$
	3_2		$\frac{2}{3}$
Screw tetrads	4_1		$\frac{1}{4}$
	4_2		$\frac{2}{4} = \frac{1}{2}$
	4_3		$\frac{3}{4}$
Screw hexads	6_1		$\frac{1}{6}$
	6_2		$\frac{2}{6} = \frac{1}{3}$
	6_3		$\frac{3}{6} = \frac{1}{2}$
	6_4		$\frac{4}{6} = \frac{2}{3}$
	6_5		$\frac{5}{6}$

The space group symbol shows that the space groups have been built up by placing a point group at each of the lattice points of the appropriate Bravais lattice. Thus *Fm3m* means the cubic face-centred lattice with the point group *m3m* associated with each lattice point and *P6₃/mmc* is a hexagonal primitive lattice derived from *P6/mmm* by replacing the sixfold rotation axis by 6_3 and one of the mirror planes by a *c* axis glide plane. The point group symmetry of *P6₃/mmc* and of *P6/mmm* would be the same. The point group symmetry of any crystal is derived immediately from the space group symbol by replacing screw axes by the appropriate rotational axes and glide planes by mirror planes in the space group symbol.

The arrangements of all of the symmetry elements in the 230 space groups are listed in a beautifully clear fashion in the *International Tables for X-Ray Crystallography*. The seventeen two-dimensional space groups (or plane groups) are illustrated in Fig. 2.42 in the way they are depicted in the *International Tables for X-Ray Crystallography*. These will now be briefly described to illustrate the points made earlier in this section. If the plane groups are fully understood the three-dimensional space group tables in the *International Tables* can be understood without difficulty.

Space groups are obtained by the application of point group symmetry to finite lattices, the possibility of translation symmetry being taken into account. There are five lattices or nets in two dimensions, shown in Fig. 1.15. We shall give them the symbol *p* for primitive and *c* for centred. The symbols for twofold, threefold, fourfold and sixfold axes and for the mirror plane will be as in Section 2.4.

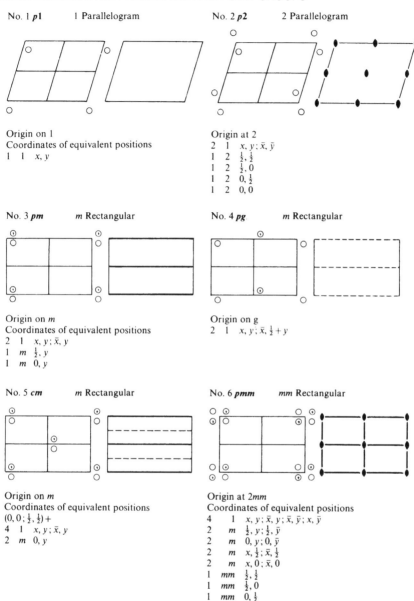

No. 1 *p*1   1 Parallelogram

Origin on 1
Coordinates of equivalent positions
1 1 *x, y*

No. 2 *p*2   2 Parallelogram

Origin at 2
2 1 *x, y*; *x̄, ȳ*
1 2 ½, ½
1 2 ½, 0
1 2 0, ½
1 2 0, 0

No. 3 *pm*   *m* Rectangular

Origin on *m*
Coordinates of equivalent positions
2 1 *x, y*; *x̄, y*
1 *m* ½, *y*
1 *m* 0, *y*

No. 4 *pg*   *m* Rectangular

Origin on g
2 1 *x, y*; *x̄*, ½ + *y*

No. 5 *cm*   *m* Rectangular

Origin on *m*
Coordinates of equivalent positions
(0, 0; ½, ½)+
4 1 *x, y*; *x̄, y*
2 *m* 0, *y*

No. 6 *pmm*   *mm* Rectangular

Origin at 2*mm*
Coordinates of equivalent positions
4   1   *x, y*; *x̄, y*; *x̄, ȳ*; *x, ȳ*
2   *m*   ½, *y*; ½, *ȳ*
2   *m*   0, *y*; 0, *ȳ*
2   *m*   *x*, ½; *x̄*, ½
2   *m*   *x*, 0; *x̄*, 0
1   *mm* ½, ½
1   *mm* ½, 0
1   *mm* 0, ½
1   *mm* 0, 0

Figure 2.42   The seventeen two-dimensional space groups arranged following the *International Tables for X-ray Crystallography* [2]. The headings for each figure read, from left to right, Number, Short Symbol (see Table 2.3), Point Group, Net. Below each figure the columns give the number of equivalent positions, the point group symmetry at that position and the coordinates of the equivalent positions. The coordinates of positions *x*, *y* are expressed in units equal to the cell edge length in these two directions

No. 7 *pmg*   *mm* Rectangular   No. 8 *pmg*   *mm* Rectangular

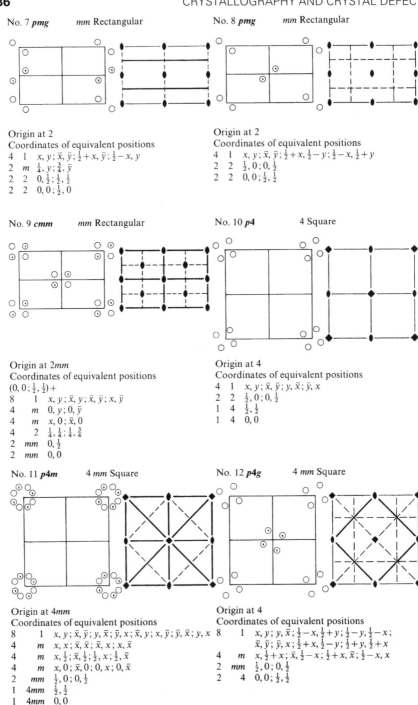

Origin at 2
Coordinates of equivalent positions
4 1 $x, y; \bar{x}, \bar{y}; \frac{1}{2}+x, \bar{y}; \frac{1}{2}-x, y$
2 m $\frac{1}{4}, y; \frac{3}{4}, \bar{y}$
2 2 $0, \frac{1}{2}; \frac{1}{2}, \frac{1}{2}$
2 2 $0, 0; \frac{1}{2}, 0$

Origin at 2
Coordinates of equivalent positions
4 1 $x, y; \bar{x}, \bar{y}; \frac{1}{2}+x, \frac{1}{2}-y; \frac{1}{2}-x, \frac{1}{2}+y$
2 2 $\frac{1}{2}, 0; 0, \frac{1}{2}$
2 2 $0, 0; \frac{1}{2}, \frac{1}{2}$

No. 9 *cmm*   *mm* Rectangular   No. 10 *p4*   4 Square

Origin at 2*mm*
Coordinates of equivalent positions
$(0, 0; \frac{1}{2}, \frac{1}{2})+$
8 1 $x, y; \bar{x}, y; \bar{x}, \bar{y}; x, \bar{y}$
4 m $0, y; 0, \bar{y}$
4 m $x, 0; \bar{x}, 0$
4 2 $\frac{1}{4}, \frac{1}{4}; \frac{1}{4}, \frac{3}{4}$
2 mm $0, \frac{1}{2}$
2 mm $0, 0$

Origin at 4
Coordinates of equivalent positions
4 1 $x, y; \bar{x}, \bar{y}; y, \bar{x}; \bar{y}, x$
2 2 $\frac{1}{2}, 0; 0, \frac{1}{2}$
1 4 $\frac{1}{2}, \frac{1}{2}$
1 4 $0, 0$

No. 11 *p4m*   4 *mm* Square   No. 12 *p4g*   4 *mm* Square

Origin at 4*mm*
Coordinates of equivalent positions
8 1 $x, y; \bar{x}, \bar{y}; y, \bar{x}; \bar{y}, x; \bar{x}, y; x, \bar{y}; \bar{y}, \bar{x}; y, x$
4 m $x, x; \bar{x}, \bar{x}; \bar{x}, x; x, \bar{x}$
4 m $x, \frac{1}{2}; \bar{x}, \frac{1}{2}; \frac{1}{2}, x; \frac{1}{2}, \bar{x}$
4 m $x, 0; \bar{x}, 0; 0, x; 0, \bar{x}$
2 mm $\frac{1}{2}, 0; 0, \frac{1}{2}$
1 4mm $\frac{1}{2}, \frac{1}{2}$
1 4mm $0, 0$

Origin at 4
Coordinates of equivalent positions
8 1 $x, y; y, \bar{x}; \frac{1}{2}-x, \frac{1}{2}+y; \frac{1}{2}-y, \frac{1}{2}-x;$
       $\bar{x}, \bar{y}; \bar{y}, x; \frac{1}{2}+x, \frac{1}{2}-y; \frac{1}{2}+y, \frac{1}{2}+x$
4 m $x, \frac{1}{2}+x; \bar{x}, \frac{1}{2}-x; \frac{1}{2}+x, \bar{x}; \frac{1}{2}-x, x$
2 mm $\frac{1}{2}, 0; 0, \frac{1}{2}$
2 4 $0, 0; \frac{1}{2}, \frac{1}{2}$

Figure 2.42 *(continued)*

No. 13 **p3**   3 Triequiangular   No. 14 **p3m1**   3 *m* Triequiangular

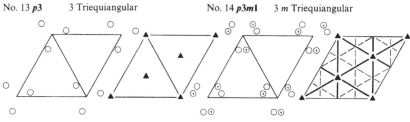

Origin at 3
Coordinates of equivalent positions

3	1	$x, y; \bar{y}, x-y; y-x, \bar{x}$
1	3	$\frac{2}{3}, \frac{1}{3}$
1	3	$\frac{1}{3}, \frac{2}{3}$
1	3	$0, 0$

Origin at 3*m*
Coordinates of equivalent positions

6	1	$x, y; \bar{y}, x-y; y-x, \bar{x};$
		$x, x-y; y-x, y; \bar{y}, \bar{x}$
3	*m*	$x, \bar{x}; x, 2x; 2\bar{x}, \bar{x}$
1	3*m*	$\frac{2}{3}, \frac{1}{3}$
1	3*m*	$\frac{1}{3}, \frac{2}{3}$
1	3*m*	$0, 0$

No. 15 **p31m**   3 *m* Triequiangular   No. 16 **p6**   6 Triequiangular

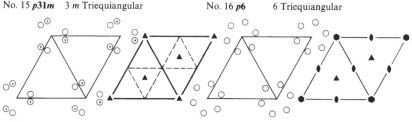

Origin at 31*m*
Coordinates of equivalent positions

6	1	$x, y; \bar{y}, x-y; y-x, \bar{x};$
		$y, x; \bar{x}, y-x; x-y, \bar{y}$
3	*m*	$x, 0; 0, x; \bar{x}, \bar{x}$
2	3	$\frac{1}{3}, \frac{2}{3}; \frac{2}{3}, \frac{1}{3}$
1	3*m*	$0, 0$

Origin at 6
Coordinates of equivalent positions

6	1	$x, y; \bar{y}, x-y; y-x, \bar{x};$
		$\bar{x}, \bar{y}; y, y-x; x-y, x$
3	2	$\frac{1}{2}, 0; 0, \frac{1}{2}; \frac{1}{2}, \frac{1}{2}$
2	3	$\frac{1}{3}, \frac{2}{3}; \frac{2}{3}, \frac{1}{3}$
1	6	$0, 0$

No. 17 **p6m**   6 *mm* Triequiangular

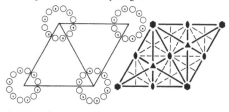

Origin at 6*mm*
Coordinates of equivalent positions

12	1	$x, y; \bar{y}, x-y; y-x, \bar{x}; y, x; \bar{x}, y-x;$
		$x-y, \bar{y}; \bar{x}, \bar{y}; y, y-x; x-y, x;$
		$\bar{y}, \bar{x}; x, x-y; y-x, y$
6	*m*	$x, \bar{x}; x, 2x; 2\bar{x}, \bar{x}; \bar{x}, x; \bar{x}, 2\bar{x}; 2x, x$
6	*m*	$x, 0; 0, x; \bar{x}, \bar{x}; \bar{x}, 0; 0, \bar{x}; x, x$
3	*mm*	$\frac{1}{2}, 0; 0, \frac{1}{2}; \frac{1}{2}, \frac{1}{2}$
2	3*m*	$\frac{1}{3}, \frac{2}{3}; \frac{2}{3}, \frac{1}{3}$
1	6*mm*	$0, 0$

Figure 2.42   (*continued*)

Table 2.3 Special forms in the crystal classes. The orientation of the axes is that used in this chapter. Monoclinic crystals are in the second setting of Fig. 2.24 and trigonal crystals are referred to hexagonal axes. There are no special forms in the triclinic system

2 $\{0\,1\,0\}$, $\{0\,\bar{1}\,0\}$, $\{h\,0\,l\}$	**32** $\{0\,0\,0\,1\}$, $\{1\,0\,\bar{1}\,0\}$, $\{2\,\bar{1}\,\bar{1}\,0\}$, $\{1\,1\,\bar{2}\,0\}$, $\{h\,k\,i\,0\}$, $\{h\,0\,\bar{h}\,l\}$, $\{0\,k\,\bar{k}\,l\}$, $\{h\,h\,2\,\bar{h}\,l\}$, $\{2\,h\,\bar{h}\,\bar{h}\,l\}$	**6mm** $\{0\,0\,0\,1\}$, $\{0\,0\,0\,\bar{1}\}$, $\{1\,0\,\bar{1}\,0\}$, $\{1\,1\,\bar{2}\,0\}$, $\{h\,k\,i\,0\}$, $\{h\,0\,\bar{h}\,l\}$, $\{h\,h\,2\,\bar{h}\,l\}$
m and **2/m** $\{0\,1\,0\}$, $\{h\,0\,l\}$	**4** $\{0\,0\,1\}$, $\{0\,0\,\bar{1}\}$, $\{h\,k0\}$	**$\bar{6}$m2** $\{0\,0\,0\,1\}$, $\{1\,0\,\bar{1}\,0\}$, $\{0\,1\,\bar{1}\,0\}$, $\{1\,1\,\bar{2}\,0\}$, $\{h\,k\,i\,0\}$, $\{h\,0\,\bar{h}\,l\}$, $\{0\,k\,\bar{k}\,l\}$, $\{h\,h\,2\,\bar{h}\,l\}$
mm2 $\{0\,0\,1\}$, $\{0\,0\,\bar{1}\}$, $\{1\,0\,0\}$, $\{0\,1\,0\}$, $\{h\,k\,0\}$, $\{h\,0\,l\}$, $\{0\,k\,l\}$	**$\bar{4}$** and **4/m** $\{0\,0\,1\}$, $\{h\,k\,0\}$	**622** and **6/mmm** $\{0\,0\,0\,1\}$, $\{1\,0\,\bar{1}\,0\}$, $\{1\,1\,\bar{2}\,0\}$, $\{h\,k\,i\,0\}$, $\{h\,0\,\bar{h}\,l\}$, $\{h\,h\,2\,\bar{h}\,l\}$
222 and **mmm** $\{1\,0\,0\}$, $\{0\,1\,0\}$, $\{0\,0\,1\}$, $\{h\,k\,0\}$, $\{h\,0\,l\}$, $\{0\,k\,l\}$	**4mm** $\{0\,0\,1\}$, $\{0\,0\,\bar{1}\}$, $\{1\,0\,0\}$, $\{1\,1\,0\}$, $\{h\,k\,0\}$, $\{h\,0\,l\}$, $\{h\,h\,l\}$	**23** $\{1\,0\,0\}$, $\{1\,1\,0\}$, $\{h\,k\,0\}$, $\{k\,h\,0\}$, $\{1\,1\,1\}$, $\{1\,\bar{1}\,1\}$, $\{h\,l\,l\}$, $\{h\,\bar{l}\,l\}$, $\{h\,h\,l\}$, $\{h\,\bar{h}\,l\}$
3 $\{0\,0\,0\,1\}$, $\{0\,0\,0\,\bar{1}\}$, $\{h\,k\,i\,0\}$	**$\bar{4}$2m** $\{0\,0\,1\}$, $\{1\,0\,0\}$, $\{1\,1\,0\}$, $\{h\,k\,0\}$, $\{h\,0\,l\}$, $\{h\,h\,l\}$, $\{h\,\bar{h}\,l\}$	**m3** $\{1\,0\,0\}$, $\{1\,1\,0\}$, $\{h\,k\,0\}$, $\{k\,h\,0\}$, $\{1\,1\,1\}$, $\{h\,l\,l\}$, $\{h\,h\,l\}$
$\bar{3}$ $\{0\,0\,0\,1\}$, $\{h\,k\,i\,0\}$	**42** and **4/mmm** $\{0\,0\,1\}$, $\{1\,0\,0\}$, $\{1\,1\,0\}$, $\{h\,k\,0\}$, $\{h\,0\,l\}$, $\{h\,h\,l\}$	**$\bar{4}$3m** $\{1\,0\,0\}$, $\{1\,1\,0\}$, $\{h\,k\,0\}$, $\{1\,1\,1\}$, $\{1\,\bar{1}\,1\}$, $\{h\,l\,l\}$, $\{h\,l\,l\}$, $\{h\,h\,l\}$, $\{h\,\bar{h}\,l\}$
3m $\{0\,0\,0\,1\}$, $\{0\,0\,0\,\bar{1}\}$, $\{1\,0\,\bar{1}\,0\}$, $\{0\,1\,\bar{1}\,0\}$, $\{1\,1\,\bar{2}\,0\}$, $\{h\,k\,i\,0\}$, $\{h\,0\,\bar{h}\,l\}$, $\{0\,k\,\bar{k}\,l\}$, $\{h\,h\,2\,\bar{h}\,l\}$	**6** $\{0\,0\,0\,1\}$, $\{0\,0\,0\,\bar{1}\}$, $\{h\,k\,i\,0\}$	**432** $\{1\,0\,0\}$, $\{1\,1\,0\}$, $\{h\,k\,0\}$, $\{1\,1\,1\}$, $\{h\,l\,l\}$, $\{h\,h\,l\}$
$\bar{3}$m $\{0\,0\,0\,1\}$, $\{1\,0\,\bar{1}\,0\}$, $\{1\,1\,\bar{2}\,0\}$, $\{h\,k\,i\,0\}$, $\{h\,0\,\bar{h}\,l\}$, $\{0\,k\,\bar{k}\,l\}$, $\{h\,h\,2\,\bar{h}\,l\}$	**$\bar{6}$** and **6/m** $\{0\,0\,0\,1\}$, $\{h\,k\,i\,0\}$	**m3m** $\{1\,0\,0\}$, $\{1\,1\,0\}$, $\{h\,k\,0\}$, $\{1\,1\,1\}$, $\{h\,l\,l\}$, $\{h\,h\,l\}$

The only additional symmetry element in two dimensions is the glide reflection line — symbol g and denoted by $---$ in Fig. 2.42. It involves reflection and a translation of one-half of the repeat distance parallel to the line.

The two-dimensional lattices and the two-dimensional point groups are combined in Table 2.4 to show the space groups which can arise. The space groups are shown in Fig. 2.42. In all the diagrams the x axis is down the page and the y axis runs across the page, the positive direction being towards the right. In each of the diagrams the left-hand one shows the equivalent general positions of the space group, i.e. the complete set of positions produced by the operation of the symmetry elements of the space group upon one initial position chosen at random. The total number of general positions is the number falling within the cell, but surrounding positions are also shown to illustrate the symmetry. The right-hand diagram is that of the group of spatially distributed symmetry operators, i.e. the true space (plane) group.

Below each of the diagrams in Fig. 2.42, besides the equivalent general positions special positions are also indicated. These are positions located on some symmetry operator so that repetition of an initial point produces fewer equivalent positions than in the general case. The symmetry at each special position is also given.

Table 2.4   Two-dimensional lattices, point groups and space groups

System and lattice symbol	Point group	Space group symbols Full	Short	Space group number
Parallelogram	1	$p1$	$p1$	1
p (primitive)	2	$p211$	$p2$	2
Rectangular	m	$p1m1$	pm	3
p and c (centred)		$p1g1$	pg	4
		$c1m1$	cm	5
	$2mm$	$p2mm$	pmm	6
		$p2mg$	pmg	7
		$p2gg$	pgg	8
		$c2mm$	cmm	9
Square p	4	$p4$	$p4$	10
	$4mm$	$p4mm$	$p4m$	11
		$p4gm$	$p4g$	12
Triequiangular	3	$p3$	$p3$	13
(hexagonal) p	$3m$	$p3m1$	$p3m1$	14
		$p31m$	$p31m$	15
	6	$p6$	$p6$	16
	$6mm$	$p6mm$	$p6m$	17

Note. The two distinct space groups $p3m1$ and $p31m$ correspond to different orientations of the point group relative to the lattice. This does not lead to distinct groups in any other case.

The group $p1$ is obtained by combining the parallelogram net and a onefold axis of rotational symmetry. There are no special positions in the cell. The group $p2$ arises by combining the parallelogram net and a diad. A mirror plane *requires* the rectangular net (Section 1.5) and if this net is combined with a *single* mirror the space group pm, No. 3, results. Points ◯ and ⊙ are in mirror relationship to one another. If the mirror in pm is replaced by a glide reflection line the space group pg results. In No. 4 the glide reflection line runs normal to the x axis. The centred rectangular lattice necessarily shows a glide reflection line as in cm, No. 5, but only one of these need be present corresponding to the plane point group m. Since, in this case, the net is multiply primitive the motif associated with the lattice point at $(\frac{1}{2}, \frac{1}{2})$ necessarily arises from that at $(0, 0)$. Hence the coordinates of equivalent positions if added to $(\frac{1}{2}, \frac{1}{2})$ give additional equivalent positions. If two mirror planes at right angles (point group mm) are combined with the rectangular lattice we get diads at the intersections of the mirrors as in pmm, No. 6. If one of the mirrors is replaced by g the diads lie half-way between the intersections (see pmg, No. 7). The group cmm necessarily involves the presence of two sets of glide reflection lines whilst $p4$ denotes the square lattice and point group 4, which together necessarily involve the presence of diads (No. 10). However, mirror planes are not *required*. If 4 lies at the intersection of two sets of mirrors we have $p4m$, the diagonal glide reflection line necessarily being present. However, 4 can also lie at the intersection of two sets of glide reflection lines, in which case again two sets of mirrors arise but the mirrors intersect in diads, giving point group symmetry mm at these points (No. 12). The triequiangular net and the point group 3 give the space group $p3$. If mirror planes are combined with the triad axis, i.e. the combination of the point group $3m$ and the triequiangular net, it is found that the mirrors can be arranged in two different ways with respect to the points of the net, yielding p31m and $p3m$1. With the hexagonal point group $6mm$, which necessarily has two sets of mirrors, this duality does not arise and the two space groups are $p6$ and $p6mm$.

PROBLEMS

For problems following 2.6, material in Sections A1.1 to A1.3 in Appendix 1 may be necessary and in problem 2.17 the material in Appendix 4 is required.

2.1 (a) Make a stereographic projection of the earth (of the same radius as in Fig. 2.13b) with the North Pole at the centre and the equator coincident with the primitive. Insert on the projection the meridians of longitude at $20°$ intervals and the lines of latitude $20°, 40°, 60°, 80°$, North and South. (Note that the lines of longitude are at $20°, 40°$, etc., from Greenwich meridian $(0°)$ but that the lines of latitude are at $70°, 50°, \cdots$ from the North Pole.)

   (b) Mark on your projection the poles of the following places: New Orleans $(30°N, 90°W)$ and Calcutta $(22°N, 90°E)$

Quito (0°N, 79°W) and Delhi (29°N, 77°E)

Perth (32°S, 116°E) and Peking (40°N, 116°E)

(Note that the pole of Perth, being in the Southern Hemisphere, will project outside the primitive. Alternatively, it can be projected from the North Pole so as to lie within the primitive.)

(c) Draw on the stereographic projection the great circles representing the shortest air routes between these pairs of places.

2.2 (a) Make a stereographic projection of the earth with the point where the Greenwich meridian crosses the equator at the centre. Insert on the projection the same lines of latitude and longitude as those in problem 2.1.

(b) Mark on your projection the poles of the same places as in 2.1. (Note that the poles of Perth and Peking will project outside the primitive.)

(c) Draw on your stereographic projection the great circles representing the shortest air routes between these pairs of places and compare with the corresponding great circles on the polar projection made in problem 2.1.

(d) Locate the pole of the great circle from Quito to Delhi and hence measure the arc of the great circle between these places.
(i) Graphically and (ii) by using a stereographic net, calculate the distance between Quito and Delhi, assuming the radius of the earth to be 4000 miles.

(e) Determine the bearing on which an aircraft must leave Delhi to fly on the great circle route to Quito (i.e. the angle between the great circle route and the meridian through Delhi) by (i) measuring the angle between the tangents to the two great circles at Delhi and (ii) measuring the angle between the poles of these great circles graphically and also by means of the stereographic net. Compare the values of the bearing obtained by the different methods.

2.3 α-Sulphur forms orthorhombic holosymmetric crystals with $a = 10.48$ Å, $b = 12.92$ Å and $c = 24.55$ Å.

(a) Draw a sketch stereogram showing the symmetry elements shown by α-sulphur.

(b) Calculate the angles $(0\,0\,1)^{\wedge}(0\,1\,1)$ and $(1\,0\,0)^{\wedge}(1\,1\,0)$. Insert these poles on a stereogram and hence:

(c) Draw an accurate stereogram of sulphur showing all the faces of the forms $\{1\,0\,0\}$, $\{0\,1\,0\}$, $\{0\,0\,1\}$, $\{1\,0\,1\}$, $\{1\,1\,0\}$, $\{1\,1\,1\}$, $\{0\,1\,1\}$ and $\{1\,1\,3\}$. Index all the faces in the upper hemisphere. Which of these are general and which are special forms?

2.4 (a) Sketch stereograms showing the operation of the undermentioned symmetry elements on *two* distinct poles, each in a general position, and insert the symmetry elements themselves, except the centre of symmetry. Centre of symmetry

Vertical plane of symmetry
Horizontal plane of symmetry
Vertical axes 2, 3, 4, 6
Vertical inversion axes $\bar{2}, \bar{3}, \bar{4}, \bar{6}$
Horizontal axes 2,4
Horizontal inversion axis $\bar{4}$

(b) The symbol $\bar{X}/m$ is not used in conventional point group notation. Draw sketch stereograms of the poles of one general form for each possible value of X and hence derive the conventional symbols that are actually used.

2.5 (a) Sketch stereograms of the symmetry elements and of the poles of one general form in the trigonal classes:

$$3, \bar{3}, 3m, 32, \bar{3}m$$

(b) In which of these classes does a rhombohedron occur as a special form? In what class is the rhombohedron the general form?

(c) Which of these classes has a centre of symmetry?

2.6 (a) Sketch stereograms of the symmetry elements and of the poles of one general form in the tetragonal classes:

$$4, \bar{4}, 4/m, 4mm, \bar{4}2m, 422, 4/mmm$$

(b) Which of these classes has a centre of symmetry?

(c) Which of these classes have closed forms as their general forms?

2.7 Prove formula (2.5).

2.8 In an orthorhombic crystal (topaz) with axial ratios $a:b:c = 0.529:1:0.477$, the following angles were measured to the face P of the general form: $(1\,0\,0)\char`^P = 67.85°/(0\,1\,0)\char`^P = 66.5°$. Determine the indices of P.

2.9 In a holosymmetric cubic crystal the angle between $(1\,1\,0)$ and a face P which lies in the zone $[0\,1\,0]$ is 53.97°. Find the indices of P and calculate the angle between P and the face P' related to P by the mirror plane parallel to $(1\,\bar{1}\,0)$. Determine the indices $[U\,V\,W]$ of the zone containing P and P' and calculate the interzone angle $[U\,V\,W]\char`^[1\,\bar{1}\,0]$.

2.10 (a) Draw sketch stereograms showing the symmetry elements and the poles of one general form in each of the *cubic* classes $23, m3, 432, \bar{4}3m$ and $m3m$.

(b) To which of the classes could a crystal belong which had the shape of (i) an octahedron, (ii) a tetrahedron and (iii) an icositetrahedron? (The icositetrahedron is the name given to the $\{1\,1\,2\}$ form.)

2.11 A wooden model of a cube is having its corners cut off to make equal faces of the form $\{1\,1\,1\}$. The first four new faces are cut in positions $(1\,1\,1)$, $(1\,1\,\bar{1})$, $(\bar{1}\,1\,1)$ and $(\bar{1}\,1\,\bar{1})$, in that order. Draw sketch stereograms to show the symmetry elements shown by the model after each new face is formed. What is the crystal system of the model in each case?

2.12 Calculate the angle between $[0\,0\,0\,1]$ and $[1\,1\,\bar{2}\,3]$ in beryllium (hexagonal, $a = 2.28$ Å, $c = 3.57$ Å). What face lies at the intersection of the zones $[1\,1\,\bar{2}\,3]$ and $[\bar{2}\,1\,1\,3]$?

2.13 In a crystal of calcite (trigonal; $c : a = 0.8543$), a face lies in the zone between $(1\,0\,\bar{1}\,1)$ and $(\bar{1}\,1\,0\,1)$ at a distance of 16.50 from $(1\,0\,\bar{1}\,1\,1)$. Determine the indices of this face.

2.14 Calculate the axial ratios $a : b : c$ and the axial angle β for a monoclinic crystal (gypsum) given that $(1\,1\,0)\overset{\frown}{}(1\,\bar{1}\,0) = 68.5°$, $(0\,0\,1)\overset{\frown}{}(1\,\bar{1}\,0) = 82.33°$, $(0\,0\,1)\overset{\frown}{}(1\,\bar{1}\,0) = 33.13°$.

2.15 Calcite has the trigonal holosymmetric point group $\bar{3}m$. Sketch the distribution of symmetry elements on a stereogram. Choose the diad axes as parallel to two crystal axes and use a hexagonal cell.

Crystals of calcite cleave parallel to $\{h\,0\,\bar{h}\,l\}$. Insert the poles of the form $\{h\,0\,\bar{h}\,l\}$ on your sketch stereogram and mark the angle between $\{h\,0\,\bar{h}\,l\}$ and $(0\,h\,\bar{h}\,l)$. Given that the angle between $(1\,0\,\bar{1}\,1)$ $(0\,1\,\bar{1}\,\bar{1})$ in calcite is $105.08°$, draw an accurate stereogram of calcite including the forms $\{1\,0\,\bar{1}\,1\}$, $\{0\,0\,0\,1\}$, $\{1\,0\,\bar{1}\,0\}$, $\{1\,1\,\bar{2}\,0\}$, $\{1\,0\,\bar{1}\,2\}$, $\{0\,1\,\bar{1}\,2\}$ and $\{2\,1\,\bar{3}\,1\}$. Determine the axial ratio c/a calcite.

2.16 Sketch stereograms showing the distribution of symmetry elements and of the pole of the faces of two different general forms in each of the non-centrosymmetric classes of the monoclinic and orthorhombic systems.
(a) Given the point group symbol for each class.
(b) In which of the classes is enantiomorphism possible?

2.17 In dealing with some imperfect crystals with a face-centred cubic lattice it is convenient to use a hexagonal unit cell. The face-centred cubic lattice can be referred to a hexagonal cell of which the z axis is parallel to [111] and of magnitude $\sqrt{3}a$, where a is the lattice parameter of the conventional cubic unit cell and the x and y axes are parallel to the $\langle 1\,0\,1\rangle$ directions and of magnitude $a/\sqrt{2}$.
(a) Draw a diagram showing the relation between the two unit cells.
(b) Write the hexagonal lattice vectors in terms of the cubic lattice vectors. Thence derive the matrix for transforming the indices of lattice planes.
(c) Find the ratio of the volumes of the two unit cells by the matrix method and check by direct calculation. How many lattice points does each contain?
(d) Obtain the hexagonal indices of the planes with indices $(1\,1\,2)$, $(1\,0\,0)$ and $(1\,\bar{1}\,0)$ referred to the conventional cubic cell.

SUGGESTIONS FOR FURTHER READING

See the suggestions for Chapter 1 as well as the following:
Coxeter, H. S. M., *An Introduction to Geometry*, Wiley, New York (1961).
Koster, G. F., *Space Groups and Their Representations in Solid State Physics*, (eds. F. Seitz and D. Turnbull) Vol. 5, p. 173, Academic Press (1957).
International Tables for X-Ray Crystallography (ed. Theo Hahn), 4th revised edn., Vol. A: *Space-Group Symmetry*, published for the International Union of Crystallography by Kluwer Academic, London (1995).
Philips, F. C., *An Introduction to Crystallography*, 4th edn, Oliver and Boyd, Edinburgh (1971).
Terpstra, P. and Codd, L. W., *Crystallometry*, Academic Press, New York (1961).

REFERENCES

1. B. D. Cullity, *Elements of X-Ray Diffraction*, Addison-Wesley.
2. *International Tables for X-Ray Crystallography* (ed. Theo Halun), 4th revised edn, published for the International Union of Crystallography by Kluwer Academic, London (1995).
3. W. F. de Jong, *General Crystallography: A Brief Compendium*, W. H. Freeman, New York (1959).

3

Crystal Structures

3.1 INTRODUCTION

The relative positions of the atoms in crystals have been deduced from measurements of the relative intensities of the diffraction spectra using X-rays, electrons or neutrons. X-rays have been most useful in this regard because the intensities of the diffracted waves are not so greatly affected by crystalline perfection as are those obtained with electrons or neutrons. Careful measurements of the spectra enable the electron density to be found throughout the unit cell of a crystal. Since only the outermost electrons are involved in binding the atoms together in a crystal, most of the electrons in the crystal reside in orbits that are the same as those in an isolated atom. The crystal structure can therefore be described by stating the relative positions of *atoms* within the unit cell. In many crystals of both elements and compounds the charge distribution of the outermost electrons, which are not involved in the chemical binding, are more or less spherically symmetric and therefore the crystal structure is sometimes regarded as being composed of spheres of the same size (if the crystal structure of an element is being considered) or of different sizes (corresponding to the different atoms in a chemical compound) packed together.

The crystal structures of the stable elements and of most simple compounds have been determined. These are listed in the books given at the end of this chapter. We shall now describe a number of the structures emphasizing the geometrical features of the relative arrangement of the atoms. The imperfections in these crystals will be described in Part II. When describing each crystal structure we shall state the space group in the style given in the *International Tables for X-Ray Crystallography* [1] because from this symbol the Bravais lattice and the point group of the crystal are immediately known (see Section 2.14).[†] We shall first describe the structures of some of the elements and then those of simple compounds. In many cases a given element or compound possesses more than one crystal structure, each being the thermodynamically stable form

---

[†] Other ways of describing crystal structure using different designations are given in the books listed at the end of this chapter. They are referred to as the Strukturbericht symbol and the Pearson symbol.

in a given regime of temperature and pressure. When this is the case an element is said to show *allotropy* and a compound is spoken of as *polymorphous*. In describing the crystal structures of some compounds we shall sometimes refer to the chemical formulae as of the type MX, M denoting a metal ion or atom and X an electro-negative element, e.g. O, F, Cl, etc. Crystal structures can be described by stating the lattice and the coordinates of the atoms in terms of the cell sides as units of length, referred to the lattice point at the origin. In multiply-primitive unit cells the same arrangement of atoms must, of course, occur around every lattice point.

3.2 COMMON METALLIC STRUCTURES

The vast majority of the elements are metallic. The metallic elements with the notable exceptions of manganese, gallium, indium, tin, protoactinium, mercury, uranium and plutonium (see Table A5.1, Appendix 5) all possess one of the three structures: either the face-centred cubic, close packed hexagonal or body-centred cubic.

FACE-CENTRED CUBIC (FM3M)

The noble metals (Cu, Ag, Au), the metals of higher valence (Al and Pb), the later transition metals (Co, Ni, Rh, Pd, Ir, Pt) and the inert gases (Ne, Ar, Kr, Xe) all possess this structure. The lattice is cubic face-centred and there is one atom at each lattice point. Figure 3.1a shows a conventional unit cell of the structure with lattice parameter a and Fig. 1.31 shows the relationship of this cell to a primitive unit cell. The coordinates of the atoms in the conventional cell are thus $(0,0,0)$, $(\frac{1}{2}, \frac{1}{2}, 0)$, $(\frac{1}{2}, 0, \frac{1}{2})$ and $(0, \frac{1}{2}, \frac{1}{2})$. There are four atoms per unit cell. Each atom possesses twelve nearest neighbours at a distance of $a/\sqrt{2}$. The *coordination number* is thus 12. This structure is the one obtained if equal spheres are placed in contact at the lattice points of a face-centred cubic (f.c.c.) lattice. A diagram illustrating this is given in Fig. 3.2a. If the radius of the spheres is R then $R = a/2\sqrt{2}$. The proportion of space filled by the spheres is $\pi\sqrt{2}/6$ or 74%. This is described by saying that the packing fraction is 0.74. A packing fraction of 0.74 is probably the closest packing of equal spheres that can be achieved, and it is certainly the closest packing that can be achieved with equal spheres at the lattice points of a Bravais lattice (see Section 3.6). Along $\langle 1\,1\,0 \rangle$ directions in the lattice, rows of spheres are in contact along the line joining their centres. Such directions are the closest packed. There are twelve such directions in all if account is taken of change in sign. The atomic centres lie at the lattice points and so Fig. 1.27 shows that in the $\{1\,1\,1\}$ planes, which lie normal to triad axes, the atom centres form a triequiangular net of points. If the atoms are spheres of radius $a/2\sqrt{2}$ the appearance of a $\{1\,1\,1\}$ plane is shown by the full circles in

Fig. 3.3. Each sphere is in contact with six equidistant spheres with centres in the plane. Since the arrangement shown in Fig. 3.3 is the closest packing of circles in a plane, the {1 1 1} planes are spoken of as being closest packed. There are eight such planes in the lattice if distinction is drawn between parallel normals of opposite sense (so that, for example, $(1\bar{1}1)$ and $(\bar{1}1\bar{1})$ are counted separately) and each contains six closest packed directions. The spacing of the {1 1 1} planes

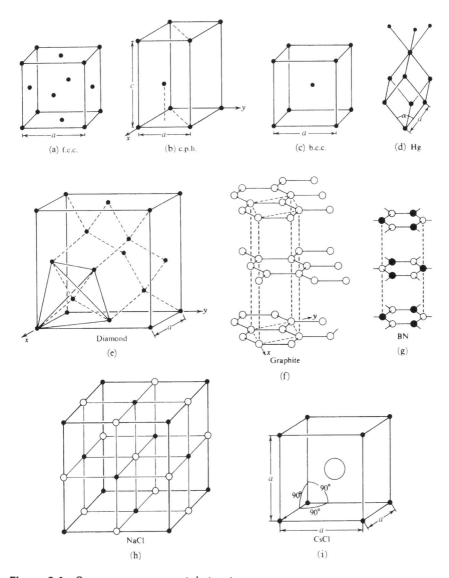

(a) f.c.c.   (b) c.p.h.   (c) b.c.c.   (d) Hg

Diamond
(e)

Graphite
(f)

BN
(g)

NaCl
(h)

CsCl
(i)

Figure 3.1   Some common crystal structures

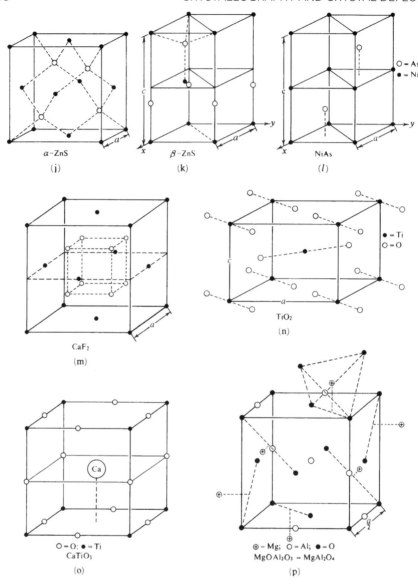

Figure 3.1 (*continued*)

is $a/\sqrt{3} = 2\sqrt{2/3}R$ and so is equal to $\sqrt{2/3}$ of the atomic spacing in $\{1\,1\,1\}$. The centres of the atoms in a given $(1\,1\,1)$ plane occupy points such as A in Fig. 3.3. If the positions of the centres of the atoms in adjacent $(1\,1\,1)$ planes are projected on to the given $(1\,1\,1)$ plane they occupy positions such as those marked B or C. The projections of the spheres centred on points B are shown

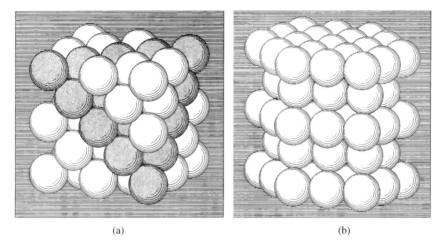

(a)   (b)

Figure 3.2   Close packing of equal spheres. (a) Face-centred cubic (f.c.c). The closest packed layers are parallel to the darker balls. (b) Close packed hexagonal (c.p.h.). Closest packed layers are horizontal. (From Bragg and Claringbull [2])

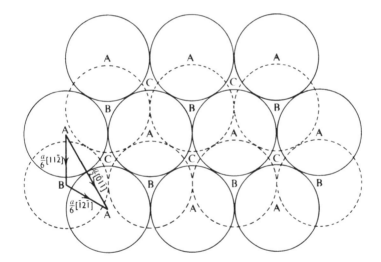

Figure 3.3

dotted in the figure. A given crystal can be considered as made up by stacking, one above the other, successive planar rafts of closest packed spheres so that proceeding in the [1 1 1] direction the centres of the atoms in adjacent rafts follow the sequence ABCABCABC ... (Fig. 3.4a). The same crystal structure, but in a different orientation, would be described by the sequence of rafts ACBACBACB

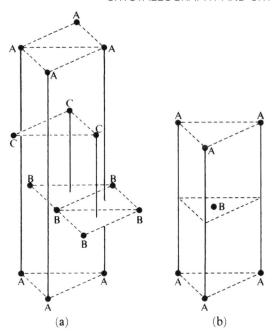

Figure 3.4　The stacking of closest packed planes in (a) the f.c.c. structure and (b) the c.p.h. structure

. . . . Any atom in a face-centred cubic crystal has twelve nearest neighbours at a distance of $a/\sqrt{2}(= 2R)$, six second nearest at $a(= \sqrt{2}2R)$, twenty-four third nearest at $\sqrt{3/2}a(= \sqrt{3}2R)$ and twelve fourth nearest at $\sqrt{2}a(= \sqrt{4}2R)$.

When the structure is regarded as being made up of spheres in contact the size of the interstices between spheres is important because many other crystal structures contain at least one set of atoms in a face-centred cubic arrangement. The largest interstice occurs at positions in the unit cell with coordinates $(\frac{1}{2}, \frac{1}{2}, \frac{1}{2})$ and equivalent positions (i.e. $0, \frac{1}{2}, 0$; $0, 0, \frac{1}{2}$ and $\frac{1}{2}, 0, 0$). There are four of these per unit cell and, hence one per lattice point; one is illustrated in Fig. 3.5. The largest sphere which can be placed in this position without disturbing the arrangement of spheres at the lattice points has radius $r = (\sqrt{2} - 1)R = 0.414R$. This sphere would have octahedral coordination with six nearest neighbours (Fig. 3.5). The sites of the largest interstices themselves form a face-centred cubic lattice.

The second largest interstice occurs at points with coordinates $(\frac{1}{4}, \frac{1}{4}, \frac{1}{4})$ and equivalent positions (Fig. 3.6). The largest sphere which can be placed here has radius $r = (\sqrt{3/2} - 1)R = 0.225R$ and possesses tetrahedral coordination. There are eight such points in the unit cell and hence two per lattice point. The sites of the second largest interstice lie on a primitive cubic lattice.

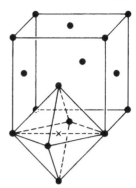

Figure 3.5 The largest interstice in the f.c.c. structure

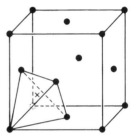

Figure 3.6 The second largest interstice in the f.c.c. structure

CLOSE-PACKED HEXAGONAL (P6₃/MMC)

This structure is exhibited by the early transition metals scandium, titanium, yttrium and zirconium, by the divalent metals beryllium, magnesium, zinc and cadmium and by most of the rare earths (Section A5.1 in Appendix 5). The hexagonal primitive unit cell contains two atoms with coordinates $(0,0,0)$ and $(\frac{2}{3}, \frac{1}{3}, \frac{1}{2})$ (Fig. 3.1b). There are thus two atoms associated with each lattice point. There is no pure rotational hexagonal axis but 6_3 axes located at $00z$, i.e. at the origin of the unit cell running parallel to the z axis. This structure can be produced by packing together equal spheres as shown in Fig. 3.2b. If each sphere has twelve nearest neighbours then the axial ratio c/a must equal $\sqrt{8/3} = 1.633$. The packing fraction is then 0.74 as in the face-centred cubic structure. The values of the axial ratio for a number of metals with this structure are given in Table 3.1. Cobalt has an axial ratio very close to the ideal. If the atomic centres are projected on to a plane parallel to $(0\,0\,0\,1)$, the basal plane, then the structure can be regarded as made up by stacking rafts of spheres in the sequence ABABAB or ACACAC (Figs. 3.3 and 3.4b). If c/a is equal to the ideal value for sphere packing there are six closest packed directions of the type $\langle 1\,1\,\bar{2}\,0 \rangle$ in the basal plane and these are the only closest packed directions. Again if $c/a = \sqrt{8/3}$ then the largest interstices have coordinates $(\frac{1}{3}, \frac{2}{3}, \frac{1}{4})$ and $(\frac{1}{3}, \frac{2}{3}, \frac{3}{4})$ (Fig. 3.7a) and the largest sphere that can be inserted without disturbing the spheres of radius

Table 3.1 Axial ratios at room temperature of some materials with a hexagonal Bravais lattice

Material	Structure	c/a	Material	Structure	c/a
Cd		1.886	BeO		1.63
Zn		1.856	ZnO		1.60
Co		1.628	β-ZnS		1.63
Mg		1.624	β-CdS	wurtzite	1.62
Re	c.p.h.	1.615	CdSe		1.63
Sc	metals	1.594	AlN		1.60
Zr		1.593	SiC		1.65
Ti		1.587	TaN		1.62
Y		1.571			
Be		1.567	NiAs		1.40
			CrS	NiAs	1.64
NH_4F	wurtzite	1.61	FeS		1.66
α-AgI		1.63	CoS		1.52
			NiS		1.55

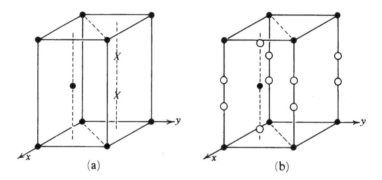

Figure 3.7 Interstices in the c.p.h. arrangement. The largest interstices are denoted by X and the second largest by open circles

R packed in contact has radius $r = 0.414R$. There are two such interstices per cell and each has octahedral coordination. The second largest interstices lie at $(0, 0, \frac{3}{8})$, $(0, 0, \frac{5}{8})$, $(\frac{2}{3}, \frac{1}{3}, \frac{1}{8})$ and $(\frac{2}{3}, \frac{1}{3}, \frac{7}{8})$ (Fig. 3.7b). These have $r/R = 0.225$ and there are four per cell, each with tetrahedral coordination.

BODY-CENTRED CUBIC (LM3M)

This structure is shown by the alkali metals lithium, sodium, potassium, rubidium and caesium at room temperature, by the transition metals vanadium, chromium, niobium, molybdenum, tantalum and tungsten and by iron, titanium and zirconium in certain temperature ranges. The lattice is body-centred cubic (b.c.c.) with one atom at each lattice point, so the atomic coordinates are $(0,0,0)$ and $(\frac{1}{2}, \frac{1}{2}, \frac{1}{2})$

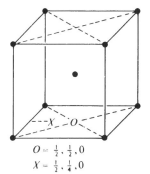

$O = \frac{1}{2}, \frac{1}{2}, 0$

Figure 3.8 $X = \frac{1}{2}, \frac{1}{4}, 0$

(Fig. 3.1c). Each atom has eight nearest neighbours at a separation of $\sqrt{3}a/2$, where a is the lattice parameter. The nearest neighbours of one atom are not nearest neighbours of one another. If the structure is made up with equal spheres these have radius $R = \sqrt{3}a/4$ and the eight $\langle 1\,1\,1 \rangle$ directions are closest packed. The packing fraction is $\pi\sqrt{3}/8 = 0.68$. The second nearest neighbours of an atom are closer than in the face-centred cubic structure. There are six second nearest neighbours at a distance $4R/\sqrt{3}(= 2.309R)$ (in f.c.c. they are at $2\sqrt{2}R = 2.828R$) and twelve third nearest at $4\sqrt{2}/3R$. There are no closest packed planes of atoms in this structure. If made up of equal spheres this structure has largest holes at coordinates $(\frac{1}{2}, \frac{1}{4}, 0)$ and equivalent positions (Fig. 3.8). There are twelve such positions per cell, i.e. six per lattice point. The largest sphere fitting in such an interstice has radius $= R(\sqrt{5/3} - 1) = 0.288R$. This interstice is smaller than the largest one in f.c.c. It has four nearest neighbours equidistant from it but the tetrahedron that these neighbours form is not regular. The second largest interstice is at $(\frac{1}{2}, \frac{1}{2}, 0)$ and equivalent positions (Fig. 3.8). There are six such sites per cell ($6 \times \frac{1}{2}$ at the centres of cube faces and $12 \times \frac{1}{4}$ at the midpoints of cube edges) and so three per lattice point. Each can accommodate a sphere of radius $r = (2/\sqrt{3} - 1)R = 0.15R$; such a sphere is at the centre of a distorted octahedron.

3.3 RELATED METALLIC STRUCTURES

The crystal structure of *indium* is very similar to the f.c.c. structure. The space group is *I*4/*mmm*. It is tetragonal with $c/a = 1.52$. This axial ratio is such that if referred to a face-centred tetragonal lattice instead of the body-centred one the axial ratio $c/a = 1.08$. Thus the structure can be described as face-centred tetragonal (*F*4/*mmm*) with one atom at each lattice point and an axial ratio of 1.08. Mercury also has a structure that can be described as distorted f.c.c. The space group is $R\bar{3}m$ and the primitive unit cell is rhombohedral with one atom at each lattice point (Fig. 3.1d). The axial angle α is 70.73° and $a = 2.993$ Å at 78 K.

The primitive unit cell of a face-centred cubic crystal has $\alpha = 60°$ (Fig. 1.31). Thus the mercury structure can be derived from the f.c.c. by squashing along a body diagonal of the cell. The atoms in the (1 1 1) planes of mercury are therefore arranged to form a triequiangular net but the spacing of these planes is too small to allow closest packing of spherical atoms. The nearest neighbours of any one atom are in adjacent (1 1 1) planes. The closest packed directions are ⟨1 0 0⟩ of the primitive rhombohedral cell of side a (Fig. 3.1d). There are six nearest neighbours at a distance a. Mercury can, of course, also be referred to a rhombohedral face-centred cell containing four atoms to emphasize the relationship to the conventional cell of an f.c.c. crystal. The larger cell for Hg has $a' = 4.581$ Å and axial angle $\alpha' = 98.22°$ at 78 K.

White tin, the form of tin stable at room temperature, has the space group $I4_1/amd$. It has a body-centred tetragonal lattice having two atoms associated with each lattice point: one at the lattice point and the other at $(0, \frac{1}{2}, \frac{1}{4})$. The value of c/a at 25° is 0.545. Each atom has four nearest neighbours at the vertices of an irregular tetrahedron and two more at a slightly greater distance (see problem 3.4). The structure is a distorted form of the diamond structure which is shown undistorted by the form of tin (grey tin) stable below room temperature (see Section 3.4.)

3.4 OTHER ELEMENTS

DIAMOND (FD3M)

The elements silicon, germanium, α-tin (the form stable below room temperature) and crystalline carbon stable at high temperature and pressure (diamond) have the structure shown in Fig. 3.1e. The Bravais lattice is face-centred cubic and there are two atoms with coordinates $(0, 0, 0)$ and $(\frac{1}{4}, \frac{1}{4}, \frac{1}{4})$ associated with each lattice point. The coordination number is four with the nearest neighbours at a distance $\sqrt{3}a/4$ arranged at the corners of a regular tetrahedron, outlined in Fig. 3.1e. The atom centres lie at the corners of triequiangular nets in {1 1 1} planes. If these are projected on to (1 1 1) the stacking sequence of successive (1 1 1) planes can be described as CA AB BC CA AB BC. Successive planes are not equally separated from one another (Fig. 3.9). The structure is very loosely packed. If equal spheres in contact have their centres at the atom centres the packing fraction is only $\sqrt{3}\pi/16 = 0.34$.

GRAPHITE (P6₃/MMC)

The stable crystal structure of carbon at room temperature is graphite, shown in Fig. 3.1f. The lattice is hexagonal with four atoms per unit cell, with coordinates $(0, 0, 0)$, $(0, 0, \frac{1}{2})$ $(\frac{1}{3}, \frac{2}{3}, 0)$ and $(\frac{2}{3}, \frac{1}{3}, \frac{1}{2})$. At room temperature $c = 6.70$ Å and $a = 2.46$ Å; thus $c/a(= 2.72)$ is large. The atoms in one $(0\,0\,0\,1)$ plane are

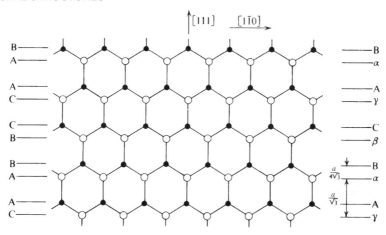

Figure 3.9 The stacking of (1 1 1) planes in the diamond and zinc-blende structures. The atomic positions are projected on to (1 1 $\bar{2}$). In diamond the filled
and open circles represent the same type of atom. The sequence of triequiangular nets is described for diamond on the left-hand side and for zinc-blende
(sphalerite) on the right-hand side

arranged at the corners of regular hexagons. Figure 1.1a shows the x and y axes
of the unit cell. The structure can be built up by stacking successive hexagonal sheets of atoms one above the other along the z axis so that the hexagons
are in the same orientation but with half the corners of hexagons in one plane
lying in the centres of the hexagons in adjacent planes (Fig. 3.1f). If the atomic
centres are all projected on to (0 0 0 1) the stacking sequence can be described
as ABABAB or ACACAC, where it must be remembered that the letters A or B
or C refer to sheets of atoms arranged at the corners of hexagons and not at the
vertices of equilateral triangles. Each atom has three nearest neighbours at a separation of $a/\sqrt{3}$ in the basal plane. Half of the atoms in any one hexagonal layer
have atoms directly above and below them at a distance of $c/2$. The separation
of the nearest neighbours in the (0 0 0 1) planes is only 1.42 Å (compared with
1.54 Å in diamond) and this is much smaller than the separation of the hexagonal
sheets — 3.35 Å. Graphite is therefore said to possess a layer structure because
the atoms are strongly bonded within a sheet and the sheets are only weakly
bound to one another.

   The structure of hexagonal boron nitride is related to that of graphite. The
atoms occur in hexagonal sheets but the sheets are stacked directly above one
another along [0 0 0 1] so that the stacking sequence would be described as
AAAAA ... , with unlike atoms above one another in consecutive layers (see
Fig. 3.1g). In any one sheet there are equal numbers of B and N atoms arranged so
that B and N alternate around any one atomic hexagon. At room temperature the
B–N separation in the sheets is 1.45 Å and the separation of the sheets is 3.33 Å.

If graphite crystals are ground or otherwise severely deformed at low temperature another form of graphite can be detected in which the atomic hexagons are arranged in the sequence ABCABCABC or alternatively CBACBACBA. Such a structure has a rhombohedral Bravais lattice (space group $R\bar{3}m$). The primitive rhombohedral cell has $a = 3.64$ Å and $\alpha = 34.5°$ at room temperature. There are two atoms per unit cell with coordinates $\pm(u, u, u)$ where $u = \frac{1}{6}(= 0.166)$.

The structures of arsenic, antimony and bismuth are also based on a primitive rhombohedral Bravais lattice (space group $R\bar{3}m$) with two atoms associated with each lattice point. The structure is shown in Fig. 3.10. The coordinates of the atoms are $\pm(u, u, u)$; u is a little less than $\frac{1}{4}$ and α somewhat less than $60°$ (Table 3.2). The structure is easily visualized as being made up of sheets of atoms lying perpendicular to [1 1 1]. Within each sheet the atoms are arranged at the corners of triequiangular nets. Referring to Fig. 3.10 and noting the values of u, the stacking sequence of these nets along [1 1 1] is BA CB AC BA .... If u is $\frac{1}{4}$, the spacing of adjacent nets would be regular. In fact, each atom in any one sheet has three nearest neighbours in the closest adjacent sheet at d_1 (Table 3.2) and three at a slightly greater distance in the sheet on the other side at d_2. The possession of three nearest neighbours fits with the positions of As, Sb and Bi in the periodic table since each atom is expected to form three covalent bonds. If $\alpha = 60°$ and $u = \frac{1}{4}$ in this structure each atom has six closest neighbours and the arrangement of the atoms is at the points of a simple cubic lattice.

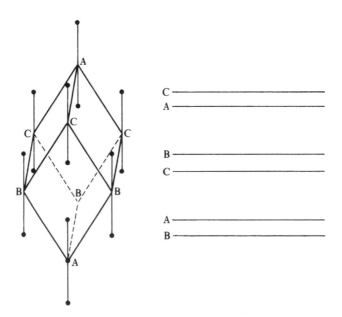

Figure 3.10 Structures of arsenic, antimony and bismuth

Table 3.2   Cell dimensions of As, Sb and Bi

	α (deg)	a (Å)	u	d_1 (Å)	d_2 (Å)
As	54.17	4.131	0.226	2.51	3.15
Sb	57.10	4.507	0.233	2.87	3.37
Bi	57.23	4.746	0.237	3.10	3.47

3.5 COMPOUNDS

A third of all compounds of the type MX crystallize in the sodium chloride structure (*Fm3m*) shown in Fig. 3.1h. The lattice is face-centred cubic with two different atoms associated with each lattice point, one, say M, at (0, 0, 0) and the other, X, at (0, 0, $\frac{1}{2}$). Each of the two types of atoms lies upon a f.c.c. lattice, and each lies at the largest interstice of the other's f.c.c. lattice. From Fig. 3.5 it is then clear that each atom has a coordination number of six, the neighbours being at the vertices of a regular octahedron. There are four formula units per conventional unit cell but there is clearly no trace of the formation of a molecule of MX in this structure.

Each {1 1 1} *lattice plane* specifies a double sheet of atoms, each of the sheets consisting entirely of atoms of one kind arranged at the points of a triequiangular net. If we denote sheets of atoms of one kind with a Greek letter and those of the other with a Roman letter then the stacking sequence along a [1 1 1] direction can be described as A γ B α C β A γ B α C β .... . The spacing of the sheets of atoms denoted by letters from the same alphabet is $a/\sqrt{3}$, where a is the lattice parameter.

All of the alkali halides with the exception of CsCl, CsBr and CsI crystallize with the sodium chloride (halite) structure and many of the sulphides, selenides and tellurides of Mg, Ca, Sr, Ba, Pb, Mn, as well as the oxides of formula MO of Mg, Ca, Sr, Ba, Cd, Ti, Zr, Mn, Fe, Co, Ni and U and some transition metal carbides and nitrides such as TiC, TiN, TaC, ZrC, ZrN, UN and UC (see Section A5.2 in Appendix 5).

When the metallic ion is of variable valence, crystals with this structure often form with some ion positions unoccupied. For example, FeO crystals would be perfect if they contained only divalent iron, Fe^{2+}. If some Fe^{3+} ions are present then for every two Fe^{3+} ions one of the sites on the iron sublattice must be empty. Such a structure is called a *defect structure* (see Section 9.1).

CsCl, CsBr, CsI (see Section A5.3 in Appendix 5) as well as many intermetallic compounds such as CuBe, CuZn, AgCd, AgMg, FeAl show the caesium chloride structure (*Pm3m*), which has a primitive cubic lattice with one atom of each kind associated with each lattice point, one, say A, at (0, 0, 0) and the other at ($\frac{1}{2}$, $\frac{1}{2}$, $\frac{1}{2}$) (Fig. 3.1i). The coordination number is 8 for each atom, the nearest neighbours being at $\sqrt{3}a/2$, where a is the lattice parameter.

The cubic form of ZnS, designated α-ZnS, shows the sphalerite structure[†] ($F\bar{4}3m$), which is also exhibited by the sulphides, selenides and tellurides of Be, Zn, Cd, Hg and the halides of Cu and AgI (Fig. 3.1j and Section A5.4 in Appendix 5). The lattice is f.c.c. with one atom of each kind associated with each lattice point, one at (0, 0, 0) and the other at $(\frac{1}{4}, \frac{1}{4}, \frac{1}{4})$, so the relationship to the structure of diamond is very close. Each type of atom lies at the lattice points of an f.c.c. lattice and each lies in the second largest interstice of the f.c.c. lattice of the other. The coordination number is four, the nearest neighbours being atoms of the other kind at a distance $\sqrt{3}a/4$ arranged at the vertices of a regular tetrahedron. The sequence of (1 1 1) planes is stacked at the same intervals as in diamond (Fig. 3.1e), but alternate planes are occupied by atoms of different chemical species. Thus, following the same nomenclature as for NaCl, the stacking sequence is γ A α B β C γ A .... .

The sphalerite structure is derived from the f.c.c. structure by placing atoms of a different kind to those at the lattice points at every other tetrahedrally coordinated interstice. A related structure, also shown by ZnS, is obtained by filling alternate tetrahedrally coordinated interstices in the close packed hexagonal structure. This is the β-ZnS or wurtzite structure ($P6_3mc$). The lattice is hexagonal with atoms of one kind at (0,0,0) and $(\frac{2}{3}, \frac{1}{3}, \frac{1}{2})$ and those of the other kind at (0,0,u) and $(\frac{2}{3}, \frac{1}{3}, \frac{1}{2} + u)$ (Fig. 3.1k). The value of u is very close to 0.375, i.e. $\frac{3}{8}$. From Fig. 3.7b the relationship to the hexagonal close packed structure is obvious. Each atom is, of course, tetrahedrally coordinated with four of the opposite kind. The stacking of planes of atoms parallel to (0 0 0 1) using the same nomenclature as for NaCl and α-ZnS is then A α B β A α B ... or equivalently A α C γ A α C γ A .... . This structure is shown by a number of compounds listed in Section A5.5 in Appendix 5. Some axial ratios at room temperature are given in Table 3.1. These ratios are quite close to the ideal, 1.633, for hexagonal close packing of one type of ion with the other in the tetrahedral interstices.

We described the sodium chloride structure as being derived from the f.c.c. structure by placing a secondset of atoms in the octahedrally coordinated largest interstices of the structure. It is not surprising therefore that a structure exists in which atoms of a different kind are placed in the octahedral interstices of the close packed hexagonal structure. This is the nickel arsenide (niccolite) structure ($P6_3/mmc$) (Fig. 3.1l). The lattice is hexagonal with atoms of one kind at (0,0,0) and (0, 0, $\frac{1}{2}$) and those of the other kind at $(\frac{2}{3}, \frac{1}{3}, \frac{1}{4})$ and at $(\frac{1}{3}, \frac{2}{3}, \frac{3}{4})$. For both kinds of atom the coordination number is six, but whilst the sites occupied by one kind of atom, those denoted by open circles in Fig. 3.1l, lie in positions corresponding to a close packed hexagonal arrangement (As), the others (Ni) lie at the lattice points of a primitive hexagonal lattice. This is obvious from Fig. 3.7a and also by stating the stacking sequence of the planes of atoms along [0 0 0 1] following the procedure for NaCl. The stacking sequence is A β A γ A β

[†] Sometimes called zinc blende.

A γ .... The atoms denoted by Greek letters lie in the centre of a trigonal prism of atoms denoted by Roman letters, whilst the atoms denoted by Roman letters are octahedrally coordinated. The axial ratios of some crystals of sulphides with this structure are listed in Table 3.1; the selenides, tellurides and antimonides of the metals listed often have the same structure (see Section A5.6 in Appendix 5). The values of c/a at room temperature depart from the ideal for hexagonal close packing of one type of ion. Usually the metal atom is situated in the type of site occupied by Ni in NiAs, but in PtB the 'anti-NiAs' structure is shown.

If all of the tetrahedral interstices in the f.c.c. structure are filled with atoms of a different kind from those at the lattice points we obtain the calcium fluoride (CaF_2) or fluorite structure ($Fm3m$) (Fig. 3.1m). The lattice is face-centred cubic with atoms of one kind, say Ca, at $(0, 0, 0)$ and the other lattice points, and with F atoms at $(\frac{1}{4}, \frac{1}{4}, \frac{1}{4})$ and $(\frac{1}{4}, \frac{3}{4}, \frac{1}{4})$ and equivalent positions. There are clearly four formula units per unit cell. The coordination number of Ca is eight, with the nearest neighbours being atoms of F arranged at the corners of a cube. The fluorine atoms lie at the lattice points of a primitive cubic lattice. F has tetrahedral coordination with four Ca atoms. The centres of atoms of one kind all form triequiangular nets of points parallel to $\{1\,1\,1\}$ lattice planes and the separation of the atoms in these places is the same for the two types of atom. The stacking sequence of these planes along $[1\,1\,1]$ can therefore be described as ... $\alpha B\gamma\ \beta C\alpha\ \gamma A\beta\ \alpha B\gamma\ \beta C\alpha$ .... The planes of atoms denoted by successive Roman letters are regularly spaced from one another; the same is true of the planes of atoms denoted by Greek letters. This structure is shown by the fluorides of Ca, Sr and Ba and by the oxides of Zr, Th, Hf and U. Oxides and sulphides of alkali metals also show this structure; in these cases it is sometimes called the antifluorite structure (see Section A5.7 in Appendix 5).

A common structure of compounds of formula AX_2, besides that of fluorite, is the rutile (TiO_2) structure ($P4_2/mmm$). It is also sometimes called the cassiterite (SnO_2) structure. This is shown by the dioxides of Ti, Sn, Pb, W, Mn, V, Nb, Ta and Ge and the fluorides of Mg, Zn, Mn, Co, Ni and Fe. It is shown in Fig. 3.1n. The lattice is primitive tetragonal with c/a about 0.65. There are titanium atoms at $(0, 0, 0)$ and $(\frac{1}{2}, \frac{1}{2}, \frac{1}{2})$ and oxygen atoms at $\pm(u, u, 0)$ and at $\pm(u + \frac{1}{2}, \frac{1}{2} - u, \frac{1}{2})$, where u is in all examples of the structure close to 0.30. Thus there are two formula units per cell. A plan of the structure is shown in Fig. 3.11. Each Ti atom is surrounded by six O atoms, but the octahedron of O atoms is not quite regular. A list of compounds with this structure is given in Section A5.8 in Appendix 5.

3.6 CLOSE PACKING

The early attempts to predict crystal structures used the notion of the close packing of spheres and as we saw in the descriptions of simple crystal structures the idea is still of great practical use.

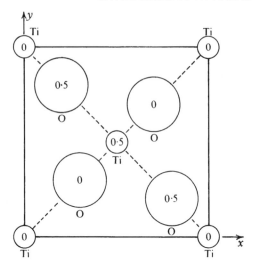

Figure 3.11 Rutile (TiO_2) structure. The numbers indicate the z coordinates of the atoms

The close packing of *circles* in a plane is defined by the mathematician as an arrangement such that each circle has at least three contacts in the plane and not all of these fall in the same semicircle. The close packing of spheres in space is such that each sphere has at least four contacts with neighbouring spheres and not all of these fall in the same hemisphere. We shall consider briefly the stacking of spheres and circles of the same diameter, following Patterson and Kasper in the *International Tables for X-Ray Crystallography*.

The hexagonal packing of circles in the plane is the closest possible planar packing (Fig. 3.3). Since each circle can have six neighbours in one and only one way, this is the largest possible number of nearest neighbours, and is the closest lattice packing in two dimensions — the centres of the circles lying at the corners of a triequiangular net. It has been proved that cubic closest packing (f.c.c. structure) is the closest *lattice* packing of spheres in three dimensions, but it has not been proved that this is the closest possible packing of spheres that fill space. The complication in a proof arises because there are an infinite number of ways in which twelve spheres can be made to contact a single sphere. In particular, it has been shown that there are special arrangements of spheres in a finite volume which have a density of packing that exceeds those of the f.c.c. and ideal hexagonal close packed arrangements. The idea of a closest packing is therefore still some-what questionable.

Among the infinite number of ways in which twelve spheres can be made to contact a single sphere there are two, the f.c.c. structure and the c.p.h. structure, which when repeated in space lead to closest packed planes of spheres. In this section we use the term closest packing for arrangements of spheres that contain closest packed planes and possess the same density of packing as does the f.c.c. structure.

A closest packed plane of spheres has the appearance shown in Fig. 3.3 and we project the centres of spheres in closest packed planes above and below on to a close packed layer. We then describe a closest packed plane by a letter A, B, C denoting the points to which the centres of the spheres in that layer project. Any sequence, e.g. ACBACBCA, is then closest packed provided any two adjacent layers are not described by the same letter. If d is the diameter of the spheres the separation of the planes is $\sqrt{2/3}d$.

About any sphere in a closest packed arrangement there are two kinds of void, illustrated in Figs. 3.12a and b, a tetrahedral void (Fig. 3.12a) and an octahedral void (Fig. 3.12b). A sphere in, say, an A layer is surrounded by six triangular voids in that layer (Fig. 3.3). The next added layer above makes three of these voids tetrahedral and three of them octahedral, and similarly for the layer below. Further, a sphere in the A layer itself covers a triangular void in the layers above and below, thus adding two tetrahedral voids. The result is that there are six octahedral voids and eight tetrahedral voids surrounding each sphere. Therefore in an infinite array there is one octahedral void and there are two tetrahedral voids, per sphere.

Since the spheres in the closest packed plane have their centres at the vertices of a triequiangular net the point group symmetry is $6mm$ (Fig. 1.15c). The pure sixfold symmetry cannot be maintained if closest packing of spheres in space is to be achieved, and the space groups of *lowest* symmetry which can show closest packing are trigonal, $P3m$ and $R3m$. Higher symmetry space groups can arise if, for example, mirror planes lie parallel to the close packed layers or a centre of symmetry is present, changing 3 to $\bar{3}$. A screw 6_3 may also be found. The space groups which can describe a sphere closest packing are then $P3m$, $R3m$, $P\bar{3}m$, $R\bar{3}m$, $P\bar{6}m2$, $P6_3mc$, $P6_3/mmc$ and $Fm3m$.

In many crystals a 'closest packed' arrangement of one type of atom or ion (not necessarily in contact) occurs with the interstices filled or partly filled with smaller atoms (or ions). There are often departures from ideal 'closest packing'.

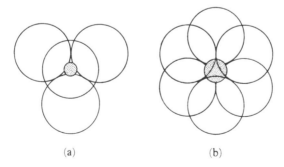

(a)   (b)

Figure 3.12

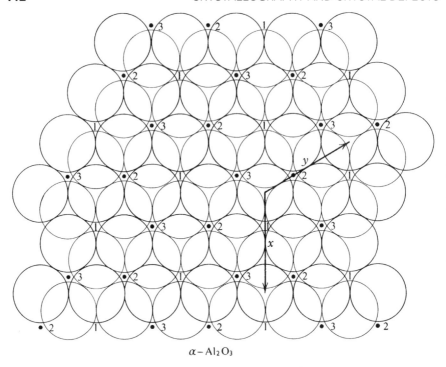

$\alpha - Al_2O_3$

Figure 3.13 The structure of α-Al_2O_3 corundum (or sapphire). The large circles represent oxygen ions in two adjacent sheets. Between the sheets shown, aluminium ions lie in the octahedral interstices γ_2 and γ_3 whilst the positions γ_1 are empty. In the next layer of aluminium ions positions γ_1 and γ_3 would be filled and γ_2 empty

Sapphire (α-Al_2O_3 — space group $R\bar{3}c$) possesses a large unit cell with the oxygen ions in a hexagonal close packed arrangement. Aluminium ions occupy the octahedral interstices. Since the formula is Al_2O_3 only two-thirds of these are filled, as shown in Fig. 3.13. The structure can be easily described by locating the 'missing' aluminium ions. If all the atomic positions are projected on to $(0\,0\,0\,1)$ of the hexagonal cell and Roman letters denote the oxygen ion positions and Greek letters the positions of missing aluminium ions, the stacking sequence is $A\gamma_1 \ B\gamma_2 \ A\gamma_3 \ B\gamma_1 \ A\gamma_2 \ B\gamma_3 \ A\gamma_1 \dots$, where the positions γ_1, γ_2, γ_3 are indicated by their suffixes in Fig. 3.13. The hexagonal unit cell is then six oxygen layers high, and contains six formula units of Al_2O_3. The lattice parameters of the hexagonal cell are $a = 4.75$ Å, $c = 12.97$ Å; Thus $c/a = 2.73$. If c/a equalled 2.816 the oxygen ions would be in an ideal c.p.h. arrangement. The lattice translations in $(0\,0\,0\,1)$ of the triply-primitive hexagonal cell are marked in Fig. 3.13. The rhombohedral primitive cell contains two formula units of Al_2O_3 and has

$a = 5.12$ Å and $\alpha = 55.28°$. If the oxygen ions were in the ideal c.p.h. arrangement the value of α would be $53.78°$. $FeTiO_3$ (ilmenite $R\bar{3}$) has the same structure as Al_2O_3 except that the Fe and Ti atoms are distributed in alternating octahedral sheets. In $LiNbO_3$ ($R3c$) the Li and Nb are distributed in the same octahedral sheet such that one species occupies the γ_2 position whilst the other occupies the γ_3 position. It can also be described as a distorted form of the perovskite structure.

In $CaTiO_3$[†] (perovskite — $Pm3m$), which is a very common structure of crystals of compounds of the type $MM'X_3$, the calcium and oxygen ions taken together form a f.c.c. arrangement with the titanium ions in the octahedral voids (Fig. 3.1o). The coordination number of Ca is 12, of Ti is 6 and of O is 2. The atomic coordinates in the primitive unit cell are Ti $(0, 0, 0)$, Ca $(\frac{1}{2}, \frac{1}{2}, \frac{1}{2})$ and O $(0, 0, \frac{1}{2})$, $(0, \frac{1}{2}, 0)$, $(\frac{1}{2}, 0, 0)$. Very accurate determination of the cell parameters indicates that the structure is in fact triclinic.

There are many slightly distorted forms of the perovskite structure which are important in solid state devices. Barium titanate, $BaTiO_3$ ($P4mm$), is a tetragonally distorted form and so are the various forms of $Pb(ZrTi)O_3$, indicating substitution of Pb by Zr or Ti. In addition are the recently discovered (1986) superconducting oxides such as $LaBa_2Cu_3xO_{7-\delta}$, where δ is a small fraction indicating an oxygen deficiency, and La may be replaced by other rare earths. The perovskite structure can be viewed as a layer structure with the layers being composed of triequiangular nets of calcium and oxygen in the ratio of 1 to 3 with Ti layers in between. Similarly, the high-temperature superconductors contain sheets of CuO.

The *spinel* structure, shown by $MgAl_2O_4$ ($Fd3m$) and by other mixed oxides of di- and tervalent metals, has an elementary cell containing 32 oxygen ions in almost perfect cubic closest packing. Eight of the 64 tetrahedral interstices are filled by the divalent Mg and sixteen of the 32 octahedral interstices are filled with the trivalent Al ions. The Mg ions form a structure of the diamond type (Fig. 3.1p). One-eighth of the face-centred cubic unit cell is shown in this figure.

Some oxides of composition MM'_2O_4 show a structure called 'inverse' spinel to distinguish it from the normal spinel which we have just described. $MgFe_2O_4$ is an example of an inverse spinel. In this, the oxygen ions are arranged in cubic closest packing and the same two types of interstice are involved. The cations are *differently* arranged. Ideally, half of the iron atoms occupy the eight tetrahedral interstices and the sixteen octahedral interstices are occupied by the remaining iron and by magnesium. Amongst the occupied octahedrally coordinated sites the Mg and Fe can occur at random. To emphasize the difference from a normal spinel the formula of this inverse spinel is sometimes written Fe $(MgFe)O_4$ or generally $M'(MM')O_4$. In fact, the structure of $MgFe_2O_4$ deviates somewhat from this ideal; the number of ions atoms in tetrahedral sites is not exactly

---

[†] $CaTiO_3$ does not in fact have this structure except above $900\,°C$; $SrTiO_3$ does.

equal to the number in octahedral sites. The olivine $(Mg_{1-x}Fe_x)SiO_4$ group of minerals includes chrysoberyl $BeOAl_2O_3$ (*Pnma*). The oxygen ions are in this case arranged in a hexagonal close packed structure but it is slightly distorted. As for the spinels, the metal ions are distributed amongst the tetrahedral and the octahedral sites; in the case of chrysoberyl half of the octahedral sites are occupied by Al ions and one-eighth of the tetrahedral ones by Be.

The concept of the occupation of the interstices in a structure by different types of ion is useful in describing the garnet structures which are so important in many solid state physics devices. Various garnets can be ferrimagnetic (e.g. YIG, see below) and form excellent laser hosts (e.g. YAG).

The garnets occur naturally as the silicates of various di- and trivalent metals. The archetype is of formula $3MnO \cdot Al_2O_3 \cdot 3SiO_2$ or $Mn_3Al_2Si_3O_{12}$. The above examples of yttrium iron garnet (YIG) of formula $Y_3Fe_5O_{12}$ and yttrium aluminium garnet (YAG) of formula $Y_3Al_5O_{12}$ contain no silicon. The substitution appears possible because in a typical natural garnet such as $3CaO \cdot Al_2O_3 \cdot 3SiO_2$ it is possible to substitute YAl for CaSi. Thus 3CaSi is replaced by 3YAl to yield $Y_3Al_5O_{12}$. This is probably easier to see and remember if we write $Al_2O_3 \cdot 3(CaO \cdot SiO_2) \equiv Al_2O_3 \cdot 3(YAlO_3)$. Because these substitutions are possible, a general formula for garnets is often used, which is $\{C_3\}[A_2](D_3)O_{12}$, where O denotes the oxygen ion or atom and C, A and D denote cations. The space group is $Ia3d$ so the structure is (essentially) cubic and there are eight formula molecules per unit cell. There are 96 so-called h sites which are occupied by oxygen. $\{C_3\}$ denotes an ion in a tetrahedral site, i.e. one surrounded by four oxygens. In the calcium aluminium garnet, $Ca_3Al_2Si_3O_{12}$, these would be occupied by silicon. There are 24 of these sites per unit cell. $[A_2]$ represents an octahedral site which is surrounded by six oxygens and there are sixteen of these per unit cell. These would be occupied by aluminium in the above example. (D_3) denotes the so-called dodecahedral site which is surrounded by eight oxygens; this would be occupied by Ca. This site is variously described as a triangular dodecahedron, hence the name, or as a distorted cube. It is illustrated in Fig. 3.14. A triangular dodecahedron is a polyhedron with twelve faces, each face of which is a triangle.

The substitution of an enormous number of cations for one another is possible and the trivalent rare earth ions have been much introduced. If strict chemical rules of valence applied then $\{C\}$ sites would be occupied by four valent cations, e.g. silicon, [A] by trivalent ions, e.g. Al, rare earth, Y or Fe^{3+}, etc., and (D) by divalent ions, e.g. Ca, Mg, Fe^{2+} or others. Because of the substitution of two cations to balance the charges of two other cations which we have mentioned, a whole host of possibilities exists and a simple valence rule cannot be followed. Further, since the substitution of large ions such as the rare earths is possible, the oxygen ions are pushed apart to accommodate the cations. The description of the structure in terms of the shape of the polyhedron surrounding a particular

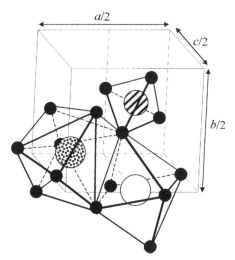

Figure 3.14 Coordination about the oxygen ions (solid black circles) in a garnet structure showing the {C_3} tetrahedral site (striped circle), the [A_2] octahedral site (solid white circle) and the (D_3) dodecahedral site (spotted circle)

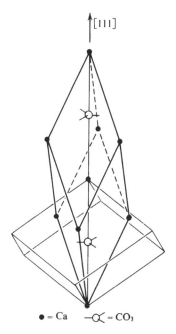

● = Ca   —◁ = CO_3

Figure 3.15 The structure of calcite ($CaCO_3$). The primitive rhombohedral unit cell is shown which contains two formula units of $CaCO_3$. The cell outlined with weak lines is the smallest 'cleavage rhombohedron' and contains four units of $CaCO_3$

type of ion and the packing together of these polyhedra to fill space is found to be very useful in many of the more complicated crystal structures.

Many complicated structures are most easily described as distorted forms of the simpler ones. For instance, the structure of calcite (one form of $CaCO_3$, with the space group $R\bar{3}c$) can be derived from that of sodium chloride by identifying the sodium ions with calcium ones and the carbonate radical CO_3 with Cl. If the sodium chloride structure is imagined to be compressed along a body diagonal ([1 1 1]) until the angle between the axes, originally 90°, becomes 101.92° we produce the doubly-primitive unit cell of calcite containing four formula units. Cleavage occurs parallel to {1 0 0} of this unit cell. The CO_3 group is triangular with the C in the centre of the triangle and the plane of the triangle normal to the direction [111] of the rhombohedral cell. The primitive unit cell of calcite contains just two formula units of $CaCO_3$, and is shown in Fig. 3.15. Calcium ions are at $(0, 0, 0)$ and $(\frac{1}{2}, \frac{1}{2}, \frac{1}{2})$ and the centre of the CO_3 radical is at $\pm(u, u, u)$. Note that u is close to $\frac{1}{4}$ in all examples of this structure; in calcite it is 0.259. The primitive cell of calcite has $a = 6.361$ Å and $\alpha = 46.10°$; the cleavage cell has $a = 6.412$ Å, $\alpha = 101.92°$.

3.7 INTERATOMIC DISTANCES

From the measured lattice parameters of simple structures the interatomic distances can be derived with the same accuracy as the parameters of the unit cell. For instance, in copper the separation of the centres of adjacent atoms is $1/\sqrt{2}$ times the cell edge. Values of the interatomic distances for most metals and for some other *elements* deduced in this way are given in Section A5.1 in Appendix 5. They are often useful in considering 'atomic radii' if the crystal structure is viewed as being made up of spheres in contact. A difference is found between the values of the atomic radii deduced from different crystal structures of the same element when the element shows allotropic forms. V. M. Goldschmidt showed that contractions of about 3 and 12% occur when a given element alters its structure from one of coordination 12 (e.g. f.c.c.) to one of coordination 8 and 4 respectively.

In crystals of compounds, interatomic distances can again be deduced from measured lattice parameters. They are useful in considering structures of compounds in terms of hard spheres in contact. However, there is a problem in dividing up the distance between two unlike atoms or ions so as to give each its own characteristic radius. This problem can only be solved by making a theoretical estimate of the size of at least one ion. Consequently, all values of ionic radii are part experimental, part theoretical in origin. Some values of ionic radii are given in Section A5.1 in Appendix 5. In crystals of compounds the state of ionization of an atom may be quite different from that in the crystal of the element, and its size will differ accordingly. The value of the ionic radius of a metal is usually less than that of the atomic radius, defined as half the distance of closest approach of atoms in the element. This is because metals form positive ions in which the electrons are drawn inwards by the excess positive charge on the nucleus. Conversely, the ionic radius of an electronegative element is usually much greater than the atomic radius.

3.8 SOLID SOLUTIONS

Many pure metals dissolve large quantities of other elements to form solid solutions. If the solute element is also metallic then the solute atom merely substitutes for the solvent atom in the crystal structure as shown in Fig. 3.16a. Such a solution is called a *substitutional solid solution*. Another type is the *interstitial solid solution* in which the solute element resides between the atoms of the solvent (Fig. 3.16b). Other elements besides the metals form solid solutions, but, since the great majority of the elements are metallic and many of them have similar chemical properties, the formation of solid solutions is most important for the metals. Solid solutions occur, rather than the formation of chemical compounds, the more similar are the chemical properties of the components. Gold dissolves silver in all proportions and NaCl dissolves KCl. Interstitial solid solutions are often formed when elements that are expected to form small atoms or ions (e.g. H, C, B, O, N) dissolve in a crystal. The two types of solution may in all cases be distinguished by density measurements and measurements of the volume of the unit cell of the solid solution. The density ρ of the crystal is given by

$$\rho = \frac{\nu M}{V}$$

where M is the molecular weight (in grams), V is the volume of the unit cell and ν is the number of formula units per unit cell. In the pure material ν is an integer. In a substitutional solid solution ν is the same integer but M alters to the

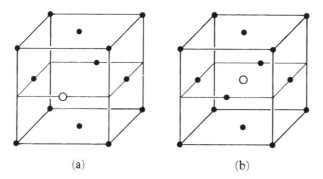

(a)   (b)

Figure 3.16   A face-centred cubic crystal showing (a) a substitutional solid solution and (b) an interstitial solid solution

average molecular weight $\overline{M}$ given by the chemical composition of the solution.[†] In an interstitial solution v is again the same for the solute but the density is increased. In a *binary* interstitial solution (two components) the value of ρ is

$$\rho = \frac{v}{V} \left(M + \frac{n_{si}}{n_s} M_{si} \right)$$

where n_{si}/n_s is the ratio of the mole fraction of solute to that of solvent and M_{si} is the molecular weight of the solute. An example of the determination of the type of solid solution is given in problem 3.9.

The various component atoms of a solid solution are usually randomly distributed amongst the sites available for them but, in some, below a certain temperature the distribution ceases to be random and what is called ordering occurs. Ordering is most easily described for a metallic solid solution. Figure 3.16a shows a part of the (1 1 1) plane of a disordered alloy of copper containing 25 at% of gold. Above a temperature of about 390 °C the copper and gold atoms occur in any of the positions at the lattice points of a face-centred cubic lattice. There is no preferred position for gold or copper. In equilibrium below 375 °C a (1 1 1) plane would appear as in Fig. 3.17b — the gold atoms are all surrounded by copper atoms. Such a structure is called an *ordered solid solution*. Figure 3.17c shows that in the fully ordered state a conventional unit cell can be chosen with the gold atoms at cube corners (0, 0, 0) and the copper atoms at the midpoints of all of the faces. This cell is the primitive unit cell of a simple cubic *superlattice*. Order–disorder changes occur in many solid solutions. The fully ordered state is always of lower symmetry than the disordered one and usually possesses a lattice with larger cell dimensions which is called a *superlattice*.

Figure 3.17c illustrates that the essential grouping in the ordered state is one with all the gold atoms surrounded by copper atoms. When ordering starts in a large crystal it may be 'out-of-step' in the various parts of the crystal. If this occurs the ordering may be perfect within various regions of the crystal (as in Figs. 3.17b and c) which are called *domains*. Where the domains are in contact the requirement of gold atoms being surrounded by copper atoms is not met, as Fig. 3.17d shows. The dotted line in Fig. 3.17d indicates the trace in (1 1 1) of what are called *anti-phase domain* boundaries. In three dimensions the boundaries are walls separating neighbouring domains. Since the neighbouring atoms are not fully ordered at the domain boundaries the boundaries represent a source of extra energy in an ordered crystal. Heating a crystal for a long time near to the ordering temperature can lead to their complete removal.

[†] $\overline{M} = \dfrac{n_1 M_1}{n} + \dfrac{n_2 M_2}{n} + \cdots$

where n_1 is the number of moles of component 1 of molecular weight M_1 in the solution and $n = n_1 + n_2 + n_3 + \cdots$.

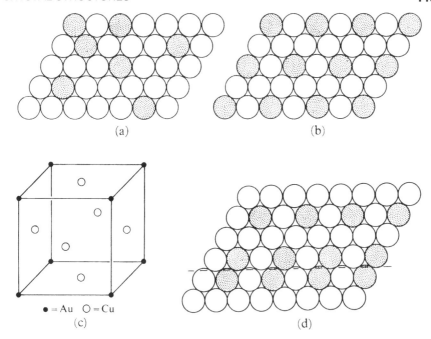

(a)   (b)

• = Au   ○ = Cu
(c)   (d)

Figure 3.17

The order we have just described is called *long-range order* because within any domain one type of lattice site is preferred for a particular atom. Many solid solutions, whilst not showing long-range order, do not show a truly random distribution of the atoms; unlike atoms occur more frequently as near neighbours than they would by pure chance. Such a state of affairs is called *short-range order*. It is very common and is shown by some solutions when they are heated above their ordering temperatures.

The structures of some ordered solid solutions are illustrated in Fig. 3.18 and examples of materials showing these structures are given in Table 3.3. The $L2_0$ or B2 order–disorder change is characterized by a body-centred cubic structure in the disordered state which changes to the caesium chloride structure on ordering. The perfectly ordered structure has the composition AB (Fig. 3.18a). The superlattice is then simple cubic. The DO_3 superlattice type with perfectly ordered composition AB_3 also has a b.c.c. structure in the disordered state. The ordered state is shown in Fig. 3.18b. It is most easily described by saying that the superlattice is composed of four equal interpenetrating f.c.c. lattices with the origins at $(0, 0, 0)$ lattice 1, $(\frac{1}{2}, 0, 0)$ lattice 3, $(\frac{1}{4}, \frac{1}{4}, \frac{1}{4})$ lattice 4 and $(\frac{3}{4}, \frac{1}{4}, \frac{1}{4})$ lattice 2. The ordered state consists of B atoms occupying the sites of lattices 2, 3 and 4 with A atoms at type 1 lattice sites. The $L1_2$ superlattice has already been described (Fig. 3.17c). The fully ordered condition requires the composition

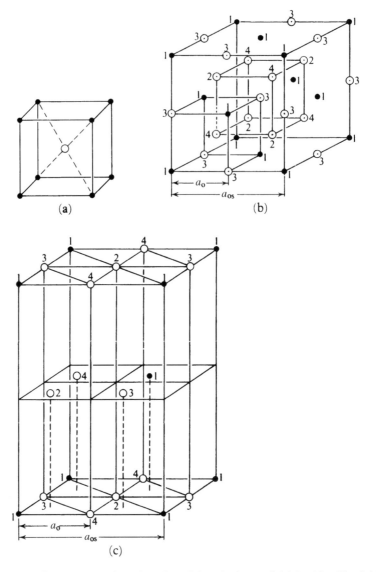

Figure 3.18 Structures of ordered solid solutions: (a) $L2_0$ (A, B), (b) DO_3, (c) DO_{19}

AB_3. A related superlattice type also of ideal composition AB_3 is called DO_{19} and is typified by Mg_3Cd. The disordered state is the close packed hexagonal structure. The ordered structure (Fig. 3.18c) can be described as four interpenetrating c.p.h. structures in parallel orientation, with the c axis the same length as in the ordered alloy but the lattice parameter a_{os} twice that of the corresponding

Table 3.3   Some examples of superlattice types

Superlattice	Examples
$L2_0$ $(CuZn)^a$	CuZn, FeCo, NiAl, CoAl, FeAl, AgMg, AuCd, NiZn
$L1_2$ (Cu_3Au)	Cu_3Au, Ni_3Mn, Ni_3Fe, Ni_3Al, Pt_3Fe, Au_3Cd, Co_3V, $TiZn_3$
DO_{19} (Mg_3Cd)	Mg_3Cd, Cd_3Mg, Ti_3Al, Ni_3Sn, Ag_3In, Co_3Mo, Co_3W, Fe_3Sn, Ni_3In, Ti_3Sn
DO_3 (Fe_3Al)	Cu_3Sb, Mg_3Li, Fe_3Al, Fe_3Si, Fe_3Be, Cu_3Al

^{a}The designation $L2_0$, etc., arises because in the literature the designation of structure type used in the early volumes of *Strukturbericht* is often followed. (See the suggestions for Further Reading at the end of the chapter).

disordered alloy. The origins of the sublattices in the ordered state are at $(0, 0, 0)$ sublattice 1, $(\frac{1}{2}, \frac{1}{2}, 0)$ sublattice 2, $(\frac{1}{2}, 0, 0)$ sublattice 3 and $(0, \frac{1}{2}, 0)$ sublattice 4. B atoms occupy the sites of sublattices 2, 3 and 4 and A atoms are found at the points of the sublattice 1.

3.9 POLYMERS

The investigation of the crystallography of long chain molecules (polymers) is relatively recent, extending only over the last quarter century, although many of the unique physical properties associated with polymers result from their ability to crystallize. In general, the synthetic polymer is built up by the repeated addition to the growing chain of one or more small chemical units called *monomers*. Polymers can typically have molecular weights between several hundred and several million. The predominant bond between atoms in a polymer molecule is the covalent bond with a dissociation energy of the order of 2 eV. The same type of strong bond is found in diamond. Many polymers have a zig-zag backbone consisting of covalently bonded carbon atoms. The angle between successive single bonds is usually a few degrees larger than the tetrahedral angle of 109.5°. A change in shape of the molecule may occur by rotation about these bonds in preference to either elongation or bending of the bonds.

If the monomers are polymerized according to a regular pattern of side groups or branches the polymer is known as a *tactic* polymer and is usually able to crystallize. If the monomers are polymerized in an irregular pattern the resulting *atactic* polymer usually cannot crystallize. The class of tactic polymers is divided into further subgroups depending upon the arrangement of side branches. For example, if an *isotactic* polymer is stretched out and viewed along its backbone, all of the branches lie along a single line, whereas in a *syndiotactic* polymer they alternate between one side of the backbone and the other.

Van der Waals forces (or 'secondary bonds') hold together molecules or portions of molecules not connected by covalent (or 'primary') bonds. These forces vary as the inverse sixth power of the separation of the atoms. The inert gases are bound by similar forces. They are at least two orders of magnitude

(a)

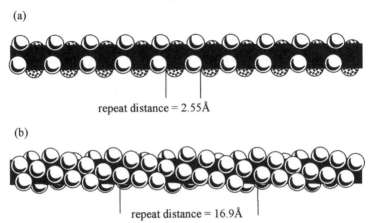

repeat distance = 2.55Å

(b)

repeat distance = 16.9Å

Figure 3.19 Molecular model of (a) polyethylene $(-CH_2-)_n$ and (b) poytetra-fluoroethylene $(-CF_2-)_n$, where the C atoms are contained in the chain (black) and the hydrogen and fluorine atoms are represented as white spheres. Note that in (b) the molecule forms an helix. Polytetrafluoroethylene can form crystals with different molecular repeat distances by changing the pitch of the helix

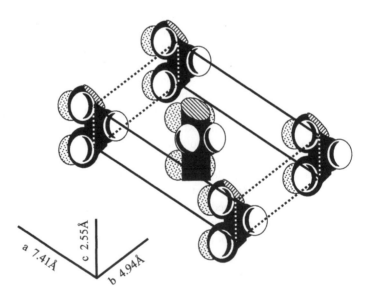

a 7.41Å c 2.55Å b 4.94Å

Figure 3.20 Unit cell of polyethylene. At the cell centre the planar zigzag backbone is rotated 90° about the c axis with respect to the chains at the cell corner where the planar zigzag backbone is parallel to (1 1 0). Note that the strongly bound polymer chain passes between many unit cells in the same crystal and may also pass unbroken and in approximately the same orientation into nearby crystals (see Fig. 3.21)

weaker than the covalent bond. Another particularly important secondary bond for determining the crystal structure of some polymers is the hydrogen bond, since its strength is greater than that of the other secondary bonds.

The attractive force between portions of polymer molecules that are in a crystalline lattice is greater than that between portions outside the crystal, so the density of a crystalline polymer is as much as 15% greater than that of the supercooled melt at the same temperature. Polymer crystals tend to be lamellar in external form. One and sometimes two dimensions are of the order of 100 Å, whereas the other dimensions can be two or more orders of magnitude larger.

The periodic placement of atoms in a tactic polymer molecule in a crystal is expressed in terms of the *repeat distance* or *repeat unit*. The repeat unit is the simplest arrangement of atoms by which the operation of linear translation (no rotation) will generate the structure of the extended molecule. It thus defines the conformation of the molecule in the crystal. The repeat unit usually contains one or more chemical repeat units. Figure 3.19 illustrates some repeat units for polyethylene $(-CH_2-)_n$ and one of the forms of polytetrafluoroethylene $(-CF_2-)_n$.

The packing of the molecules is expressed in terms of the unit cell with the familiar axes **a, b** and **c** and angles α, β, and γ, as in Chapter 1 of this book, and

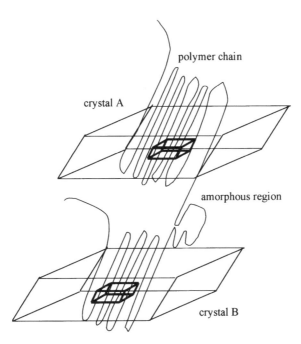

Figure 3.21 Illustration of a continuous polymer chain running through neighbouring crystals A and B. The small parallelepipeds indicate unit cells in the two crystals

contains one or more repeat units. Figure 3.20 shows an example of the unit cell of polyethylene. Note that the crystal forms with continuous covalent bonding along the c axis and van der Waals bonding along the a and b axes. Polymer crystal imperfections may result from such factors as terminal groups (ends of molecules), branches and cross-overs of the molecules as well as modifications of the numerous types of defects found in crystals composed of small molecules. The same polymer chain may run many times through the same crystal and through amorphous regions into neighbouring crystals, as illustrated in Fig. 3.21. Some common polymer crystal structures are presented in Section A5.9 in Appendix 5.

3.10 QUASICRYSTALS

The first sentence in Section 1.5 defines a crystal. This definition precludes the existence of fivefold symmetry as the argument in that section of this book shows. However, regular packing of groups of atoms in solids can show fivefold symmetry. This was observed to occur in nature in 1984 by Shechtmann, Blech, Gratias and Cahn who found that a rapidly cooled alloy of Mn–Al showed what is called *icosahedral* symmetry. An icosahedron[†] is a twenty-sided figure, each face of which is a regular triangle. The faces meet with five at each vertex, as illustrated in Fig. 3.22. Figure 3.23 shows a possible packing of spheres which demonstrates icosohedral coordination. The arrangements are called quasicrystals — a name coined by Levine and Steinhardt — and they occur in many binary and ternary intermetallic systems. Many are metastable and can only be obtained by rapid solidification. They are formed in systems where there is a strong

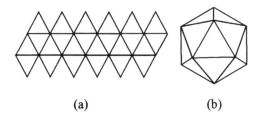

(a)   (b)

Figure 3.22 The net of (a) 20 equilateral triangles which make up (b) the solid icosahedron

---

[†] The icosahedron is one of the so-called Platonic solids — solids in which all faces and all vertices are identical. There are five such solids: cube, tetrahedron, octahedron, dodecahedron and icosahedron. Why five and only five is very simply shown and proof is given in the *13th Book of Euclid's Elements*. Each face is a regular polygon; hence the angles of the faces at any vertex must be equal and together less than 360° and must be three or more in number. Each angle in a regular triangle is 60°; hence a solid can be formed with three, four or five regular triangles as faces — the tetrahedron, octahedron and icosahedron respectively. The angle in a square is 90°; hence three will form the solid angle of a cube, but four will not. The angle in a regular pentagon is 108°; hence three will form a solid angle and four will not, yielding the (regular) dodecahedron.

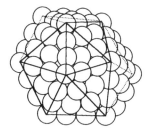

Figure 3.23 The packing of spheres into a solid icosahedron. (After Mackay [4])

tendency for pentagonal or icosahedral groupings of atoms to occur. (See the review by Steurer [3].)

PROBLEMS

3.1 Show that if a crystal has the close packed hexagonal structure and the atomic sites are occupied by equal-sized hard spheres in contact then the axial ratio c/a will be equal to 1.633.

3.2 What point group describes the symmetry of the interatomic forces acting on (a) a carbon atom in the diamond structure, (b) Cs in the CsCl structure, (c) Zn in the zinc-blende structure, (d) Zn in the wurtzite structure, (e) Ni in the NiAs structure, (f) As in the NiAs structure?
 Hint: Draw a stereogram showing the arrangement of the neighbours at various (close) distances from a given atom.

3.3 In the fluorite structure CaF_2 find the spacing along [1 1 1] of successive planes of (a) Ca atoms and (b) F atoms, in terms of the conventional cell side a.

3.4 Sketch the structure of white tin. Take the c/a ratio as 0.545 and show that the ratio of the separation of nearest neighbours to that of second nearest neighbours is 0.950.

3.5 Find the position of the 4_1 axis and of the axial glide planes in the unit cell of white tin.

3.6 Show with a diagram that if an element possessed the structure of bismuth with a value of α (cf. Table 3.2) of 60° the lattice points in Fig. 3.10 would form an f.c.c. lattice. If, in addition, $u = 0.250$ what type of lattice would the atomic positions form?

3.7 Molybdenum disulphide MoS_2 has a hexagonal primitive lattice with atoms in the following positions:

Mo at $(0, 0, 0)$ and $\left(\frac{2}{3}, \frac{1}{3}, \frac{1}{2}\right)$

S at $\left(0, 0, \frac{1}{2} + z\right) \left(0, 0, \frac{1}{2} - z\right) \left(\frac{2}{3}, \frac{1}{3}, z\right) \left(\frac{2}{3}, \frac{1}{3}, -z\right)$

(a) Draw a plan on $(0\,0\,0\,1)$ of a number of unit cells.
(b) Describe the stacking of the planes along $[0\,0\,0\,1]$ in the A, B, ..., α, β notation and compare with that in CaF_2.
(c) Locate the centres of symmetry.
(d) Re-express the coordinates of the atoms with respect to a new origin at a centre of symmetry having positive coordinates with respect to the old origin. Write the coordinates in the short form $\pm(x, y, z)$.

3.8 The intermetallic compound $CaCu_5$ has a hexagonal P lattice with $a = 5.092$ Å and $c = 4.086$ Å. The atomic coordinates are

Ca at $(0, 0, 0)$

Cu (1) at $\left(\frac{1}{3}, \frac{2}{3}, 0\right)$ and $\left(\frac{2}{3}, \frac{1}{3}, 0\right)$

Cu (2) at $\left(\frac{1}{2}, 0, \frac{1}{2}\right) \left(0, \frac{1}{2}, \frac{1}{2}\right)$ and $\left(\frac{1}{2}, \frac{1}{2}, \frac{1}{2}\right)$

Draw a projection of the structure on $(0\,0\,0\,1)$.
(a) How many formula units are there per unit cell?
(b) Find the interatomic distances: Ca–Cu (1), Ca–Cu (2), Cu (1)–Cu (2), Cu (2)–Cu (2).
(c) Calculate the density of $CaCu_5$.

At wts: Ca = 40.08, Cu 63.54. Mass of H atom = 1.66×10^{-24} g

3.9 A solution of carbon in face-centred cubic iron has a density of 8.142 g cm^{-3} and a lattice parameter of 3.583 Å. The solution contains 0.8% by weight of carbon. Is it an interstitial or a substitutional solid solution?

At wts: Fe = 55.85, C = 12.01. Mass of H atom = 1.66×10^{-24} g

3.10 A crystal of wüstite (approximate composition FeO) has the sodium chloride structure and contains 76.08% by weight of iron. The density is found to be 5.613 and the lattice parameter is 4.2816 Å. Does the crystal contain iron ion vacancies or interstitial oxygen ions? How many vacancies or interstitials are there per cm^3?

SUGGESTIONS FOR FURTHER READING

Bragg, W. L. and Claringbull, G. F., *Crystal Structures of Minerals*, G. Bell and Sons, London (1965).
Buchanan, R. C. and Park, T., *Materials Crystal Chemistry*, Marcel Dekker Inc., New York (1997).
deJong, W. F., *General Crystallography*, W. H. Freeman, San Francisco (1959).
Geil, P. H., *Polymer Single Crystals*, Wiley (1963).

Hahn, T., *International Tables for Crystallography*, Vols. II and III, 4th revised edn, Published for the International Union of Crystallography by Kluwer Academic, London (1989).

Hyde, B. G., Thompson, J. G. and Withers, R. L., 'Crystal structures of principal ceramic materials', *Structure and Properties of Ceramics*, (ed. M. Swain), of *Materials Science and Technology* (eds. R. W. Cahn, P. Haasen and E. J. Kramer), Vol. 11, Wiley–VCH, Chichester and Weinheim (1998), Ch. 1.

Janot, C., *Quasicrystals a Primer*, 2nd edn, Clarendon Press, Oxford (1994).

Pearson, W. B., *Handbook of Lattice Spacings and Structures of Metals*, Pergamon Press (1958).

Structure Reports A: Metals and Inorganic Compounds. eds. Utrecht: Bohn, Schetema and Holkama), for the International Union of Crystallography, 1965–.

Structure Reports B: Organic and Organometallic Compounds. (eds. Utrecht: Ooshoek, Scheltema and Holkema), for the International Union of Crystallography, 1965–.

Villars, P. and Calvert, L. D., *Pearson's Handbook of Crystallographic Data for Inter-metallic Phases*, American Society for Metals, c.1985.

Wyckoff, R. W. G., *Crystal Structures*, 2nd edn, Vols. 1–6, Wiley–Interscience, New York (1963–1971).

REFERENCES

1. *International Tables for X-Ray Crystallography* (ed. Theo Hahn), 4th revised edn, published for the International Union of Crystallography by Kluwer academic, London (1995).
2. W. L. Bragg and G. F. Claringbull, *Crystal Structures of Minerals*, G. Bell and Sons, London (1965).
3. W. Steurer, *Zeit. f. Kristallog.*, 190, 179 (1990).
4. A. L. Mackay, *Acta Crystallogr.*, **15**, 916 (1962).

4

Tensors

4.1 NATURE OF A TENSOR

There are a number of ways to introduce tensors. We shall regard them as quantities that relate two vectors. Each vector represents a definite physical quantity. Suppose we wish to know the relationship between the electric field in a crystal, represented by the vector $\mathbf{E}$, and the current density (i.e. current per unit area of cross-section perpendicular to the current), represented by the vector $\mathbf{J}$. In general, in a crystal the components of $\mathbf{J}$ referred to three mutually perpendicular axes (Ox_1, Ox_2, Ox_3), which we can call J_1, J_2, J_3, will be related to the components of $\mathbf{E}$, referred to the same set of axes in such a way that the components J_1, J_2, J_3 each depends linearly on *all three of the components* E_1, E_2 and E_3. It is usual to write this in the following way:

$$J_1 = \sigma_{11}E_1 + \sigma_{12}E_2 + \sigma_{13}E_3$$
$$J_2 = \sigma_{21}E_1 + \sigma_{22}E_2 + \sigma_{23}E_3$$
$$J_3 = \sigma_{31}E_1 + \sigma_{32}E_2 + \sigma_{33}E_3 \qquad (4.1)$$

The nine quantities σ_{11}, σ_{12}, σ_{13}, σ_{21}, σ_{22}, σ_{23}, σ_{31}, σ_{32}, σ_{33} are called the components of the conductivity tensor. The electrical conductivity tensor relates the vectors $\mathbf{J}$ and $\mathbf{E}$. If we write all of the relations (4.1) in the shorthand form

$$\mathbf{J} = \sigma\mathbf{E} \qquad (4.2)$$

we see that σ is a quantity that multiplies the vector $\mathbf{E}$ in order to obtain the vector $\mathbf{J}$. When a tensor relates two vectors in this way it is called a tensor of the second rank or second order.

Many physical properties are represented by tensors like the electrical conductivity tensor. Such tensors are called matter tensors. Some examples are given in Table 4.1.

In addition there are field tensors of which two are very important, namely stress and strain. The stress tensor relates the vector traction (force per unit area)

Table 4.1 Properties represented by second-rank tensors

Tensor	Vectors related	
Electrical conductivity	Electric field	Current density
Thermal conductivity	Thermal gradient (negative)	Thermal current density
Diffusivity	Concentration gradient (negative)	Flux of atoms
Permittivity	Electric field	Dielectric displacement
Dielectric susceptibility	Electric field	Dielectric polarization
Permeability	Magnetic field	Magnetic induction
Magnetic susceptibility	Magnetic field	Intensity of magnetization

and the orientation of an element of area in a stressed body. The strain tensor relates the displacement of a point in a strained body and the position of the point (see Chapter 5).

4.2 TRANSFORMATION OF COMPONENTS OF A VECTOR

If we know the components of a vector, say **P**, referred to a set of orthogonal axes (Ox_1, Ox_2, Ox_3), it is often necessary to know what are the components of the same vector referred to a different set of axes (Ox'_1, Ox'_2, Ox'_3) which are again mutually perpendicular and have the same origin as Ox_1, Ox_2 and Ox_3 (Fig. 4.1). We must first define how the two sets of axes are related to one another. We do this by setting down a table of the cosines of the angles between each new axis and the three of the old set. The table will appear as:

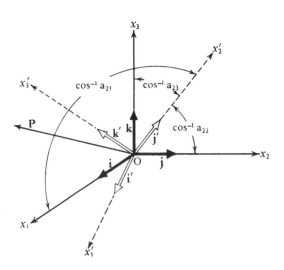

Figure 4.1

$$\begin{array}{c} \text{Old} \\ \begin{array}{c|ccc} & x_1 & x_2 & x_3 \\ \hline x'_1 & a_{11} & a_{12} & a_{13} \\ \text{New} \quad x'_2 & a_{21} & a_{22} & a_{23} \\ x'_3 & a_{31} & a_{32} & a_{33} \end{array} \end{array} \tag{4.3}$$

Here a_{32}, for example, is the cosine of the angle between the new axis 3 and the old axis 2, i.e. the angle x'_3Ox_2 in Fig. 4.1. Similarly, a_{11} is the cosine of the angle between Ox'_1 and Ox_1, etc. In an array of the type of (4.3) it should be noted that the sum of the squares of any row or column of the array is equal to one, because both sets of axes are orthogonal. Therefore, for example, $a_{11}^2 + a_{21}^2 + a_{31}^2 = 1$ and $a_{21}^2 + a_{22}^2 + a_{23}^2 = 1$, etc.

Now let $\mathbf{P}$ have components P_1, P_2, P_3 along the old axes so that

$$\mathbf{P} = P_1\mathbf{i} + P_2\mathbf{j} + P_3\mathbf{k} \tag{4.4}$$

where $\mathbf{i}$, $\mathbf{j}$, $\mathbf{k}$ are unit vectors along Ox_1, Ox_2, Ox_3 respectively. The component of $\mathbf{P}$ along the new axis Ox'_1 can be found by resolving the three old components, i.e. $P_1\mathbf{i}$, $P_2\mathbf{j}$ and $P_3\mathbf{k}$, along the new axis Ox'_1 and adding up the projections, thus obtaining

$$P'_1 = a_{11}P_1 + a_{12}P_2 + a_{13}P_3 \tag{4.4a}$$

Similarly, we find the new components of $\mathbf{P}$ along Ox'_2, Ox'_3, i.e. P'_2, P'_3, as

$$P'_2 = a_{21}P_1 + a_{22}P_2 + a_{23}P_3 \tag{4.4b}$$

and

$$P'_3 = a_{31}P_1 + a_{32}P_2 + a_{33}P_3 \tag{4.4c}$$

The vector $\mathbf{P}$ has the same magnitude and direction referred to any set of axes. If $\mathbf{i'}$, $\mathbf{j'}$, $\mathbf{k'}$ are unit vectors along Ox'_1, Ox'_2, Ox'_3 then

$$P_1\mathbf{i} + P_2\mathbf{j} + P_3\mathbf{k} = P'_1\mathbf{i'} + P'_2\mathbf{j'} + P'_3\mathbf{k'} \tag{4.5}$$

From the array (4.3) we can also write

$$\mathbf{i'} = a_{11}\mathbf{i} + a_{12}\mathbf{j} + a_{13}\mathbf{k}$$
$$\mathbf{j'} = a_{21}\mathbf{i} + a_{22}\mathbf{j} + a_{23}\mathbf{k}$$
$$\mathbf{k'} = a_{31}\mathbf{i} + a_{32}\mathbf{j} + a_{32}\mathbf{k} \tag{4.6a}$$

and, conversely,

$$\mathbf{i} = a_{11}\mathbf{i'} + a_{21}\mathbf{j'} + a_{31}\mathbf{k'}$$
$$\mathbf{j} = a_{12}\mathbf{i'} + a_{22}\mathbf{j'} + a_{32}\mathbf{k'}$$
$$\mathbf{k} = a_{13}\mathbf{i'} + a_{23}\mathbf{j'} + a_{33}\mathbf{k'} \tag{4.6b}$$

The relations (4.4a), (4.4b) and (4.4c) could also be deduced by using (4.6b) to express $\mathbf{i}, \mathbf{j}$ and $\mathbf{k}$ each in terms of $\mathbf{i}', \mathbf{j}'$ and $\mathbf{k}'$ and then substituting in (4.4), noting that P'_1 is the coefficient of the unit vector $\mathbf{i}'$. It is a useful exercise to write out the substitution in full. Proceeding in the same way we can also obtain expressions for the old components of the vector $\mathbf{P}$, i.e. $[P_1, P_2, P_3]$, in terms of the new components by substituting for $\mathbf{i}', \mathbf{j}', \mathbf{k}'$ in terms of $\mathbf{i}, \mathbf{j}, \mathbf{k}$ (using Eqns 4.6a) on the right-hand side of Eqn (4.5) and subsequently comparing terms on the two sides of the equation. Doing this we find

$$P_1 = a_{11}P'_1 + a_{21}P'_2 + a_{31}P'_3$$

$$P_2 = a_{12}P'_1 + a_{22}P'_2 + a_{32}P'_3$$

$$P_3 = a_{13}P'_1 + a_{23}P'_2 + a_{33}P'_3 \tag{4.7}$$

Relations (4.7) are the converse of relations (4.4a), (4.4b) and (4.4c). We are now in a position to relate these results on the transformation of the components of a vector to a relation such as (4.2).

A tensor is a quantity that multiplies one vector to give another, generally non-parallel, vector of different magnitude. If we return to our example of electrical conductivity in Section 4.1 we note that we could state Eqns (4.1) in the following way:

$$J_1\mathbf{i} = (\sigma_{11}E_1 + \sigma_{12}E_2 + \sigma_{13}E_3)\mathbf{i}$$

$$J_2\mathbf{j} = (\sigma_{21}E_1 + \sigma_{22}E_2 + \sigma_{23}E_3)\mathbf{j}$$

$$J_3\mathbf{k} = (\sigma_{31}E_1 + \sigma_{32}E_2 + \sigma_{33}E_3)\mathbf{k} \tag{4.8}$$

Having done this let us add up these three equations and express the right-hand side in a new form so that the equation reads

$$J_1\mathbf{i} + J_2\mathbf{j} + J_3\mathbf{k} = [\sigma_{11}\mathbf{ii} + \sigma_{12}\mathbf{ij} + \sigma_{13}\mathbf{ik} + \sigma_{21}\mathbf{ji} + \sigma_{22}\mathbf{jj} + \sigma_{23}\mathbf{jk}$$

$$+ \sigma_{31}\mathbf{ki} + \sigma_{32}\mathbf{kj} + \sigma_{33}\mathbf{kk}]{\cdot}(E_1\mathbf{i} + E_2\mathbf{j} + E_3\mathbf{k}) \tag{4.9}$$

The quantity in square brackets is to be regarded as an operator, which operates on vector $\mathbf{E}$ in order to obtain the vector $\mathbf{J}$ from $\mathbf{E}$. The form in which the operator is written here is the one that is appropriate to the orthogonal axes defined by the unit vectors $\mathbf{i}, \mathbf{j}, \mathbf{k}$.[†] We let the operator inside the square brackets 'multiply' the vector $\mathbf{E}(= E_1\mathbf{i} + E_2\mathbf{j} + E_3\mathbf{k})$ as if we were forming the scalar (or dot) product of vector multiplication (Appendix 2).

To understand Eqn (4.9) it is best to write out the right-hand side term by term. The first term $\sigma_{11}\mathbf{ii}{\cdot}E_1\mathbf{i} = \sigma_{11}E_1\mathbf{i}$, since $\mathbf{i}\cdot\mathbf{i} = 1$. The next term $\sigma_{11}\mathbf{ii}{\cdot}E_2\mathbf{j} = 0$, since $\mathbf{i}\cdot\mathbf{j} = 0$, the axes being orthogonal. Similarly, $\sigma_{11}\mathbf{ii}{\cdot}E_3\mathbf{k} = 0$ and $\sigma_{12}\mathbf{ij}{\cdot}E_1\mathbf{i} = 0$, but the fifth term $\sigma_{12}\mathbf{ij}{\cdot}E_2\mathbf{j} = \sigma_{12}E_2\mathbf{i}$, since $\mathbf{j}\cdot\mathbf{j} = 1$. If all terms

---

[†] Note that inside the square brackets we do not have $\mathbf{i} \cdot \mathbf{i}$ (which equals one) but $\mathbf{ii}$.

are written out and similar ones collected we finally obtain just Eqns (4.8), which can also be represented as in Eqns (4.1). Equation (4.9) is the expanded form of (4.2), or, put another way, the most succinct expression of (4.9) is (4.2).

Equation (4.9) contains the components of the conductivity tensor $\sigma_{11}, \sigma_{12}, \sigma_{13}, \ldots, \sigma_{33}$ referred to the axes (Ox_1, Ox_2, Ox_3) defined by the unit vectors **i, j, k.** We have seen how to transform the components of a vector so we can also transform the components of the tensor. If we take new axes represented by the unit vectors **i', j', k'**, the components of the conductivity tensor σ will be different and we can find what they are by substituting for **i, j, k** in Eqn (4.9) the values of **i', j', k'** referred to the new axes, using (4.6a). We would also have to substitute for E_1, E_2, E_3 the components of **E** along the new axes, i.e. E'_1, E'_2, E'_3, so that the new components of the tensor $\sigma'_{11}, \sigma'_{12}, \sigma'_{33}$, etc., multiplied by the new components of **E** in a form such as Eqn (4.9) will yield the components of the vector **J** referred also to the new set of axes. This is very lengthy to write out and so we shall first introduce a convenient short notation.

4.3 DUMMY SUFFIX NOTATION

The tensor T relates the vector **P** with components $[P_1 P_2 P_3]$ and $\mathbf{q}[q_1 q_2 q_3]$ so that

$$P_1 = T_{11}q_1 + T_{12}q_2 + T_{13}q_3$$
$$P_2 = T_{21}q_1 + T_{22}q_2 + T_{23}q_3$$
$$P_3 = T_{31}q_1 + T_{32}q_2 + T_{33}q_3 \qquad (4.10)$$

These three equations can be written as

$$P_1 = \sum_{j=1}^{j=3} T_{1j}q_j, \qquad P_2 = \sum_{j=1}^{j=3} T_{2j}q_j, \qquad P_3 = \sum_{j=1}^{j=3} T_{3j}q_j$$

or, even more succinctly, as

$$P_i = \sum_{j=1}^{j=3} T_{ij}q_j \qquad (i = 1, 2, 3)$$

We now leave out the summation sign and introduce a convention, called the Einstein convention, that if a suffix occurs twice in the same term summation with respect to that suffix is automatically implied. In what follows the summation will always run over the values 1, 2 and 3 for both i and j. Equations (4.10) can then be written as

$$P_i = T_{ij}q_j \qquad (4.11)$$

Here j is called a dummy suffix since it does not matter which letter except i is taken to represent it. Equation (4.11) could equally well read

$$P_i = T_{ik}q_k$$

If we apply this notation to Eqns (4.4a), (4.4b) and (4.4c) all three of them are contained in the expression

$$P'_i = a_{ij}P_j \tag{4.12a}$$

Note that the dummy suffix occurs in neighbouring places. If we apply the same notation to Eqns (4.7) we obtain

$$P_i = a_{ji}P'_j \tag{4.12b}$$

so that when writing the 'old' components in terms of the 'new' the dummy suffixes are separated.

When first learning this notation it is best to write out the sums term by term. Thus $P'_i = a_{ij}P_j$ gives, on expansion,

$$P'_1 = a_{1j}P_j$$
$$P'_2 = a_{2j}P_j$$
$$P'_3 = a_{3j}P_j$$

These three sums can then be expanded without fear of making an error.

4.4 TRANSFORMATION OF COMPONENTS OF A SECOND-RANK TENSOR

The two vectors $\mathbf{P} = [P_1 P_2 P_3]$, $\mathbf{q} = [q_1 q_2 q_3]$ are related by the tensor T_{ij}. The components of T_{ij} relate the components of $\mathbf{q}$ to those of $\mathbf{P}$ according to Eqn (4.10) or more succinctly according to Eqn (4.11). Now the *components* of the two vectors depend upon the choice of axes, because this choice determines the values $[P_1 P_2 P_3]$ and $[q_1 q_2 q_3]$. The vectors $\mathbf{P}$ and $\mathbf{q}$ themselves do not change. When the axes are changed and hence the components of $\mathbf{P}$ and $\mathbf{q}$ change, the components T_{ij} will also change.

Suppose $\mathbf{P}[P_1 P_2 P_3]$ and $\mathbf{q}[q_1 q_2 q_3]$ are related by the tensor T such that

$$\mathbf{P} = T\mathbf{q} \tag{4.13}$$

or equivalently, for a specific choice of the axes **i, j, k,**

$$P_i = T_{ij}q_j \tag{4.14}$$

If now we choose new axes **i', j', k'** so that

$$P'_i = T'_{ij}q'_j \tag{4.15}$$

we then wish to find the relation between the nine components T'_{ij} and the nine components T_{ij}.

We can find these directly by writing the relation (4.13) in the operational form of Eqn (4.9), i.e.

$$P_1\mathbf{i} + P_2\mathbf{j} + P_3\mathbf{k} = [T_{11}\mathbf{ii} + T_{12}\mathbf{ij} + T_{13}\mathbf{ik} + T_{21}\mathbf{ji} + T_{22}\mathbf{jj} + T_{23}\mathbf{jk}$$
$$+ T_{31}\mathbf{ki} + T_{32}\mathbf{kj} + T_{33}\mathbf{kk}]\cdot(q_1\mathbf{i} + q_2\mathbf{j} + q_3\mathbf{k}) \quad (4.16)$$

We now substitute for $\mathbf{i}$, $\mathbf{j}$, $\mathbf{k}$ in this equation the values of these quantities in terms of $\mathbf{i'}$, $\mathbf{j'}$ and $\mathbf{k'}$ according to (4.6b). The substitution in the left-hand side yields $P'_1\mathbf{i'} + P'_2\mathbf{j'} + P'_3\mathbf{k'}$, directly from (4.5). The quantity in round brackets on the right-hand side of (4.16) similarly becomes $q'_1\mathbf{i'} + q'_2\mathbf{j'} + q'_3\mathbf{k'}$. The tensor operator in square brackets is treated in the same way and substitution made for $\mathbf{i}$, $\mathbf{j}$ and $\mathbf{k}$ in terms of $\mathbf{i'}$, $\mathbf{j'}$ and $\mathbf{k'}$ from (4.6b). There will be 81 terms in all. If all those containing (say) the vectors $\mathbf{j'k'}$ are collected, there will be nine of these and the coefficient of $\mathbf{j'k'}$ will be the new component T'_{23}. If we write out the expression for T'_{23} it is

$$T'_{23} = (a_{21}a_{31}T_{11} + a_{21}a_{32}T_{12} + a_{21}a_{33}T_{13} + a_{22}a_{31}T_{21} + a_{22}a_{32}T_{22}$$
$$+ a_{22}a_{33}T_{23} + a_{23}a_{31}T_{31} + a_{23}a_{32}T_{32} + a_{23}a_{33}T_{33}) \quad (4.16a)$$

Similarly, the other eight components of T'_{ij} may be found by collecting the terms in $\mathbf{i'i'}$, $\mathbf{i'j'}$, $\mathbf{i'k'}$, $\mathbf{j'j'}$, $\mathbf{j'i'}$, $\mathbf{k'i'}$, $\mathbf{k'j'}$ and $\mathbf{k'k'}$ respectively.

The complete transformation scheme of the components of T_{ij} is most easily written out and memorized using the dummy suffix notation. When using the old axes $\mathbf{i}$, $\mathbf{j}$, $\mathbf{k}$ we have the old components of $\mathbf{P}$, viz. $[P_1\,P_2\,P_3]$. These are related to the old components of $\mathbf{q}$, viz. $[q_1\,q_2\,q_3]$, by Eqn (4.14). We now choose a new set of axes $\mathbf{i'}$, $\mathbf{j'}$, $\mathbf{k'}$ related to the old by the array of cosines given by (4.3). To find the new components of the tensor T_{ij} relating $[P'_1\,P'_2\,P'_3]$ with $[q'_1\,q'_2\,q'_3]$ we must perform the following operations:

(a) Write P' in terms of P.
(b) Write P in terms of q.
(c) Write q in terms of q'.

Operation (a) is carried out by (4.12a) where j is a dummy suffix. Therefore we can write (4.12a) as

$$P'_i = a_{ik}P_k$$

Now we use Eqn (4.14) to carry out operation (b) and have

$$P_k = T_{kl}q_l$$

Operation (c) is carried out by using an equation like (4.12b) for the q_l. Finally, when we combine the three operations we have

$$P'_i = a_{ik}P_k = a_{ik}T_{kl}q_l = a_{ik}T_{kl}a_{jl}q'_j$$

or

$$P'_i = a_{ik}T_{kl}a_{jl}q'_j$$

This is of the form of Eqn (4.15) and therefore, by comparing the two, we have the important formula

$$T'_{ij} = a_{ik}T_{kl}a_{jl} \tag{4.17}$$

which is the transformation formula of the components of a second-rank tensor when the axes of reference are changed.

Equation (4.17) is very important and must be fully understood. The dummy suffix notation is such that the order in which a product is written does not matter and so the right-hand side of Eqn (4.17) could be written as

$$a_{ik}a_{jl}T_{kl} \qquad \text{or as} \qquad T_{kl}a_{jl}a_{ik}$$

Equation (4.17) is most easy to remember as

$$T'_{ij} = a_{ik}a_{jl}T_{kl} \tag{4.18}$$

When fully written out Eqn (4.18) contains 81 terms, nine for each value of i and j. For each value of i and j the sum is to be taken over both k and l. It is helpful to write out the expansion summing over each of these separately. Thus, as an example,

$$T'_{23} = a_{2k}a_{3l}T_{kl}$$

so summing over k

$$T'_{23} = a_{21}a_{3l}T_{1l} + a_{22}a_{3l}T_{2l} + a_{23}a_{3l}T_{3l}$$

Now, summing over l, we have

$$T'_{23} = a_{21}a_{31}T_{11} + a_{21}a_{32}T_{12} + a_{21}a_{33}T_{13} + a_{22}a_{31}T_{21} + a_{22}a_{32}T_{22}$$
$$+ a_{22}a_{33}T_{23} + a_{23}a_{31}T_{31} + a_{23}a_{32}T_{32} + a_{23}a_{33}T_{33} \tag{4.19}$$

which is precisely the same as (4.16a).

The reverse transformation to (4.18), i.e. the transformation giving the components of the tensor T_{ij} in terms of those of the new components T'_{ij}, can be accomplished in exactly the same way. This will lead to

$$T_{ij} = a_{ki}a_{lj}T'_{kl} \tag{4.20}$$

It is a useful exercise to prove this.

4.5 DEFINITION OF A TENSOR

If the operator T relates two vectors $\mathbf{P}$ and $\mathbf{q}$ so that T may be written in the form of the operator given in Eqn (4.16), then T is known as a dyadic or tensor of the second rank. Alternatively, we can define a tensor of the second rank as a physical quantity which, with respect to a set of axes x_i, has nine components which transform, according to Eqn (4.18), when the axes of reference are changed.[†]

A tensor T_{ij} is said to be symmetric if $T_{ij} = T_{ji}$ and to be skew symmetric or antisymmetric if $T_{ij} = -T_{ji}$. Thus the tensor with components

$$\begin{bmatrix} 9 & -4 & 1 \\ -4 & 7 & 2 \\ 1 & 2 & 6 \end{bmatrix}$$

is symmetric. A skew symmetric tensor necessarily has all the diagonal terms T_{ij} equal to zero. Thus the tensor

$$\begin{bmatrix} 0 & -\gamma & \beta \\ \gamma & 0 & -\alpha \\ -\beta & \alpha & 0 \end{bmatrix} \tag{4.21}$$

is antisymmetric.

Whether or not a tensor is symmetric or antisymmetric is independent of the choice of axes. The condition for two tensors to be equal is that each component of one shall be equal to the corresponding component of the other.

Any second-rank tensor or dyadic may be expressed as the sum of a symmetric tensor and of a skew symmetric tensor. This is because any component T_{ij} can always be written as

$$T_{ij} = \tfrac{1}{2}(T_{ij} + T_{ji}) + \tfrac{1}{2}(T_{ij} - T_{ji})$$

The first term then gives the component of a symmetric tensor and the second term that of an antisymmetric tensor.

Any *symmetric* tensor S_{ij} can be transformed by a suitable choice of axes so that it takes on the simple form

$$\begin{bmatrix} S_{11} & 0 & 0 \\ 0 & S_{22} & 0 \\ 0 & 0 & S_{33} \end{bmatrix} \tag{4.22}$$

i.e. all $S_{ij} = 0$ unless $i = j$. Such a tensor when expressed in this form is said to be referred to its *principal axes*. When referred to its principal axes the components S_{11}, S_{22}, S_{33} are called the *principal components* and are often written

---

[†] According to this definition a scalar is a tensor of zero rank, because it transforms according to the law $\rho' = \rho$ when the axes are changed. A vector is a tensor of the first rank, since when the axes are changed its components transform according to the rule (Eqn 4.12a) $P'_i = a_{ij}P_j$. The components of the second-rank tensor transform according to $T'_{ij} = a_{ik}a_{jl}T_{kl}$; those of the third-rank tensor as $T'_{ijk} = a_{il}a_{jm}a_{kn}T_{lmn}$ and so on. Just as a second-rank tensor relates two vectors so a third-rank tensor relates a second-rank tensor and a vector and a fourth-rank tensor relates two second-rank tensors and so on. Sometimes the word 'order' is used instead of rank to describe the number of suffixes appropriate to a tensor quantity.

simply as S_1, S_2, S_3 respectively. The corresponding dyadic is of the form

$$S_1\mathbf{ii} + S_2\mathbf{jj} + S_3\mathbf{kk} \qquad (4.23)$$

We shall not prove any of the above statements. Proofs can be found in the books listed at the end of this chapter.

4.6 TENSOR REFERRED TO PRINCIPAL AXES

When referred to its principal axes the symmetric tensor (S_{ij}) relating vectors **P** and **q** has only its diagonal components S_{11}, S_{22} and S_{33} not equal to zero. Then the equations

$$P_i = S_{ij}q_j$$

have the simple form

$$P_1 = S_{11}q_1, \qquad P_2 = S_{22}q_2, \qquad P_3 = S_{33}q_3$$

Now let us return to the simple example of electrical conductivity. The conductivity tensor σ_{ij} is symmetric and when referred to its principal axes, we have all the $\sigma_{ij} = 0$ except σ_{11}, σ_{22} and σ_{33}. Then suppose we apply an electrical field **E** and let it have components $[E_1\,E_2\,O]$ along these principal axes. Then

$$J_1 = \sigma_{11}E_1 + \sigma_{12}E_2 + \sigma_{13}E_3 = \sigma_{11}E_1$$

because σ_{12} and σ_{13} both equal zero.

Similarly, $J_2 = \sigma_{22}E_2$ and $J_3 = \sigma_{33}E_3$, but we have taken $E_3 = 0$ and therefore $J_3 = 0$. We can represent this on a diagram as in Fig. 4.2. This diagram can be

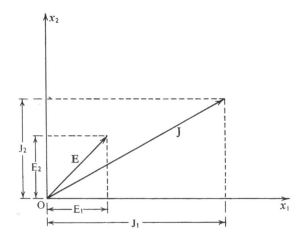

Figure 4.2

constructed by drawing $\mathbf{E}$, finding E_1 and E_2 and multiplying E_1 by σ_{11} to get J_1 and E_2 by σ_{22} to get J_2. We then construct J from its components along the axes, i.e. J_1 and J_2. It should be noted carefully that $\mathbf{E}$ and $\mathbf{J}$ are not parallel. If $\mathbf{E}$ were directed along Ox_1, we would have $J_1 = \sigma_{11}E_1$ because then both J_2 and J_3 equal zero. Thus, if $\mathbf{E}$ is directed along a principal axis, $\mathbf{J}$ will be parallel to $\mathbf{E}$. The conductivity will be different, in general, along the three principal axes, because σ_{11}, σ_{22} and σ_{33} will not have the same value.

When we speak of the conductivity in a particular direction we mean that if $\mathbf{E}$ is applied in that direction and the current density is measured *in the same direction*, to give a value $J_{\parallel}$, then the conductivity in this direction is $J_{\parallel}$ divided by the magnitude of $\mathbf{E}$, i.e. $J_{\parallel}/|\mathbf{E}|$. We can find an expression for this by resolving $\mathbf{J}$ parallel to $\mathbf{E}$. Suppose $\mathbf{E}$ is applied in a direction such that its direction cosines with respect to the principal axes of the conductivity tensor are $\cos\alpha$, $\cos\beta$, $\cos\gamma$. Then we have

$$J_1 = \sigma_{11}E_1 = \sigma_{11}E\cos\alpha$$

$$J_2 = \sigma_{22}E_2 = \sigma_{22}E\cos\beta$$

$$J_3 = \sigma_{33}E_3 = \sigma_{33}E\cos\gamma$$

where E is the magnitude of $\mathbf{E}$ (i.e. $|\mathbf{E}|$).
Then

$$J_{\parallel} = J_1\cos\alpha + J_2\cos\beta + J_3\cos\gamma$$

$$= E(\sigma_{11}\cos^2\alpha + \sigma_{22}\cos^2\beta + \sigma_{33}\cos^2\gamma)$$

Therefore the conductivity in the direction considered is

$$\sigma = J_{\parallel}/E$$

$$= \sigma_{11}\cos^2\alpha + \sigma_{22}\cos^2\beta + \sigma_{33}\cos^2\gamma \qquad (4.24)$$

The steps in deriving Eqn (4.24) are illustrated diagrammatically in Fig. 4.3 for the simple case where $\mathbf{E}$ is normal to the principal axis Ox_3 of the conductivity tensor so that $\cos\gamma = 0$.

It is instructive to derive the result given in Eqn (4.24) in a different way. Suppose we consider the meaning of the component σ'_{11} of the conductivity tensor whether or not it is referred to principal axes. Component σ'_{11} relates the electric field along axis Ox'_1 to the component of the current along the same axis Ox'_1. If, therefore, we wish to find the value of the conductivity in a particular direction, having been given the components of the conductivity tensor referred to its principal axes, we can proceed as follows. Choose a new set of axes such that Ox'_1 is along the direction we are interested in. Then the component σ'_{11} of the conductivity tensor referred to this new set of axes gives us the conductivity along this particular direction. We are only interested in σ'_{11}, so in writing out

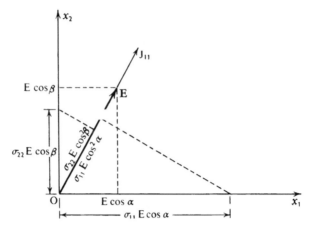

Figure 4.3 Derivation of the magnitude of the conductivity in a particular direction. The figure is drawn for $\sigma_{11} = 2.5$ and $\sigma_{22} = 0.75(\text{cm})^{-1}$

the array of the a_{ij} for this transformation we need know only the values of the cosines of the angles between Ox'_1 and the principal axes of the conductivity tensor, Ox_1, Ox_2 and Ox_3. Following the scheme of the expressions (4.3) we need to know only a_{11}, a_{12} and a_{13}. These quantities are $\cos \alpha$, $\cos \beta$ and $\cos \gamma$ respectively. Using the transformation formula (4.18), we have

$$\sigma'_{ij} = a_{ik} a_{jl} \sigma_{kl}$$

and hence

$$\sigma'_{11} = a_{1k} a_{1l} \sigma_{kl} \tag{4.25}$$

Since σ_{ij} is given referred to its principal axes, all the terms in Eqn (4.25) are zero unless $k = l$, so

$$\sigma'_{11} = a_{11} a_{11} \sigma_{11} + a_{12} a_{12} \sigma_{22} + a_{13} a_{13} \sigma_{33}$$

Substituting for a_{11}, a_{12}, a_{13}, we obtain

$$\sigma'_{11} = \sigma_{11} \cos^2 \alpha + \sigma_{22} \cos^2 \beta + \sigma_{33} \cos^2 \gamma$$

as we had in Eqn (4.24).

It is clear that we could have proceeded in exactly the same way to find the conductivity in a particular direction, even if the values of the components of σ_{ij}, the conductivity tensor, had not been given to us referred to principal axes. In this case there would have been, in general, nine terms in the expansion of Eqn (4.25). The derivation of (4.25) did not assume that we were given the components of σ_{ij} referred to principal axes. Thus, we can say that to find the value of a property of a crystal in a particular direction we proceed as follows. Let the components of the tensor representing this property be given as T_{ij} referred to axes (Ox_1, Ox_2, Ox_3). Choose an axis along the direction of interest. Let this axis have direction cosines referred to (Ox_1, Ox_2, Ox_3) of a_1, a_2, a_3 respectively. Then the value of the property T in the direction we are interested in is given by

$$T = a_i a_j T_{ij} \qquad (4.26)$$

This relation holds for all second-rank tensors whether or not they are symmetrical.

4.7 LIMITATIONS IMPOSED BY CRYSTAL SYMMETRY

The discussion in this section applies to tensors used to represent *physical properties* of crystals and strictly only to crystals without defects.

Suppose two vector quantities **P** and **q** are related by the tensor T_{ij} in a *crystal* so that this tensor represents a property of the crystal. The crystal then determines the relationship between components of **P** and those of **q** and we choose to represent this relationship by means of the tensor T_{ij}. We state the components of T_{ij} with respect to certain axes. If we now take a different set of axes in the crystal, the components of **P** and of **q** along these new axes will in general be different and, further, the relationship between the components of **P** and those of **q** along the new axes will alter.

However, if we select new axes in the crystal which are related to the old ones by some symmetry operation, the values of the components of **P** and of **q** with respect to the new set of axes will be different from those referred to the old ones, but the *relationship* between the components of **P** and those of **q** will be the same, so that

$$T'_{ij} = T_{ij}$$

This is because the properties of the crystal are the same along the new axes as along the old ones. In order for this to be true for particular symmetry elements, limitations are imposed upon the components of T_{ij} when representing a property of a crystal. The same general type of argument will also impose restrictions upon the components of tensors of the third and higher orders; an example occurs in Section 5.5.

Physical properties characterized by a second-order tensor are necessarily *centrosymmetric*. This is implicit in the linear relations

$$P_i = T_{ij} q_j$$

because if we substitute $-P_i$ for P_i and $-q_j$ for q_j (i.e. we reverse the directions of **P** and of **q**) the relation is still satisfied by the same values of T_{ij}.

It will assist in understanding what follows if we look at the preceding statement in another way. Suppose that for one set of axes (Ox_1, Ox_2, Ox_3) the relation between the P_i and the q_j are given by T_{ij}. If we now reverse the axes of reference, leaving **P** and **q** the same as before, this corresponds to choosing a new set of axes such that the array of the a_{ij} relating the axes is

$$\begin{matrix} -1 & 0 & 0 \\ 0 & -1 & 0 \\ 0 & 0 & -1 \end{matrix}$$

We now apply the transformation formula (4.18)

$$T'_{ij} = a_{ik}a_{jl}T_{kl}$$

All a_{ij} are equal to zero unless $i = j$. Therefore

$$T'_{ij} = a_{ii}a_{jj}T_{ij} = T_{ij} \text{ since } a_{11} = a_{22} = a_{33} = -1$$

What we have done here is to leave the measured quantities **P** and **q** the same and to imagine the crystal inverted through a centre of symmetry. We have obtained the same result as if we had reversed the directions of **P** and **q**.

Now suppose the crystal contains a diad axis of symmetry. If we measure a certain property along a certain direction with respect to this axis and then rotate the crystal 180° about this axis and remeasure the property, we will get the same value as before. This imposes restrictions on the values of the components of the *symmetric* second-rank tensor S_{ij} which represents this property. To see what these restrictions are, take axes (Ox_1, Ox_2, Ox_3) and suppose there is a diad axis along Ox_2. Initially we suppose the tensor S_{ij} to be symmetric and so to have six independent components. Now, if we take new axes related to the old by a rotation of 180° about Ox_2 the physical property must remain the same as before. The new axes are related to the old by the array of a_{ij} as

	Ox_1	Ox_2	Ox_3
Ox'_1	-1	0	0
Ox'_2	0	1	0
Ox'_3	0	0	-1.

The components of the tensor S_{ij} with respect to the new axes are given in terms of the old components by

$$S'_{ij} = a_{ik}a_{jl}S_{kl}$$

and we must have $S'_{ij} = S_{ij}$ for all i and j. If we work out the components S'_{ij} one by one we find

$$S'_{11} = a_{11}a_{11}S_{11} = S_{11}, \qquad S'_{22} = a_{22}a_{22}S_{22} = S_{22}$$

$$S'_{33} = a_{33}a_{33}S_{33} = S_{33}, \qquad S'_{13} = a_{11}a_{33}S_{13} = S_{13}$$

but

$$S'_{23} = a_{22}a_{33}S_{23} = -S_{23} \quad \text{and} \quad S'_{12} = a_{11}a_{22}S_{12} = -S_{12}$$

We insisted, however, that $S'_{23} = S_{23}$ and therefore, if the property represented by S_{ij} is to be the same under this transformation, we must have $S_{23} = 0$. Similarly, S_{12} must be zero. Thus a symmetrical second-rank tensor representing a physical property of a crystal with a twofold axis of rotational symmetry must have the components S_{23} and S_{12} equal to zero when referred to axes so that Ox_2 corresponds to the diad axis.

Table 4.2 summarizes the limitations on the components of a symmetrical second-rank tensor representing a physical property of a crystal for each of the crystal systems. In the cubic, tetragonal, hexagonal and trigonal systems the principal axes of the tensor coincide with the crystal axes and in the mono-clinic system one of each coincides. The derivation of this table is dealt with in Section 4.8 by a rapid method. However, we could deduce part of it using the procedure we have just outlined for a diad axis and applying this to the defining symmetry elements of each of the crystal classes. Table 4.2 applies to any of the physical properties of a crystal listed in Table 4.1 (see the footnote following Eqn (4.27)).

4.8 REPRESENTATION QUADRIC

Section 4.6 has demonstrated that the measured value of a property that can be represented by a second-order tensor will vary with the direction of measurement

Table 4.2 Number of independent components of physical properties represented by second-rank (order) tensors

Crystal system	Orientation of principal axes with respect to the crystal axes	Form of tensor	Number of independent components
Cubic	Any; representation quadric is a sphere	$\begin{bmatrix} S & 0 & 0 \\ 0 & S & 0 \\ 0 & 0 & S \end{bmatrix}$	1
Tetragonal Hexagonal Trigonal[a]	x_3 parallel to 4, 6, 3 or $\bar{3}$	$\begin{bmatrix} S_1 & 0 & 0 \\ 0 & S_1 & 0 \\ 0 & 0 & S_3 \end{bmatrix}$	2
Orthorhombic	x_1, x_2, x_3 parallel to the diads along the x, y, z axes	$\begin{bmatrix} S_1 & 0 & 0 \\ 0 & S_2 & 0 \\ 0 & 0 & S_3 \end{bmatrix}$	3
Monoclinic	x_2 parallel to the diad along the y axis	$\begin{bmatrix} S_{11} & 0 & S_{13} \\ 0 & S_{22} & 0 \\ S_{13} & 0 & S_{33} \end{bmatrix}$	4
Triclinic	Not fixed	$\begin{bmatrix} S_{11} & S_{12} & S_{13} \\ S_{12} & S_{22} & S_{23} \\ S_{13} & S_{23} & S_{33} \end{bmatrix}$	6

[a]A hexagonal cell is used.

and that this variation can be found by the procedure of that section, leading to Eqn (4.26), which is

$$T = a_i a_j T_{ij}$$

If we expand this equation we obtain

$$T = a_i(a_1 T_{i1} + a_2 T_{i2} + a_3 T_{i3})$$

$$= a_1^2 T_{11} + a_1 a_2 T_{12} + a_1 a_3 T_{13} + a_2 a_1 T_{21} + a_2^2 T_{22}$$

$$+ a_2 a_3 T_{23} + a_3 a_1 T_{31} + a_3 a_2 T_{32} + a_3^2 T_{33} \qquad (4.27)$$

Here a_1, a_2, a_3 are the direction cosines of the direction we are considering with respect to the same axes to which T_{ij} is referred. We shall now restrict our discussion to symmetrical second-rank tensors so that $T_{ij} = T_{ji}$. When second-order tensors are used to represent physical properties then in nearly every case the tensor can be shown to be symmetrical (a proof always involves thermodynamic considerations).[†] When second-rank tensors are used to represent stresses and small strains then we shall show in Chapter 5 that they are always symmetric. When the tensor is symmetric, terms such as $a_1 a_2 T_{12}$ and $a_2 a_1 T_{21}$ are equal. If we denote the tensor now by S_{ij}, S being chosen to remind ourselves that it is symmetric, we have, from Eqn (4.27),

$$S = a_1^2 S_{11} + a_2^2 S_{22} + a_3^2 S_{33} + 2a_1 a_2 S_{12} + 2a_1 a_3 S_{13} + 2a_2 a_3 S_{23} \qquad (4.28)$$

If now S_{ij} were referred to principal axes as well as being symmetric, Eqn (4.27) would reduce to

$$S = a_1^2 S_1 + a_2^2 S_2 + a_3^2 S_3 \qquad (4.29)$$

because $S_{12} = S_{13} = S_{23} = 0$. Equation (4.28) is of the same form as the equation to the general surface of the second degree (called a quadric) written in polar coordinates and referred to its centre as origin. The general equation is

$$\frac{1}{r^2} = A \cos^2 \alpha + B \cos^2 \beta + C \cos^2 \gamma + 2D \cos \alpha \cos \beta + 2E \cos \alpha \cos \gamma$$

$$+ 2F \cos \beta \cos \gamma \qquad (4.30)$$

where r is the radius vector and $\cos \alpha$, $\cos \beta$, $\cos \gamma$ the direction cosines of r with respect to a set of orthogonal axes. When the general surface of the second degree is referred to its principal axes it takes the form

$$\frac{1}{r^2} = A \cos^2 \alpha + B \cos^2 \beta + C \cos^2 \gamma \qquad (4.31)$$

[†] All of the properties listed in Table 4.1 can be shown to be represented by *symmetric* second-rank tensors. However, in the case of the transport properties, e.g. electrical or thermal conductivity and diffusion, appeal to experiment is necessary to complete the proof. To the authors' knowledge the requisite experimental data are lacking in the case of the diffusivity tensor.

which is of the same form as Eqn (4.29). When the axes, to which a general surface of the second degree is referred, are altered, the coefficients in Eqn (4.30) transform in the same way as do the components of a symmetrical second-rank tensor. Because of this the variation of a given *property* of a crystal with direction, given by Eqns (4.28) and (4.29), can be represented by a figure in three-dimensional space which gives the variation of the property S with direction.

The general surface of the second degree can be an ellipsoid or a hyperboloid of one or two sheets. To explain the procedure for representing the variation of the property S with direction in a crystal we shall confine ourselves to the case when the values of S_1, S_2 and S_3 in Eqn (4.29) and A, B and C in Eqn (4.31) are all positive. In this case the second-degree surface, which we shall call the representation quadric, is an ellipsoid.[†]

Fixing attention on Eqns (4.29) and (4.31), we see that if we construct an ellipsoid of semi-principal axes $1/\sqrt{S_1}$, $1/\sqrt{S_2}$, $1/\sqrt{S_3}$, as in Fig. 4.4, then the length r of any radius vector of the ellipsoid (representation quadric) is equal to the reciprocal of the square root of the magnitude of the property S in that direction. Thus, if we return to the example of electrical conductivity, and we know the values σ_1, σ_2, σ_3 of the components of the electrical conductivity referred to principal axes, then we can construct the conductivity quadric (which, if σ_{ii} are all positive, will be an ellipsoid). If now a field **E** is applied in any direction the magnitude of the conductivity in that direction can be found by drawing a radius vector r in the direction of **E**, measuring the value of r and taking the reciprocal of the square root of r to find the conductivity in that direction.

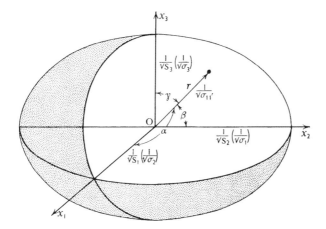

Figure 4.4 Representation ellipsoid

[†] If two of the coefficients S_1, S_2, S_3 are positive and the other negative the quadric is a hyperboloid of one sheet and if two are negative and the other positive it is a hyperboloid of two sheets. If all three values are negative, the quadric is an imaginary ellipsoid.

We are now in a position to deal rapidly with the limitations imposed by crystal symmetry on any physical property of a crystal which can be described by a *symmetric* second-rank tensor. This symmetry of physical properties of a crystal is governed by *Neumann's principle*, which states: 'The symmetry elements of a physical property of a crystal must include the symmetry elements of the point group.'[†] The representation quadric shows how the given property varies with direction, and its symmetry is that of the physical property. The number of independent coefficients in the equation to the representation surface (Eqn 4.30) equals the number of independent components of the tensor representing the physical property (Eqn 4.28). If we look for the symmetry of the representation quadric of a property of a crystal belonging to each of the Laue groups,[‡] we find that the symmetry of the representation quadric for a symmetrical second-rank tensor is governed only by the symmetry of the crystal system. The results are collected in Table 4.2.

A *cubic* crystal must possess four triad axes. The representation quadric is then a sphere and so the axes, to which the symmetrical second rank tensor representing any of the properties shown in Table 4.1 is referred, are then of no consequence. The crystal is *isotropic* with regard to this property. *Hexagonal, trigonal and tetragonal* crystals have *two* independent components for the properties indicated in Table 4.1. This is because the representation quadric must be a surface of revolution about the hexad, triad or tetrad respectively. A general quadric shows three mutually perpendicular diad axes. In the *orthorhombic* system these must lie along the crystal diads. There are then just *three* independent components of the tensor necessary to describe the physical property. In the monoclinic system one of the diad axes of the quadric must lie parallel to the crystallographic diad. There are *four* independent components of the symmetrical second-rank tensor describing a property. These can be thought of as three necessary to specify the length of the semi-axes of the quadric and one to fix the angle in the plane, normal to the crystal diad, between a crystal axis and a principal axis of the quadric. In the *triclinic system*, since a symmetrical second-rank tensor is centrosymmetric and the holosymmetric class in this system possesses just a centre of symmetry, there are six independent components possible for any property that may be represented by a symmetrical second-rank tensor.

PROBLEMS

4.1 Define a tensor of the second rank. Write down two physical properties of crystals that can be represented by a second-rank tensor, and for each state the two physical quantities that are related by the tensor. For one

---

[†] The principle does not state that the symmetry of the physical property is the same as that of the point group. The symmetry of physical properties is often higher than that of the point group.

[‡] The Laue groups are important here because we showed in Section 4.7 that all properties represented by a second-rank tensor are necessarily centrosymmetric.

of your examples write down in full the equations relating the components of the two physical quantities and explain your notation. Explain the physical significance of the tensor component D_{12}, where D is the diffusivity tensor.

4.2 Does the array of the a_{ij} in Eqn (4.3) represent the components of a second-rank tensor?

4.3 Express the following tensor as the sum of a symmetric tensor and of an antisymmetric tensor:

$$\begin{pmatrix} 12 & 6 & 0 \\ 4 & 7 & 0 \\ 0 & 0 & 3 \end{pmatrix}$$

4.4 If σ is a symmetric dyadic and $\mathbf{n}$ and $\mathbf{b}$ are vectors, show that

$$(\sigma \cdot \mathbf{n}) \cdot \mathbf{b} = (\sigma \cdot \mathbf{b}) \cdot \mathbf{n}.$$

4.5 A crystal possesses a single fourfold rotational axis of symmetry parallel to the z axis. Find the necessary relations between the components of a second-rank tensor representing a physical property of the crystal when the tensor is referred to axes parallel to the crystal axes. How do you reconcile your result with the entry for tetragonal crystals in Table 4.2?

4.6 Prove that whether a tensor is symmetric or antisymmetric is independent of the choice of axes.
Hint. Show that if $T_{ij} = T_{ji}$, then $T'_{ij} = T'_{ji}$.

4.7 (a) The electrical conductivity tensor of a crystal has the components

$$\sigma_{ij} = \begin{pmatrix} 18.25 & -\sqrt{3} \times 2.25 & 0 \\ -\sqrt{3} \times 2.25 & 22.75 & 0 \\ 0 & 0 & 9 \end{pmatrix} \times 10^{+8}\Omega^{-1}m^{-1}$$

Take new axes rotated $60°$ about x_3 in a clockwise direction looking along negative x_3 and write down a table of the direction cosines between the new and old axes as in Section 4.2. Check that the sum of the squares of the a_{ij} in each row and column equals 1.

(b) Write down the components of the conductivity tensor referred to the new axes. Check that σ_{ij} has not altered. The tensor is now referred to principal axes.

(c) Draw a section of the representation quadric (i.e. the conductivity ellipsoid in this case) in the plane $x'_3 = 0$. Draw radius vectors of the resulting ellipse at angles of $30°$ and $60°$ to x'_1 and hence find the conductivity in these directions.

(d) Check your results in (c) by direct calculations.

(e) Suppose an electric field $\mathbf{E}$ of $100 Vm^{-1}$ is established in the direction OE at $60°$ to x'_1 in the plane $x'_3 = 0$. Write down the components of this electric field along x'_1 and x'_2 and thence find the electric current densities in these two directions. Finally, find the component of the resultant current density along **E** and hence the electrical conductivity in this direction.

(f) Find the direction of the resultant current vector **J** in (e).

(g) Draw the direction of the resultant current density vector on your diagram prepared in (c) and note that **J** is parallel to the normal to the surface of the representation quadric at the point where OE meets the representation quadric.

This last result is general and is known as the *radius–normal* property. It can be stated as follows. If S_{ij} are the components of a symmetrical second-rank tensor relating the vectors **p** and **q** so that $p_i = S_{ij}q_j$ then the direction of **p** for a given **q** can be found by drawing a radius vector OQ of the representation quadric parallel to **q** and finding the normal to the quadric at Q.

SUGGESTIONS FOR FURTHER READING

Lovett, D. R., *Tensor Properties of Crystals*, Adam Hilger, Bristol and New York (1989).
Page, L., *Introduction to Theoretical Physics*, Van Nostrand, New York (1928).
Nye, J. F., *Physical Properties of Crystals*, Clarendon Press, Oxford (1985).
Sands, D. E., *Vectors and Tensors in Crystallography*, Addison-Wesley, Reading, Massachusetts and London (1982).
Wooster, W. A., *Tensors and Group Theory for the Physical Properties of Crystals*, Clarendon Press Oxford, (1973).

PART II
Imperfect Crystals

5

Strain, Stress and Elasticity

5.1 STRAIN: INTRODUCTION

When forces are applied to crystals or when imperfections are formed inside them the atoms change their *relative* positions. This change in the relative positions is called strain, and in this section we describe a way of specifying it.

The basic ideas of strain are those of *extension* and of *shear*. The first is easily understood. Consider a very thin rod of length l (Fig. 5.1a). Let the rod be stretched so that l' is its length in the strained state. Then the extension, e, is defined as

$$e = \frac{l' - l}{l} \tag{5.1}$$

that is

$$l' = l(1 + e)$$

Thus e is the ratio of the change in length to the original length. It is positive in tension and negative in compression. Shear is a measure of the change in angle between two lines in a body when the body is distorted. It is defined as follows (Fig. 5.1b). If OP, OR are two perpendicular straight lines in the unstrained state and O'P', O'R' are the positions of the corresponding lines in the strained state, then the shear strain γ associated with these two directions at the point O is defined as

$$\gamma = \tan\left(\frac{\pi}{2} - \theta\right) \tag{5.2}$$

where θ is the angle between O'P' and O'R' in the strained state.

The description of general distortions, involving both extensions and shears, can be very complicated. We shall consider two important special cases.

(a) *Infinitesimal strain*, where it is assumed that the quantities e and γ are so small that their squares and products can be neglected. For elastic strains within a crystal, this assumption is a good one, except at points which are only a very few atomic distances from a defect.

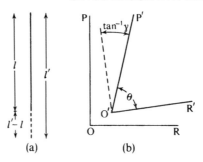

Figure 5.1   (a) Definition of extension, (b) definition of shear

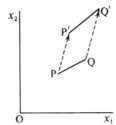

Figure 5.2

(b) *Homogeneous strain*, where every part of the strained region suffers the same distortion. Large homogeneous strains will be encountered in Chapters 6, 10 and 11.

5.2 INFINITESIMAL STRAIN

The distortion of a body can be described by giving the displacement of each point from its location in the undistorted state. Any displacements which do *not* correspond to a translation or rotation of the body as a whole will produce a strain.

We now confine ourselves to two dimensions and choose an origin fixed in space such as O in Fig. 5.2. Let P be a point with coordinates (x_1, x_2) in the unstrained state, which after distortion of the body moves to the point P'. (Then the displacement of the point P is the vector PP'.) Let the coordinates of P' be $(x_1 + u_1, x_2 + u_2)$.

Now consider a point Q with coordinates $(x_1 + dx_1, x_2 + dx_2)$ lying infinitesimally close to P in the unstrained state. After deformation Q moves to Q'. Now in a strained body the displacement of Q will not be exactly the same as that of P. The displacement of Q to Q' has components $(u_1 + du_1, u_2 + du_2)$. We can write

$$du_1 = \frac{\partial u_1}{\partial x_1} dx_1 + \frac{\partial u_1}{\partial x_2} dx_2 \qquad (5.3)$$

and

$$du_2 = \frac{\partial u_2}{\partial x_1}dx_1 + \frac{\partial u_2}{\partial x_2}dx_2 \qquad (5.4)$$

Defining the four quantities at the point P,

$$e_{11} = \frac{\partial u_1}{\partial x_1}, \qquad e_{12} = \frac{\partial u_1}{\partial x_2}, \qquad e_{22} = \frac{\partial u_2}{\partial x_2}, \qquad e_{21} = \frac{\partial u_2}{\partial x_1}$$

Eqns (5.3) and (5.4) can be compactly written as

$$du_i = e_{ij}\, dx_j \qquad (j = 1, 2) \qquad (5.5)$$

Since du_i and dx_j are vectors, then according to the definition given in Chapter 4, e_{ij} is a tensor. We shall now demonstrate the physical meaning of the various e_{ij} when each is very small compared to unity.

Let us take two special positions of the vector PQ first parallel to $Ox_1(PQ_1)$ and then parallel to $Ox_2(PQ_2)$ and find out how a rectangular element at P is distorted (Fig. 5.3). For PQ_1 we put $dx_2 = 0$ and obtain

$$du_1 = \frac{\partial u_1}{\partial x_1}dx_1 = e_{11}\, dx_1 \qquad (5.6)$$

$$du_2 = \frac{\partial u_2}{\partial x_1}dx_1 = e_{21}\, dx_1 \qquad (5.7)$$

From Fig. 5.3 it is clear that e_{11} measures the extension per unit length of PQ_1 resolved along Ox_1, while e_{21} measures the anticlockwise rotation of PQ_1, provided that e_{11} and e_{21} are small. Similarly, e_{22} measures the change in length per unit length of PQ_2, resolved along Ox_2 and e_{12} measures the small clockwise rotation of PQ_2.

The tensor e_{ij} is not completely satisfactory as a measure of strain, because it is possible to have non-zero components of e_{ij} without there being any distortion of the body. Thus, consider a rigid-body rotation through a small angle ω, illustrated in Fig. 5.4. Evidently we have $e_{11} = e_{22} = 0$, but $e_{12} = -\omega$ and $e_{21} = \omega$. To

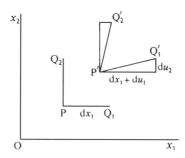

Figure 5.3

remove the component of rotation from a general e_{ij} we express it as the sum of an antisymmetrical tensor and a symmetrical tensor, so that

$$e_{ij} = \tfrac{1}{2}(e_{ij} - e_{ji}) + \tfrac{1}{2}(e_{ij} + e_{ji}) \tag{5.8}$$

Then $\omega_{ij} = \tfrac{1}{2}(e_{ij} - e_{ji})$ measures the rotation (as in Fig. 5.4) and $\varepsilon_{ij} = \tfrac{1}{2}(e_{ij} + e_{ji})$ is defined as the *pure strain*. Figure 5.5 shows the geometrical interpretation of Eqn (5.8). It should be noted that the change in angle between the two lines originally at right angles in Fig. 5.5 is $2\varepsilon_{12}$. Thus the shear component of the pure strain tensor ε_{12} is half the shear strain γ defined in Eqn (5.2).

  In specifying infinitesimal strain in three dimensions the method is the same as for two dimensions. The variation of displacement $\mathbf{u}(= u_1, u_2, u_3)$ with variation in position $\mathbf{x}(= x_1, x_2, x_3)$ is used to define nine components of a tensor e_{ij},

$$e_{ij} = \frac{\partial u_i}{\partial x_j} \qquad (i, j = 1, 2, 3) \tag{5.9}$$

The strain tensor ε_{ij} is then defined as the symmetrical part of e_{ij},

$$\varepsilon_{ij} = \tfrac{1}{2}(e_{ij} + e_{ji}) \tag{5.10}$$

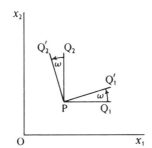

Figure 5.4

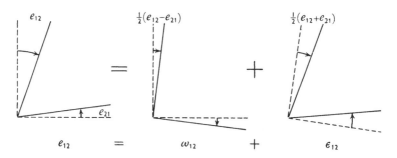

Figure 5.5

In full,

$$\begin{bmatrix} \varepsilon_{11} & \varepsilon_{12} & \varepsilon_{13} \\ \varepsilon_{12} & \varepsilon_{22} & \varepsilon_{23} \\ \varepsilon_{13} & \varepsilon_{23} & \varepsilon_{33} \end{bmatrix} = \begin{bmatrix} e_{11} & \frac{1}{2}(e_{12}+e_{21}) & \frac{1}{2}(e_{13}+e_{31}) \\ \frac{1}{2}(e_{21}+e_{12}) & e_{22} & \frac{1}{2}(e_{23}+e_{32}) \\ \frac{1}{2}(e_{31}+e_{13}) & \frac{1}{2}(e_{32}+e_{23}) & e_{33} \end{bmatrix}$$

The diagonal components of ε_{ij} are the changes in length per unit length of lines parallel to the axes and are called the tensile strains. The off-diagonal components measure shear strains and, for instance, ε_{13} is one-half the change in angle between two lines originally parallel to the Ox_1 and Ox_3 axes.

Since pure strain is a symmetrical second-rank tensor it can be referred to principal axes (Chapter 4). The shear components then vanish and we have

$$\varepsilon = \begin{bmatrix} \varepsilon_1 & 0 & 0 \\ 0 & \varepsilon_2 & 0 \\ 0 & 0 & \varepsilon_3 \end{bmatrix} \tag{5.11}$$

The geometrical meanings of the principal strains ε_1, ε_2 and ε_3 are shown in Fig. 5.6, which shows that a unit cube whose edges are parallel to the principal axes is changed into a rectangular-sided figure with edges $(1 + \varepsilon_1)$, $(1 + \varepsilon_2)$ and $(1 + \varepsilon_3)$. The change in volume per unit volume is called the dilatation, and is given by

$$\Delta = (1+\varepsilon_1)(1+\varepsilon_2)(1+\varepsilon_3) - 1 = \varepsilon_1 + \varepsilon_2 + \varepsilon_3 \tag{5.12}$$

where terms such as $\varepsilon_1\varepsilon_2$ and $\varepsilon_1\varepsilon_2\varepsilon_3$ have been neglected, since the strains are small. When the strain is referred to any other axes, the dilatation is always given by the invariant sum $(\varepsilon_{11} + \varepsilon_{22} + \varepsilon_{33})$. The components of either e_{ij} or ε_{ij} can be transformed to new axes by means of the standard transformation rule for second-rank tensors (Chapter 4).

In many books on elasticity the strain is described by the quantities

$$\begin{bmatrix} \varepsilon_x & \gamma_{xy} & \gamma_{zx} \\ \gamma_{xy} & \varepsilon_y & \gamma_{yz} \\ \gamma_{zx} & \gamma_{yz} & \varepsilon_z \end{bmatrix}$$

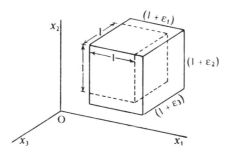

Figure 5.6

Then we see that, for instance, $\gamma_{xy} \equiv 2\varepsilon_{12}$. The γ values are usually called shear strains and it should be noted that they are equal to twice the corresponding tensor shear strains defined in this book. This array of the values of ε and γ does *not* form a tensor.

5.3 HOMOGENEOUS STRAIN

A body is said to be homogeneously strained if the distortion is the same everywhere. Then the e_{ij} in Eqns (5.5) and (5.9) are constants, independent of position in the body. The equation

$$du_i = \frac{\partial u_i}{\partial x_j} dx_j = e_{ij} \, dx_j \tag{5.13}$$

can then be integrated immediately to obtain

$$u_i = (u_0)_i + e_{ij}x_j \qquad (i, j = 1, 2, 3) \tag{5.14}$$

Equation (5.14) shows that in homogeneous strain, the displacements are linear functions of the position coordinates.

The constants $(u_0)_i$ represent the translation of the body as a whole, and are of no further interest. Subtracting this translation from the displacement, we obtain the residual displacement

$$x'_i - x_i = e_{ij}x_j \tag{5.15}$$

where x'_i are the new coordinates (referred to axes that have been translated by $(u_0)_i$ but not rotated) of the point which was originally at x_i. Equation (5.15) written out in full is

$$x'_1 = (1 + e_{11})x_1 + e_{12}x_2 + e_{13}x_3$$
$$x'_2 = e_{21}x_1 + (1 + e_{22})x_2 + e_{23}x_3$$
$$x'_3 = e_{31}x_1 + e_{32}x_2 + (1 + e_{33})x_3$$

The shape into which a line or surface whose equation is

$$f(x_1, x_2, x_3) = 0 \tag{5.16}$$

is deformed can be determined by solving Eqn (5.15) for x_1, x_2 and x_3 in terms of x'_1, x'_2 and x'_3, and substituting these values into Eqn (5.16). In this way it can be shown that any homogeneous strain, large or small, has the following properties:

(a) Straight lines remain straight lines and in general are rotated and stretched or contracted, all straight lines in the same direction being rotated through the same angle and stretched or contracted in the same ratio. Similarly, planes are deformed into planes.

(b) A sphere is deformed into an ellipsoid. The ellipsoid into which a sphere of unit radius is deformed is called the strain ellipsoid.

(c) The axes of the strain ellipsoid are derived from three mutually perpendicular diameters of the unit sphere, which are called the principal axes. These axes form the only set of three orthogonal directions which remain orthogonal after the strain.

(d) In the unstrained state there is a particular ellipsoid, called the reciprocal strain ellipsoid, which deforms into the unit sphere. Its axes are the principal axes.

In general, the principal axes are rotated by the strain. A deformation which leaves the principal axes unrotated is called a pure strain. A general homogeneous strain can be accomplished in two stages: a pure strain in which the principal axes receive their correct extensions, followed by a rotation in which they are brought to their final position. The above results are applied in discussing the geometry of martensitic transformations in Chapter 11.

We shall now mention some very simple types of homogeneous strain.

SIMPLE EXTENSION

$$
\begin{array}{l}
x'_1 = kx_1, \\
x'_2 = x_2, \\
x'_3 = x_3,
\end{array}
\qquad
\begin{pmatrix}
e_{11} = k - 1 & e_{12} = 0 & e_{13} = 0 \\
e_{21} = 0 & e_{22} = 0 & e_{23} = 0 \\
e_{31} = 0 & e_{32} = 0 & e_{33} = 0
\end{pmatrix}
\qquad (5.17)
$$

The principal axes Ox_1, Ox_2 and Ox_3 are not rotated; therefore simple extension is a pure strain.

SIMPLE SHEAR

$$
\begin{array}{l}
x'_1 = x_1 + gx_2, \\
x'_2 = x_2, \\
x'_3 = x_3,
\end{array}
\qquad
\begin{pmatrix}
e_{11} = 0 & e_{12} = g & e_{13} = 0 \\
e_{21} = 0 & e_{22} = 0 & e_{23} = 0 \\
e_{31} = 0 & e_{32} = 0 & e_{33} = 0
\end{pmatrix}
\qquad (5.18)
$$

where g is the magnitude of the shear strain, as defined in Eqn (5.2). A more detailed description of simple shear will be developed in Chapters 6 and 10, where it is required for the description of slip and twinning. Simple shear is not a pure strain; its non-rotational part is described below.

PURE SHEAR

$$
\begin{array}{l}
x'_1 = kx_1, \\
x'_2 = k^{-1}x_2, \\
x'_3 = x_3,
\end{array}
\qquad
\begin{pmatrix}
e_{11} = k - 1 & e_{12} = 0 & e_{13} = 0 \\
e_{21} = 0 & e_{22} = 1/k - 1 & e_{23} = 0 \\
e_{31} = 0 & e_{32} = 0 & e_{33} = 0
\end{pmatrix}
\qquad (5.19)
$$

In this strain, the principal axes are Ox_1, Ox_2 and Ox_3, and they are not rotated. The relationship between a pure shear and a simple shear is illustrated by Fig. 5.7. A simple shear of strength g, referred to axes Ox_1^0, Ox_2^0 distorts a circle into the ellipse shown in Fig. 5.7. The same ellipse can be produced by a strain which is a pure shear, referred to axes Ox_1, Ox_2, with Ox_1 at an angle of $+\pi/4 - \frac{1}{2}\tan^{-1} g/2$ to Ox_1^0. The required magnitude of the pure shear is given by

$$k - 1/k = g \qquad\qquad (5.20)$$

In order for the *displacements* produced by this pure shear to be identical to those produced by the simple shear, it must be preceded by a rotation of $\tan^{-1} g/2$. Thus point P, for example, is rotated to Q, and the pure shear then carries Q to P'. Alternatively, the displacements of the simple shear can be produced by a pure shear referred to axes at $\pm\pi/4 + \frac{1}{2}\tan^{-1} g/2$ to Ox_1^0, followed by a rotation of $\tan^{-1} g/2$. These axes are principal axes of the simple shear; Ox_1, Ox_2 and Ox_3 (which remains unchanged) are the principal axes of the simple shear in their final, rotated position.

When the simple shear is of very small magnitude ($g = \gamma$, where $\gamma \ll 1$), it is equivalent to a pure shear referred to axes rotated through an angle of $\pi/4$ together with a rotation of $\gamma/2$. This can also be seen by carefully studying Fig. 6.8.

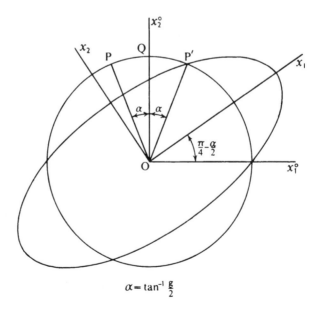

$$\alpha = \tan^{-1}\frac{g}{2}$$

Figure 5.7

5.4 STRESS

When atoms in a crystal are displaced relative to one another, forces act on them, tending to restore their normal spatial relationship. If a plane is passed through a strained region of crystal, it will be found that the atoms on one side of the plane are exerting forces upon the atoms on the other side. At any point on the plane, the force acting per unit area of plane is defined as the traction on the plane at that point. The force is customarily resolved into components normal and parallel to the plane, f_n and f_s respectively, and corresponding normal and shear stress components are defined as follows:

$$\sigma_n = \operatorname*{Lt}_{\delta A \to 0} \frac{f_n}{\delta A} \tag{5.21}$$

$$\tau = \operatorname*{Lt}_{\delta A \to 0} \frac{f_s}{\delta A} \tag{5.22}$$

The area δA is shown in Fig. 5.8; it surrounds the point P, so that by finding the limit of the series of average stresses $f/\delta A$ as δA decreases, the stress at the point itself can be defined. For this definition to be meaningful, the force at P must not be changing too sharply on an atomic scale.

The procedure outlined above could be repeated for any plane passing through the point P. To define the stress at P completely, it is sufficient to specify the stresses acting on three mutually perpendicular planes passing through P. It is then possible to calculate the stress acting on any other plane by means of a force balance. Figure 5.9a shows the stresses acting on the faces of an infinitesimal cube located at P, using two alternative notations. In the numerical suffix notation, the stress component σ_{ij} is the component in the direction of the x_j axis of the traction acting on the face whose normal is parallel to x_i. The sign convention is that a stress is positive if the force component and the outward-going normal of the face on which it acts have the same sense, relative to the axes to which they are respectively parallel. The letter-suffix notation follows the same pattern, except

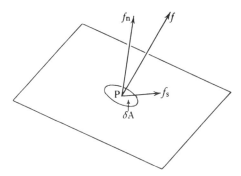

Figure 5.8

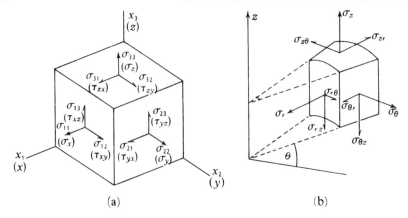

Figure 5.9 Definition of stress components (a) in Cartesian coordinates and (b) in cylindrical polar coordinates

that shear stresses are distinguished by the letter τ. It is sometimes convenient to work with cylindrical polar coordinates instead of Cartesian coordinates, and the stress components for this case are shown in Fig. 5.9b.

Since the cube in Fig. 5.9a is supposed to be infinitesimally small, the forces acting on opposite faces must be equal and opposite. If the cube were of side δ, the shear stresses τ_{xy} and τ_{yx} would produce a couple $(\tau_{xy} - \tau_{yx})\delta^3$, tending to rotate the cube about the z axis. Since the moment of inertia of the cube varies as δ^5, it vanishes as $\delta \to 0$ more rapidly than does the couple $(\tau_{xy} - \tau_{yx})\delta^3$, so that, to avoid an infinite angular acceleration,

$$\tau_{xy} = \tau_{yx}$$

and in general

$$\sigma_{ij} = \sigma_{ji} \tag{5.23}$$

In order to calculate the stress acting upon any given plane at P, an infinitesimal body is constructed having one face, ABC, parallel to the given plane and all its other faces parallel to the cube faces (Fig. 5.10). A balance of the forces acting on the tetrahedron ABCO then determines the required stress. Let $\mathbf{P} = [P_1 P_2 P_3]$ be the traction exerted on the face ABC by the material outside the tetrahedron. Setting the sum of forces in the Ox_1 direction equal to zero

$$P_1(\text{ABC}) = \sigma_{11}(\text{BOC}) + \sigma_{21}(\text{AOC}) + \sigma_{31}(\text{AOB})$$

or

$$P_1 = \sigma_{11}l_1 + \sigma_{21}l_2 + \sigma_{31}l_3 \tag{5.24}$$

where l_i are the direction cosines of the normal to the plane ABC. We obtain similar equations for P_2 and P_3, and all three equations can be written

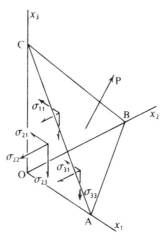

Figure 5.10   Calculation of the stress acting on the plane ABC

compactly as

$$P_i = \sigma_{ji}l_j \tag{5.25}$$

Resolving **P** into two components, one normal and one parallel to ABC, we can obtain the normal stress σ_n and shear stress τ, where σ_n is given by the sum of the components P_i, each resolved normal to ABC,

$$\sigma_n = P_i l_i$$

Therefore, from Eqn (5.25),

$$\sigma_n = \sigma_{ji}l_j l_i \tag{5.26}$$

By resolving the forces,

$$\tau = [P^2 - \sigma_n^2]^{1/2} \tag{5.27}$$

Equation (5.25) shows that the components of the vector **P** representing the traction on a plane are linearly related to those of the unit vector which is normal to that plane. It follows that the relating coefficients σ_{ij} form a tensor of the second rank. When the axes of reference are rotated, the components of a given stress transform according to the general transformation law

$$\sigma_{ij} = a_{ik}a_{jl}\sigma_{kl} \tag{5.28}$$

where a_{ij} is the cosine of the angle between the new Ox_i axis and the old Ox_j axis. Equation (5.28) can also be derived directly from Eqn (5.25) by setting the normal to the plane ABC parallel to one of the new axes.

Because the stress tensor is symmetrical ($\sigma_{ij} = \sigma_{ji}$), it is always possible to find a set of axes, the principal axes, such that a cube with its edges parallel to them has no shear stresses acting upon its faces. Referred to the principal axes, the stress takes the form

$$\begin{pmatrix} \sigma_1 & 0 & 0 \\ 0 & \sigma_2 & 0 \\ 0 & 0 & \sigma_3 \end{pmatrix}$$

where σ_1, σ_2 and σ_3 are called the principal stresses.

In very special cases, a body may be under a homogeneous stress, i.e. the stress may be the same at every point. If the stress is homogeneous, parallel cubes of any size and at any location have the same tractions on their faces. Such a condition obtains in the tensile test of a uniform rod, the stress at every point being

$$\begin{pmatrix} F/A & 0 & 0 \\ 0 & 0 & 0 \\ 0 & 0 & 0 \end{pmatrix}$$

where F is the tensile force, A the cross-sectional area of the rod and the tensile axis is parallel to Ox_1.

More generally, the stress in a body varies from point to point. However, the different components of stress cannot vary in an entirely arbitrary way, independently of one another. The variations in the different stress components must be related in such a way that there is a balance of the forces acting on any internal body (assuming that the body as a whole is in static equilibrium). Figure 5.11 shows one face of a cube of side 2δ in a varying stressfield. By setting the sum of forces in the Ox_1 direction equal to zero we obtain

$$\frac{\partial \sigma_{11}}{\partial x_1} + \frac{\partial \sigma_{21}}{\partial x_2} + \frac{\partial \sigma_{31}}{\partial x_3} + f_1 = 0 \tag{5.29}$$

where f_1 is the component of body force per unit volume (e.g. due to gravity) acting in the Ox_1 direction. Two similar equations are obtained with respect to the Ox_2 and Ox_3 directions, and all three equations of equilibrium can be written compactly as

$$\frac{\partial \sigma_{ij}}{\partial x_i} + f_j = 0 \tag{5.30}$$

The repetition of the suffix i in Eqn (5.30) implies the summation of the differential coefficients obtained by setting $i = 1, 2$ and 3 in turn.

We shall now mention some important types of stress. A pure shear stress is of the form $\sigma_{12} = S$, with all other σ_{ij} zero, or

$$\begin{pmatrix} 0 & S & 0 \\ S & 0 & 0 \\ 0 & 0 & 0 \end{pmatrix}$$

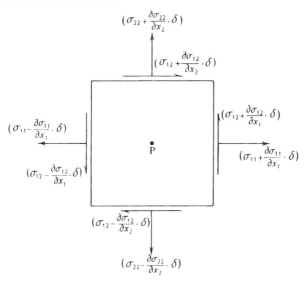

Figure 5.11

When the axes of reference are rotated through $45°$ about Ox_3, in an anticlockwise sense, the stress components transform to

$$\begin{pmatrix} S & 0 & 0 \\ 0 & -S & 0 \\ 0 & 0 & 0 \end{pmatrix}$$

The principal stresses of a pure shear are therefore equal tensile and compressive stresses acting on planes at $45°$ to those on which the shear stress acts alone (Fig. 5.12).

In a general state of stress, shear stress components exist on all planes other than those normal to the principal axes. It can be shown that the largest shear stress occurs on the plane which bisects the angle between the planes on which the greatest and least of the principal stresses act (Fig. 5.13). If $\sigma_1 < \sigma_2 < \sigma_3$, the magnitude of the largest shear stress is $\frac{1}{2}(\sigma_3 - \sigma_1)$.

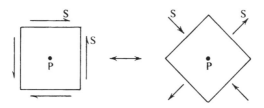

Figure 5.12   A shear stress at the point P

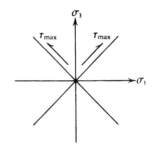

Figure 5.13 If σ_3 is the largest principal stress and σ_1 is the smallest, the largest shear stress acts in the direction shown, on a plane that is normal to the plane of the figure

A hydrostatic pressure p takes the form

$$\begin{pmatrix} -p & 0 & 0 \\ 0 & -p & 0 \\ 0 & 0 & -p \end{pmatrix}$$

which remains the same, no matter what axes of reference are chosen. In other words, a hydrostatic pressure produces no shear stress anywhere.

It is possible to produce any state of stress by superimposing a hydrostatic pressure or tension upon three pure shear stresses. To prove this, we note that a general stress can be written as the sum of a hydrostatic component and a component called the deviatoric stress. Referred to principal axes, the stress can be written as

$$\begin{pmatrix} \sigma_1 & 0 & 0 \\ 0 & \sigma_2 & 0 \\ 0 & 0 & \sigma_3 \end{pmatrix} = \frac{1}{3}\begin{pmatrix} \sigma_{ii} & 0 & 0 \\ 0 & \sigma_{ii} & 0 \\ 0 & 0 & \sigma_{ii} \end{pmatrix} + \begin{pmatrix} \sigma_1 - \frac{1}{3}\sigma_{ii} & 0 & 0 \\ 0 & \sigma_2 - \frac{1}{3}\sigma_{ii} & 0 \\ 0 & 0 & \sigma_3 - \frac{1}{3}\sigma_{ii} \end{pmatrix}$$
$$(5.31)$$

where $\sigma_{ii} = \sigma_1 + \sigma_2 + \sigma_3$. The hydrostatic component $\frac{1}{3}\sigma_{ii}$ is invariant, i.e. the sum $(\sigma_{11} + \sigma_{22} + \sigma_{33})$ remains the same whatever axes of reference are chosen. The deviatoric stress can be expressed as the sum of three pure shear stresses, as follows:

$$\begin{pmatrix} \sigma_1 - \frac{1}{3}\sigma_{ii} & 0 & 0 \\ 0 & \sigma_2 - \frac{1}{3}\sigma_{ii} & 0 \\ 0 & 0 & \sigma_3 - \frac{1}{3}\sigma_{ii} \end{pmatrix} = \frac{1}{3}\begin{pmatrix} \sigma_1 - \sigma_2 & 0 & 0 \\ 0 & \sigma_2 - \sigma_1 & 0 \\ 0 & 0 & 0 \end{pmatrix}$$

$$+ \frac{1}{3}\begin{pmatrix} \sigma_1 - \sigma_3 & 0 & 0 \\ 0 & 0 & 0 \\ 0 & 0 & \sigma_3 - \sigma_1 \end{pmatrix} + \frac{1}{3}\begin{pmatrix} 0 & 0 & 0 \\ 0 & \sigma_2 - \sigma_3 & 0 \\ 0 & 0 & \sigma_3 - \sigma_2 \end{pmatrix} \quad (5.32)$$

5.5 ELASTICITY OF CRYSTALS

In a strained crystal, the stress at any point is linearly related to the strain at that point, provided that the strain is very small. This relationship is known as Hooke's law. The physical basis for Hooke's law is merely that at small strains the relative displacements of the atoms are small, so that the interatomic force–separation relationship operates over such a small range as to be sensibly linear. In most solids, strains do in fact remain small over a quite useful range of stress.

We have seen that a stress and a small strain can each be represented by a symmetrical second-rank tensor, having six independent components. The most general linear dependence of stress on strain has the form

$$\sigma_{11} = c_{1111}\varepsilon_{11} + c_{1112}\varepsilon_{12} + c_{1113}\varepsilon_{13} + c_{1121}\varepsilon_{21} + c_{1122}\varepsilon_{22} + c_{1123}\varepsilon_{23}$$
$$+ c_{1131}\varepsilon_{31} + c_{1132}\varepsilon_{32} + c_{1133}\varepsilon_{33} \tag{5.33}$$

together with similar equations for the remaining stress components. All these equations can be written in the concise form

$$\sigma_{ij} = c_{ijkl}\varepsilon_{kl} \tag{5.34}$$

The constants c_{ijkl} are called stiffness constants. The existence of non-zero values of certain of the c_{ijkl} has rather surprising consequences. For example, if c_{1112} is not zero, then the occurrence of only the shear strain ε_{12} implies the existence of a proportional tensile stress σ_{11}. By symmetry arguments, it can be shown that constants of the type c_{1112} are zero in an isotropic medium, but they are in general not zero in a single crystal. The existence of a single strain component in a crystal may require that there be non-zero values of all the stress components.

The elastic strain produced by an applied stress is given by a set of equations similar to Eqns (5.34):

$$\varepsilon_{ij} = S_{ijkl}\sigma_{kl} \tag{5.35}$$

The constants S_{ijkl} are called compliances. Equation (5.35) shows that, in general, one component of stress will produce non-zero values of all of the strain components. Earlier experimenters usually determined compliances directly, by applying a stress and measuring the resulting strain. More recent determinations of the elastic properties of crystals have been based on measurements of the velocities of elastic waves in a crystal, and these measurements yield stiffness constants. The compliances can be calculated if the stiffness constants are known, through Eqns (5.34) and (5.35).

Since the sets of Eqns (5.35) and (5.34) both consist of nine equations, each containing nine terms on the right-hand side, it appears at first that 81 compliances or stiffness constants must be specified. However, the number of independent constants can always be reduced from 81 to 21.

Because $\varepsilon_{12} = \varepsilon_{21}$, the constants c_{ij12} and c_{ij21} always occur together in $\sigma_{ij} = (c_{ij12} + c_{ij21})\varepsilon_{12}$ in Eqn (5.34). It is therefore permissible to set $c_{ij12} = c_{ij21}$, and, in general, $c_{ijkl} = c_{ijlk}$. Next, suppose that only the strain component ε_{11} exists. We have

$$\sigma_{12} = c_{1211}\varepsilon_{11}$$

and

$$\sigma_{21} = c_{2111}\varepsilon_{11}$$

Since $\sigma_{12} = \sigma_{21}$, $c_{1211} = c_{2111}$ and in general $c_{ijkl} = c_{jikl}$. Similar arguments apply to the S_{ijkl}. The number of independent constants is now reduced to 36, and at this stage it becomes possible to introduce a contracted notation in which two suffixes are replaced by one. The notation for the stress σ_{ij} is contracted as follows:

$$\begin{pmatrix} \sigma_{11} & \sigma_{12} & \sigma_{13} \\ \sigma_{12} & \sigma_{22} & \sigma_{23} \\ \sigma_{13} & \sigma_{23} & \sigma_{33} \end{pmatrix} \rightarrow \begin{pmatrix} \sigma_1 & \sigma_6 & \sigma_5 \\ \sigma_6 & \sigma_2 & \sigma_4 \\ \sigma_5 & \sigma_4 & \sigma_3 \end{pmatrix}$$

A corresponding contraction is applied to c_{ijkl}, so that, for example, $c_{1233} \rightarrow c_{63}$, $c_{2323} \rightarrow c_{44}$, $c_{2332} \rightarrow c_{44}$, etc. In order that Eqn (5.34) may now be written in the contracted form,

$$\sigma_i = c_{ij}\varepsilon_j \tag{5.36}$$

it is necessary to introduce factors of $\frac{1}{2}$ into the contracted form of ε_{ij}:

$$\begin{pmatrix} \varepsilon_{11} & \varepsilon_{12} & \varepsilon_{13} \\ \varepsilon_{12} & \varepsilon_{22} & \varepsilon_{23} \\ \varepsilon_{13} & \varepsilon_{23} & \varepsilon_{33} \end{pmatrix} \rightarrow \begin{pmatrix} \varepsilon_1 & \frac{1}{2}\varepsilon_6 & \frac{1}{2}\varepsilon_5 \\ \frac{1}{2}\varepsilon_6 & \varepsilon_2 & \frac{1}{2}\varepsilon_4 \\ \frac{1}{2}\varepsilon_5 & \frac{1}{2}\varepsilon_4 & \varepsilon_3 \end{pmatrix}$$

In other words, the contracted shear strains ε_4, ε_5 and ε_6 are shear strains (γ), as defined in Section 5.2, and *not* tensor shear strains. Factors of 2 and 4 must also be introduced into the definition of S_{ij}, as follows:

$$s_{mn} = 2s_{ijkl} \qquad \text{when one only of either } m \text{ or } n \text{ is 4, 5 or 6}$$

$$s_{mn} = 4s_{ijkl} \qquad \text{when neither } m \text{ nor } n \text{ is 1, 2 or 3}$$

or, more succinctly,

$$s_{ijpq} = \tfrac{1}{2}(1 + \delta_{ij})\tfrac{1}{2}(1 + \delta_{pq})s_{mn}$$

For example, $s_{11} = s_{1111}$, but $s_{14} = 2s_{1123}$ and $s_{44} = 4s_{2323}$. With these definitions, Eqns (5.35) can be written as

$$\varepsilon_i = s_{ij}\sigma_j \tag{5.37}$$

It will now be shown that the fact that the energy stored in an elastically strained crystal depends on the strain, and not on the path by which the strained state is reached, implies that $c_{ij} = c_{ji}$. When the strain in a body is increased by $d\varepsilon_{ij}$, the tractions acting upon a unit cube within the body do work dw, where

$$dw = \sigma_{11}\,d\varepsilon_{11} + \sigma_{12}\,d\varepsilon_{12} + \sigma_{13}\,d\varepsilon_{13} + \sigma_{21}\,d\varepsilon_{21} + \sigma_{22}\,d\varepsilon_{22} + \sigma_{23}\,d\varepsilon_{23}$$

$$+ \sigma_{31}\,d\varepsilon_{31} + \sigma_{32}\,d\varepsilon_{32} + \sigma_{33}\,d\varepsilon_{33} \tag{5.38}$$

or, in contracted notation,

$$dw = \sigma_i \, d\varepsilon_i \tag{5.39}$$

(The quantity dw must be positive, otherwise the crystal would not be stable in the unstrained state.) It must be possible to integrate Eqn (5.39) to obtain the energy due to a finite strain. Suppose that a strain ε_1 is first imposed, keeping all other ε_i zero. The work done, per unit volume, is

$$
\begin{aligned}
w_1 &= \int_0^{\varepsilon_1} \sigma_1 \, d\varepsilon_1 \\
&= \int_0^{\varepsilon_1} c_{11}\varepsilon_1 \, d\varepsilon_1 \\
&= \tfrac{1}{2}c_{11}\varepsilon_1^2
\end{aligned}
\tag{5.40}
$$

Suppose that ε_2 is then increased. The work done is

$$
\begin{aligned}
w_2 &= \int_0^{\varepsilon_2} \sigma_2 \, d\varepsilon_2 \\
&= \int_0^{\varepsilon_2} c_{22}\varepsilon_2 \, d\varepsilon_2 + \int_0^{\varepsilon_2} c_{21}\varepsilon_1 \, d\varepsilon_2 \\
&= \tfrac{1}{2}c_{22}\varepsilon_2^2 + c_{21}\varepsilon_1\varepsilon_2
\end{aligned}
\tag{5.41}
$$

The total stored energy is

$$w = w_1 + w_2 = \tfrac{1}{2}c_{11}\varepsilon_1^2 + \tfrac{1}{2}c_{22}\varepsilon_2^2 + c_{21}\varepsilon_1\varepsilon_2 \tag{5.42}$$

The same state of strain can be reached by imposing the strains ε_1 and ε_2 in the reverse order, giving

$$w = \tfrac{1}{2}c_{22}\varepsilon_2^2 + \tfrac{1}{2}c_{11}\varepsilon_1^2 + c_{12}\varepsilon_2\varepsilon_1 \tag{5.43}$$

It follows from Eqns (5.42) and (5.43) that $c_{12} = c_{21}$ and, in general,

$$c_{ij} = c_{ji} \tag{5.44}$$

By writing Eqn (5.39) in the form

$$dw = \sigma_i s_{ij} \, d\sigma_j \tag{5.45}$$

and obtaining the energy increase in terms of the applied stress, it can be shown that

$$s_{ij} = s_{ji} \tag{5.46}$$

The most general compliance and stiffness constants can now be written in the form of matrices with 21 independent terms; for example,

$$
\begin{matrix}
c_{11} & c_{12} & c_{13} & c_{14} & c_{15} & c_{16} \\
c_{12} & c_{22} & c_{23} & c_{24} & c_{25} & c_{26} \\
c_{13} & c_{23} & c_{33} & c_{34} & c_{35} & c_{36} \\
c_{14} & c_{24} & c_{34} & c_{44} & c_{45} & c_{46} \\
c_{15} & c_{25} & c_{35} & c_{45} & c_{55} & c_{56} \\
c_{16} & c_{26} & c_{36} & c_{46} & c_{56} & c_{66}
\end{matrix}
$$

When arbitrary axes of references are chosen, any crystal has 21 different elastic constants although, because of crystal symmetry, there may be some relations between the constants. However, if the axes of reference are related to the crystal structure, the requirements of symmetry may ensure that some constants are zero. In a cubic crystal, the axes of reference are normally chosen to be parallel to the crystal axes. We may consider the least symmetrical class of cubic symmetry, 23. The result which we shall obtain will then apply to all classes of cubic symmetry. The axes Ox_1, Ox_2 and Ox_3 are parallel to $[1\,0\,0]$, $[0\,1\,0]$ and $[0\,0\,1]$ respectively. The threefold axes then imply the equivalence of the Ox_1, Ox_2 and Ox_3 directions, so that

$$
\begin{aligned}
c_{11} &= c_{22} = c_{33}, & s_{11} &= s_{22} = s_{33} \\
c_{12} &= c_{23} = c_{31}, & s_{12} &= s_{23} = s_{31} \\
c_{44} &= c_{55} = c_{66}, & s_{44} &= s_{55} = s_{66}
\end{aligned}
\tag{5.47}
$$

The fact that Ox_1, Ox_2 and Ox_3 are axes of twofold symmetry implies that the remaining constants are zero. For example, the production of a strain ε_3 by means of a stress σ_6 would be inconsistent with the twofold symmetry about Ox_1, as Fig. 5.14 shows. In Fig. 5.14b the crystal has been rotated through 180° about Ox_1. The shear stress must produce the same response in Fig. 5.14b as in Fig. 5.14a, and in particular

$$
\varepsilon'_3 = \varepsilon_3
\tag{5.48}
$$

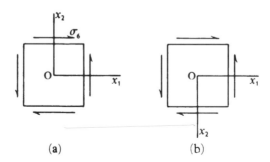

(a)   (b)

Figure 5.14

However, the sign of the shear stress has changed (the axes of reference rotate with the crystal). Therefore,

$$\varepsilon_3 = s_{36}\sigma_6 \tag{5.49}$$

$$\varepsilon'_3 = -s_{36}\sigma_6 \tag{5.50}$$

It follows that $\varepsilon_3 = 0$ and $s_{36} = 0$.

The restrictions on the elastic constants imposed by crystal symmetry could be derived formally by developing the condition that the elastic constants must remain unchanged when the axes of reference are operated on by each of the symmetry operations of the point group of the crystal (see the book by Nye [1]).

The elastic properties of a cubic crystal are completely defined by the three constants c_{11}, c_{12} and c_{44}. The compliances are related to the stiffness constants through Eqns (5.36) and (5.37). For cubic crystals,[†]

$$s_{11} = \frac{c_{11} + c_{12}}{(c_{11} - c_{12})(c_{11} + 2c_{12})}$$

$$s_{44} = \frac{1}{c_{44}} \tag{5.51}$$

and

$$s_{12} = \frac{-c_{12}}{(c_{11} - c_{12})(c_{11} + 2c_{12})}$$

The stiffness constants are obtained in terms of the compliances merely by interchanging c and s in Eqns (5.51). The elastic constants of various cubic crystals are listed in Table 5.1.

Symmetry relations reduce the number of independent elastic constants in all crystal systems other than the triclinic. The forms which the matrix takes when a coordinate axis lies along a given symmetry axis are shown in Table 5.2. When one coordinate axis is parallel to one axis of symmetry and a second coordinate axis is parallel to another axis of symmetry, the matrix must obey the restrictions on its form given by *both* the corresponding entries in Table 5.2. With the aid of Table 5.2, the number of independent elastic constants for each class of symmetry is then readily obtained. Because stress and strain are inherently centrosymmetric, the presence or absence of a centre of symmetry in the crystal does not affect the number of elastic constants. In other words, all classes within the same *Laue group* (Table 2.1) behave alike. It follows that all monoclinic classes have the same number of constants, which from the first row of Table 5.2 is seen to be thirteen. By superposing the forms in all three entries of the first row of Table 5.2, in accordance with the fact that the orthorhombic lattice has three mutually perpendicular diad axes, it is seen that all orthorhombic classes have nine constants. The Laue group $4/m$ of the tetragonal system (see Table 2.1) has

---

[†] See Problems 5.17 and 5.18.

Table 5.1 Elastic constants of cubic crystals at room temperature. (Axes of reference parallel to cube axes: $c_{11} = c_{22} = c_{33}$, $c_{44} = c_{55} = c_{66}$, $c_{12} = c_{23} = c_{13}$. Units of c_{ij} an GPa; units of are S_{ij}, $(GPa)^{-1}$

Material	c_{11}	c_{44}	c_{12}	s_{11}	s_{44}	s_{12}	Ref.
Ag	124.0	46.1	93.4	0.0229	0.0217	−0.00983	2
Al	108.2	28.5	61.3	0.0157	0.0351	−0.00568	2
Au	186	42.0	157	0.0233	0.0238	−0.01065	2
Cu	168.4	75.4	121.4	0.01498	0.01326	−0.00629	2
Ni	246.5	124.7	147.3	0.00734	0.00802	−0.00274	2
Pb	49.5	14.9	42.3	0.0951	0.0672	−0.0438	3
Fe	228	116.5	132	0.00762	0.00858	−0.00279	4
Mo	46	110	176	0.0028	0.0091	−0.00078	2
Na	7.32	4.19	6.25	0.64	0.239	−0.295	5
Nb(Cb)	245.5	29.3	139	0.0069	0.0342	−0.00249	6
Ta	267	82.5	161	0.00685	0.0121	−0.00258	7
V	228	42.6	119	0.00683	0.0235	−0.00234	7
W	501	151.4	198	0.00257	0.0066	−0.00073	2
C(diamond)	1076	575.8	125.0	0.000953	0.00174	−0.000099	2
Ge	128.9	67.1	48.3	0.00978	0.0149	−0.00266	2
Si	165.7	79.6	63.9	0.00768	0.01256	−0.00214	2
NaCl	48.7	12.6	12.4	0.0229	0.0794	−0.00465	2
LiF	111.2	62.8	42.0	0.01135	0.0159	−0.0031	2
MgO	289.2	154.6	88.0	0.00403	0.0647	−0.00094	8
TiC	500	175	113	0.00218	0.0572	−0.00040	9

seven independent constants, as can be seen from the third row of Table 5.2. The second Laue group of the tetragonal system, 4/mmm, has six constants. This can be seen by superposing the entry for a fourfold axis parallel to Ox_3, say, upon the entry for a mirror plane normal to Ox_2 (i.e. $\bar{2}$ parallel to Ox_2), which shows that the constant 16 must be zero. In the trigonal system, the Laue group $\bar{3}$ has seven independent constants and the Laue group $\bar{3}m$ has six independent constants. Both of these results can be verified by inspecting Table 5.2.

A hexagonal crystal evidently has five independent elastic constants. When the x_3 axis is chosen to be parallel to the sixfold c axis, as is conventional, the constants c_{11}, c_{33}, c_{44}, c_{12} and c_{13} must be specified. The compliances can be derived from the stiffness constants by applying the equations listed in Table 5.3. The corresponding equations for tetragonal and trigonal crystals are also given in Table 5.3. The elastic constants of some hexagonal crystals are given in Table 5.4.

In an isotropic medium, the elastic constants must be independent of the choice of coordinate axes. This requirement imposes an extra condition in addition to the conditions of cubic symmetry, which can be shown to be

$$c_{44} = \tfrac{1}{2}(c_{11} - c_{12}) \qquad (5.52)$$

Axis of	Symmetry axis parallel to Ox₁						Symmetry axis parallel to Ox₂						Symmetry axis parallel to Ox₃					
2 or 2̄	11	12	13	14	0	0	11	12	13	0	15	0	11	12	13	0	0	16
	12	22	23	24	0	0	12	22	23	0	25	0	12	22	23	0	0	26
	13	23	33	34	0	0	13	23	33	0	35	0	13	23	33	0	0	36
	14	24	34	44	0	0	0	0	0	44	0	46	0	0	0	44	45	0
	0	0	0	0	55	56	15	25	35	0	55	0	0	0	0	45	55	0
	0	0	0	0	56	66	0	0	0	46	0	66	16	26	36	0	0	66
3 or 3̄	11	12	12	0	0	0	11	12	13	−e	0	−d	11	12	13	14	15	0
	12	22	23	0	25	26	12	22	12	0	0	0	12	11	13	−14	−15	0
	12	23	22	0	−25	−26	13	12	11	e	0	d	13	13	33	0	0	0
	0	0	0	α	−a	b	−e	0	e	44	0	f	14	−14	0	44	0	−15
	0	25	−25	−a	55	0	0	0	0	0	β	0	15	−15	0	0	44	14
	0	26	−26	b	0	55	−d	0	d	f	0	44	0	0	0	−15	14	γ
4 or 4̄	11	12	12	0	0	0	11	12	13	0	−c	0	11	12	13	0	0	16
	12	22	23	c	0	0	12	22	12	0	0	0	12	11	13	0	0	−16
	12	23	22	−c	0	0	13	12	11	0	c	0	13	13	33	0	0	0
	0	c	−c	44	0	0	0	0	0	44	0	0	0	0	0	44	0	0
	0	0	0	0	55	0	−c	0	c	0	55	0	0	0	0	0	44	0
	0	0	0	0	0	55	0	0	0	0	0	44	16	−16	0	0	0	66
6 or 6̄	11	12	12	0	0	0	11	12	13	0	0	0	11	12	13	0	0	0
	12	22	23	0	0	0	12	22	12	0	0	0	12	11	13	0	0	0
	12	23	22	0	0	0	13	12	11	0	0	0	13	13	33	0	0	0
	0	0	0	α	0	0	0	0	0	44	0	0	0	0	0	44	0	0
	0	0	0	0	55	0	0	0	0	0	β	0	0	0	0	0	44	0
	0	0	0	0	0	55	0	0	0	0	0	44	0	0	0	0	0	γ

	a	b	c	d	e	f	α	β	γ
Stiffness	c_{26}	c_{25}	c_{16}	c_{14}	c_{15}	c_{14}	$\tfrac{1}{2}(c_{22} - c_{23})$	$\tfrac{1}{2}(c_{11} - c_{13})$	$\tfrac{1}{2}(c_{11} - c_{12})$
Compliance	$2s_{26}$	$2s_{25}$	$2s_{16}$	$2s_{14}$	$2s_{15}$	$2s_{14}$	$2(s_{22} - s_{23})$	$2(s_{11} - s_{13})$	$2(s_{11} - s_{12})$

Table 5.3 The equations giving the compliances s_{ij} in terms of the stiffness constants c_{ij} for the more symmetrical crystal systems. Stiffness constants are expressed in terms of compliances by interchanging c_{ij} and s_{ij} in the equations

Crystal system	Equations
Cubic	$s_{11} = \dfrac{c_{11} + c_{12}}{(c_{11} - c_{12})(c_{11} + 2c_{12})}$
	$s_{12} = \dfrac{-c_{12}}{(c_{11} - c_{12})(c_{11} + 2c_{12})}$
	$s_{44} = \dfrac{1}{c_{44}}$
Hexagonal	$s_{11} + s_{12} = \dfrac{c_{33}}{c}$
	$s_{11} - s_{12} = \dfrac{1}{c_{11} - c_{12}}$
	$s_{13} = \dfrac{-c_{13}}{c}$
	$s_{33} = \dfrac{c_{11} + c_{12}}{c}$
	$s_{44} = \dfrac{1}{c_{44}}$
	$c = c_{33}(c_{11} + c_{12}) - 2c_{13}^2$
Tetragonal	$s_{11} + s_{12} = \dfrac{c_{33}}{c}$
	$s_{11} - s_{12} = \dfrac{1}{c_{11} - c_{12}}$
	$s_{13} = \dfrac{-c_{13}}{c}$
	$s_{33} = \dfrac{c_{11} + c_{12}}{c}$
	$s_{44} = \dfrac{1}{44}$
	$s_{66} = \dfrac{1}{c_{66}}$
	$c = c_{33}(c_{11} + c_{12}) - 2c_{13}^2$
Trigonal	$s_{11} + s_{12} = \dfrac{c_{33}}{c}$
	$s_{11} - s_{12} = \dfrac{c_{44}}{c'}$
	$s_{13} = \dfrac{-c_{13}}{c}$
	$s_{14} = \dfrac{-c_{14}}{c'}$
	$s_{33} = \dfrac{c_{11} + c_{12}}{c}$
	$s_{44} = \dfrac{c_{11} - c_{12}}{c'}$
	$c = c_{33}(c_{11} + c_{12}) - 2c_{13}^2$
	$c = c_{44}(c_{11} - c_{12}) - 2c_{14}^2$

Table 5.4 Elastic constants of hexagonal crystals at room temperature. x_3 axis parallel to $[0\,0\,0\,1]$, x_1 and x_2 axes anywhere in basal plane; $c_{11} = c_{22}$, $c_{44} = c_{55}$, $c_{13} = c_{23}$, $c_{66} = \frac{1}{2}(c_{11} - c_{12})$, $s_{66} = 2(s_{11} - s_{12})$. Units of c_{ij}, GPa; units of s_{ij}, $(\text{GPa})^{-1}$

Material	c_{11}	c_{33}	c_{44}	c_{12}	c_{13}	s_{11}	s_{33}	s_{44}	s_{12}	s_{13}	Ref.
Be	292.3	336.4	162.5	26.70	14.00	0.00348	0.00298	0.00616	−0.00030	−0.00013	12
C(graphite)	1160	46.6	2.3	290	109	0.00111	0.0332	0.435	−0.00005	−0.00249	13
Cd	115.8	51.4	20.4	39.8	40.6	0.0124	0.0352	0.0498	−0.00076	−0.00920	14
Co	307	358.1	78.3	165	103	0.00472	0.00319	0.01324	−0.00231	−0.00069	2
Hf	181.1	196.9	55.7	77.2	66.1	0.00715	0.00613	0.018	−0.00247	−0.00157	15
Mg	59.7	61.7	16.4	26.2	21.7	0.022	0.0197	0.061	−0.00785	−0.0050	2
Re	612.5	682.7	162.5	270	206	0.00212	0.0017	0.00616	−0.00080	−0.00040	16
Ti	162.4	180.7	46.7	92.0	69.0	0.00958	0.00698	0.0214	−0.00462	−0.00189	15
Zn	161	61.0	38.3	34.2	50.1	0.00838	0.02838	0.0261	−0.00053	−0.00731	2
ZnO	209.7	210.9	42.5	121.1	105.1	0.00787	0.00694	0.0235	−0.00344	−0.00221	16
Zr	143.4	164.8	32.0	72.8	65.3	0.01013	0.00799	0.0313	−0.00404	−0.00241	15

Equivalently,

$$s_{44} = 2(s_{11} - s_{12}) \tag{5.53}$$

These conditions for isotropy can also be written down immediately from the form of the matrixes for a hexagonal crystal (Table 5.2).

A cubic crystal fulfils all of the conditions for isotropy except Eqn (5.52) and its degree of anisotropy can be measured by the departure from unity of the ratio A, where

$$A = \frac{2c_{44}}{c_{11} - c_{12}} \tag{5.54}$$

The ratio A measures the relative resistance of the crystal to two types of shear strain, because c_{44} measures the resistance to shear on the plane $(0\,1\,0)$ in the direction $[0\,0\,1]$, while $\frac{1}{2}(c_{11} - c_{12})$ is the stiffness with respect to shear on $(1\,1\,0)$ in $[1\,\bar{1}\,0]$. Of the crystals listed in Table 5.1 only W and Al come close to being isotropic.

Isotropic materials are often encountered, however, because metals are ordinarily found in the form of an aggregate of crystals which are oriented at random, unless a preferred orientation has developed during solidification or subsequent deformation. The elastic properties of an isotropic material are commonly specified by the two constants, Young's modulus E and Poisson's ratio v. These are defined from the effect of a simple tensile stress σ as follows:

$$E = \frac{\sigma}{\varepsilon} \tag{5.55}$$

$$v = -\frac{\varepsilon'}{\varepsilon} \tag{5.56}$$

where ε is the tensile strain in the direction of the tensile force and ε' is the tensile strain in directions normal to this. In terms of the compliances,

$$E = \frac{1}{s_{11}} \tag{5.57}$$

$$v = -\frac{s_{12}}{s_{11}} \tag{5.58}$$

Other constants, dependent on E and v, are also used. For example, the rigidity modulus or shear modulus μ is defined as

$$\mu = \tau/\gamma \tag{5.59}$$

where γ is the elastic shear strain produced by a shear stress τ. Evidently

$$\mu = \frac{1}{s_{44}} = c_{44} \tag{5.60}$$

and using Eqns (5.53), (5.57) and (5.58),

$$\mu = \frac{E}{2(1+v)} \quad (5.61)$$

Using the constants E, v and μ, Hooke's law for an isotropic solid can be written as

$$\varepsilon_{11} = \frac{\sigma_{11}}{E} - \frac{v}{E}(\sigma_{22} + \sigma_{33})$$

$$\varepsilon_{22} = \frac{\sigma_{22}}{E} - \frac{v}{E}(\sigma_{11} + \sigma_{33})$$

$$\varepsilon_{33} = \frac{\sigma_{33}}{E} - \frac{v}{E}(\sigma_{11} + \sigma_{22})$$

$$\varepsilon_{12} = \frac{\sigma_{12}}{2\mu} \quad (5.62)$$

$$\varepsilon_{23} = \frac{\sigma_{23}}{2\mu}$$

$$\varepsilon_{31} = \frac{\sigma_{31}}{2\mu}$$

Two other constants in common use are the Lamé constants λ and μ. The Lamé constant μ is identical to the shear modulus, while λ is identical to the stiffness constant c_{12} of an isotropic material:

$$\lambda = \frac{vE}{(1+v)(1-2v)}$$

In terms of the Lamé constants, Hooke's law can be written as

$$\sigma_{ij} = \lambda\varepsilon_{ii}\delta_{ij} = 2\mu\varepsilon_{ij} \quad (5.63)$$

(where $\varepsilon_{ii} = \varepsilon_{11} + \varepsilon_{22} + \varepsilon_{33}$ and $\delta_{ij} = 1$ when $i = j$ but $\delta_{ij} = 0$ when $i \neq j$.)

The reduction of the number of independent elastic constants to two for an isotropic material, three for a cubic crystal and so on, follows purely from symmetry. Further relationships among the constants have been derived in the past by making the specific assumptions that the solid consists of a lattice of atoms and that the atoms exert forces upon one another that act along the lines joining them. These relationships are called the Cauchy relations. They express the equality of constants in four-suffix notation, which are related by an exchange of one suffix from the first pair of suffixes with one from the second pair. In the general case, this reduces the number of constants from 21 to 15. For an isotropic material the Cauchy relation is

$$v = \tfrac{1}{4} \quad (5.64)$$

for a cubic crystal

$$c_{12} = c_{44} \quad (5.65)$$

and for a hexagonal or trigonal crystal

$$c_{13} = c_{44}$$
$$c_{11} = 3c_{12}$$

(5.66)

As can be seen from Tables 5.1 and 5.3, the Cauchy relations are very seldom obeyed, even approximately. Amongst cubic crystals, a number of alkali halides obey the Cauchy relation quite well.

In solving problems of elasticity in crystals it is sometimes necessary to use coordinate axes which are not aligned with respect to the crystal axes in the standard way. The elastic constants may be transformed to the new axes by applying the equations

$$c'_{ijkl} = a_{im}a_{jn}a_{ko}a_{lp}c_{mnop}$$

(5.67)

where a_{ij} is the cosine of the angle between the new axis x'_i and the old axis x_j. This transformation requires the use of the non-contracted notation. A tabulated form of these equations, using the contracted notation c_{ij} and s_{ij}, has been presented by Hearmon [10]. If one of the new axes is an axis of symmetry, or is normal to a mirror plane, the calculation is simplified by symmetry relations, which can be written down immediately from Table 5.2.

PROBLEMS

5.1 Derive expressions for the displacements u_1 and u_2 parallel to fixed axes Ox_1 and Ox_2 of a point (x_1, x_2) in a body which is rotated about the origin through an angle θ. Hence show that, when θ is very small, its magnitude is given by $\frac{1}{2}(\partial u_1/\partial x_2 - \partial u_2/\partial x_1)$.

5.2 A body is subjected to an elastic pure shear strain by stretching it along the x_1 axis and compressing it along the x_2 axis. The displacements are $u_1 = ex_1$, parallel to Ox_1, and $u_2 = -ex_2$, parallel to Ox_2, where e is very small. Write down the strain tensor referred to the axes Ox_1, Ox_2 and Ox_3 and obtain the strain tensor referred to axes Ox'_1, Ox'_2 and Ox'_3, which are derived by rotating the axes Ox_1 and Ox_2 through $45°$ anticlockwise about Ox_3.

5.3 A body is subjected to an elastic tensile strain along the x_1 axis. The displacements are $u_1 = ex_1$, $u_2 = -vex_2$, $u_3 = -vex_3$, where v is Poisson's ratio (a material constant). Write down the strain tensor referred to the axes Ox_1, Ox_2 and Ox_3 and obtain the tensor referred to the axes Ox'_1, Ox'_2, Ox_3 whose orientation is given in problem 5.2.

5.4 Prove that the homogeneous simple shear $e_{12} = g$ transforms a plane which is symmetrically inclined to Ox_1, Ox_2 and Ox_3, i.e. whose normal is parallel to the vector $(1, 1, 1)$, into another plane, whose normal is parallel to the vector $(1, 1 - g, 1)$.

5.5 Prove that the homogeneous simple shear $e_{12} = g$ transforms the circle $x_1^2 + x_2^2 = 1$ into an ellipse whose major axis makes an angle θ with Ox_1, where $\tan 2\theta = 2/g$.

5.6 When a body is subjected to a uniaxial tensile stress σ, prove that the planes on which the largest shear stress exists lie at an angle of $45°$ to the tensile axis, and determine the magnitude of the largest shear stress.

5.7 The cylindrical coordinates of a point (r, θ, z) are related to a set of mutually perpendicular axes Ox_1, Ox_2, Ox_3 as follows: the z axis is parallel to Ox_3 and the angle θ is measured anticlockwise from Ox_2, looking along the positive Ox_3 direction. Show that

$$\sigma_{rz} = \sigma_{13} \sin \theta + \sigma_{23} \cos \theta$$

$$\sigma_{\theta z} = \sigma_{13} \cos \theta - \sigma_{23} \sin \theta$$

5.8 Show that a uniaxial compressive stress is equivalent to a hydrostatic pressure superimposed on two pure shear stresses. Write down the equation in terms of the stress tensors that demonstrate this result and illustrate the equation with sketches showing the stresses.

5.9 The only non-zero stress components at a point P are σ_{11}, σ_{22} and σ_{12}, referred to axes Ox_1, Ox_2, Ox_3. Consider a plane through P, parallel to Ox_3 and making an angle of θ with the x_2 axis, measured anticlockwise from Ox_2, looking along Ox_3. Show that the normal stress σ_n acting on this plane at the point is given by

$$\sigma_n = \sigma_{11} \cos^2 \theta + \sigma_{22} \sin^2 \theta - 2\sigma_{12} \sin \theta \cos \theta$$

and that the shear stress τ is given by

$$\tau = \sigma_{11} \sin \theta \cos \theta - \sigma_{22} \sin \theta \cos \theta + \sigma_{12}(\cos^2 \theta - \sin^2 \theta)$$

5.10 An elastically isotropic body of Young's modulus E and Poisson's ratio ν is elastically strained by a tensile stress σ. Determine the dilatation of the body, i.e. the change in volume per unit volume. The bulk modulus of a solid is defined as $K = -P/\Delta$, where Δ is the dilatation produced by a stress whose hydrostatic pressure component is P. Show that $K = E/3(1 - 2\nu)$.

5.11 In a cubic crystal, the stiffness constant c_{44} is by definition the shear modulus applicable to shear on a cube plane, $\{0\,0\,1\}$ in a $\langle 1\,0\,0 \rangle$ direction. Show that the same modulus, c_{44}, applies to shear on a cube plane in any direction.

5.12 Show that, in a cubic crystal, the shear modulus applicable to shear on a $\{1\,1\,0\}$ plane in a $\langle 1\,\bar{1}\,0 \rangle$ direction is $\frac{1}{2}(c_{11} - c_{12})$.

Hint. Consider the principal strains of a shear strain on $\{1\,1\,0\}$ in a $\langle 1\,\overline{1}\,0 \rangle$ direction.

5.13 In a single crystal, Young's modulus in a particular crystallographic direction is defined as follows. When a tensile stress is applied in that direction, the ratio of the stress to the tensile strain in that direction is Young's modulus. Show that, in a cubic crystal, the reciprocal of Young's modulus is given by

$$1/E = s_{11} - 2(s_{11} - s_{12} - \tfrac{1}{2}s_{44})(a_{11}^2 a_{12}^2 + a_{11}^2 a_{13}^2 + a_{12}^2 a_{13}^2)$$

where a_{11}, a_{12} and a_{13} are the cosines of the angles between the direction in question and the Ox_1, Ox_2, Ox_3 axes respectively.

Hint. Transform the compliance s_{1111} to new axes, with Ox'_1 parallel to the direction in question. Take great care in replacing the tensor compliances with the corresponding components in contracted notation.

5.14 By symmetry arguments, deduce the form of the elastic constant matrix of an orthorhombic crystal when the axes of reference are chosen to be parallel to the crystal axes. Check your result from Table 5.2.

5.15 Consider a sphere of radius R_0 which is subject to a homogeneous simple shear of amount s. Show that it is transformed into an ellipsoid of semi-principal axes R_1, R_2 and R_3, where

$$\frac{R_1^2}{R_0^2} = \frac{2}{2 + s^2 - s\sqrt{s^2 + 4}}$$

$$\frac{R_2^2}{R_0^2} = \frac{2}{2 + s^2 + s\sqrt{s^2 + 4}}$$

$$R_3^2 = R_0^2$$

and that if θ is the angle between the major axis and the shear direction,

$$\theta = \tan^{-1}\left[\frac{+\sqrt{s^2 + 4} - s}{2}\right]$$

5.16 Write down expressions for Young's modulus measured along $[1\,0\,0]$ for a cubic crystal measured under the following conditions:
(a) Uniaxial tension along $[1\,0\,0]$.
(b) Uniaxial tension along $[1\,0\,0]$ with no strains allowed to occur along $[0\,1\,0]$ or $[0\,0\,1]$.
(c) Biaxial tension along $[1\,0\,0]$ and $[0\,1\,0]$ with no strain allowed to occur along $[0\,0\,1]$.
(d) Triaxial tension.

5.17 By considering a cubic crystal subjected to certain simple states of stress and strain, e.g. a simple tension parallel to a cube axis, a simple shear on a cube face parallel to a cube edge, a tension parallel to a cube edge with lateral contraction prevented, derive the relationship between the c_{ij} and the s_{ij} (Eqns 5.51).

5.18 Using the method of finding the inverse of a matrix given in Section A 1.4 in appendix 1 or otherwise, show that the equations giving the stiffness constants in terms of the compliances for a cubic crystal are

$$c_{11} = \frac{s_{11} + s_{12}}{(s_{11} - s_{12})(s_{11} + 2s_{12})}$$

$$c_{12} = \frac{-s_{12}}{(s_{11} - s_{12})(s_{11} + 2s_{12})}$$

$$c_{44} = \frac{1}{s_{44}}$$

Hint. Since the matrix of the s_{ij} contains so many zeros it may be quicker to use the fact that a matrix multiplied by its inverse is equal to the unit matrix, than to proceed directly as in Appendix 1.

SUGGESTIONS FOR FURTHER READING

Hearmon, R. F. S., 'Equations for transforming elastic and piezoelectric constants of crystals', *Acta Cryst.*, **10**, 121 (1957).
Huntington, H. B., 'The elastic constants of crystals', *Solid State Phys*, **7**, 213 (1958).
Jaeger, J. C., *Elasticity, Fracture and Flow*, Methuen (1962).
Love, A. E. H., *The Mathematical Theory of Elasticity*, Dover (1944).
Nye, J. F., *Physical Properties of Crystals*, Clarendon Press, Oxford (1985).
Simmons, G. and Wang, H. *Single Crystal Elastic Constants and Calculated Aggregate Properties*, 2nd MIT Press, Cambridge, Massachusetts and London (1971).

REFERENCES

1. J. F. Nye, *Physical Properties of Crystals*, Clarendon Press, Oxford (1985).
2. H. B. Huntington, *Solid State Phys.*, **7**, 213 (1958).
3. D. L. Waldorf and G. A. Alers, *J. Appl. Phys.*, **33**, 3266 (1962).
4. A. E. Lord and D. N. Beshers, *J. Appl. Phys.*, **36**, 1620 (1965).
5. W. B. Daniels, *Phys. Rev.*, **119**, 1246 (1960).
6. K. J. Carroll, *Bull. Am. Phys. Soc.*, **7**, 123 (1962).
7. D. I. Bolef, *J. Appl. Phys.*, **32**, 100 (1961).
8. D. H. Chung, *Phil. Mag.*, **8**, 833 (1963).
9. J. J. Gilman and B. W. Roberts, *J. Appl. Phys.*, **32**, 1405 (1961).
10. R. F. S. Hearmon, 'Equations for transforming elastic and piezoelectric constants of crystals'. *Acta Cryst.*, **10**, 121 (1957).

11. P. C. Waterman, *Phys. Rev.*, **113**, 1240 (1959).
12. J. F. Smith and C. L. Arbogast. *J. Appl. Phys.*, **31**, 99 (1960).
13. G. B. Spence, *Proceedings of the Fifth Conference on Carbon*, **2**, 531. Pergamon Press, 1961.
14. C. W. Garland and J. Silverman, *Phys. Rev.*, **119**, 1218 (1960).
15. E. S. Fisher and C. J. Renken, *Phys. Rev.*, **135A**, 482 (1964).
16. M. L. Shepard and J. F. Smith, *J. Appl. Phys.*, **36**, 1447 (1965).
17. T. B. Bateman, *J. Appl. Phys.*, **33**, 3309 (1962).

6

Glide

6.1 TRANSLATION GLIDE

A useful distinction between a solid and a liquid is that the liquid cannot permanently resist small forces which tend to change the shape of the liquid body, but
which allow the volume to remain constant. Such forces are shear forces and a
solid can resist these, provided that they are small. However, if large shear forces
are applied to a crystal the crystal will *yield* or give way to these in a *plastic*
manner. That is to say, if the shear stresses are large enough and are applied for
a sufficient length of time the shape of the crystal will be altered permanently.

All crystals will yield plastically at a sufficiently high temperature. Crystals
of many metals and of the alkali halides and some other non-metals, such
as graphite or the crystals of inert gases, yield plastically at very low
temperatures — sometimes close to absolute zero. These temperatures are much
too low for atomic diffusion to be occurring during the time of the experiment.
At low temperatures, crystals yield plastically by a process called *glide*.[†]

Glide is the translation of one part of a crystal with respect to another without
a change in volume. The translation usually takes place upon a specific crystallographic plane and in a particular direction in that plane.

The process of glide is illustrated schematically in Fig. 6.1. When a shearing
stress too small to produce glide is applied to the crystal, the crystal is elastically
deformed and if the stress is uniform throughout the crystal, the strain within
the crystal is homogeneous (Fig. 6.1a). In discussing the change of shape due to
glide we neglect the elastic deformation. Figure 6.1b shows the appearance of
the crystal after glide has occurred in the direction β and on the plane shown

---

[†] The examples we have just given are of crystals that glide under normal confining pressures of one
atmosphere or less. Crystals that do not glide normally break under the action of sufficiently large
forces and are said to be *brittle*. The property of brittleness or that of its antonym ductility is not an
absolute property but depends upon the state of stress and in particular upon the hydrostatic pressure.
Under a sufficiently high hydrostatic pressure a superposed shear stress can always produce glide in
a crystal. For instance, sapphire crystals can be made to glide at room temperature under a pressure
of 25 000 atm. At high temperatures, permanent deformation can occur by means of diffusion in
crystals.

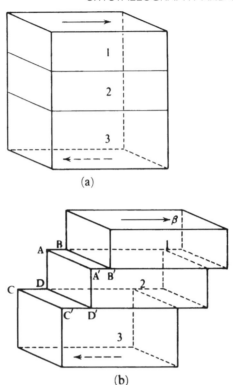

(a)

(b)

Figure 6.1

in the figure. Comparing Figs. 6.1a and b, the shape of the crystal is seen to be altered but not the volume, and the orientation of the lattice is unchanged. The two parts of the crystal on either side of the glide plane remain in an identical orientation.

Glide was first discovered in 1867 by Reusch, who recognized steps on the surface of a crystal of rock salt. Figure 6.1b shows that at the surface of a crystal undergoing glide, steps will be produced, AA'B'B and CC'D'D in Fig. 6.1b. With the optical microscope the steps are usually seen as lines and were called slip lines by Ewing and Rosenhain who made the first serious investigation of them in many metal crystals in 1899. Very careful measurement fails to detect any change in crystal orientation of the parts of a crystal on either side of a slip line and there is no evidence for any change in crystal structure as the two parts of the crystal slide over one another. It is therefore clear that the translation of part 1 of the crystal with respect to part 2 in Fig. 6.1b and of part 2 with respect to part 3 must each be equal to *an integral number of lattice translation vectors*. In view of this, the process is often called *translation glide*.

Slip lines are easily seen under an optical microscope and sometimes can be observed with the naked eye. Therefore the translation at a single step can be larger than a micrometre (10^{-4} cm), corresponding to a movement of some thousands of lattice vectors. Glide in a crystal usually occurs on a well-defined crystallographic plane of low indices, which is called the *glide plane* or slip plane, and always in a definite crystallographic direction, the slip direction.

The indices of the slip plane in a crystal can be determined easily by a two-surface analysis of the direction of the trace of the glide plane in two (or more) faces of a crystal which have been cut or polished at a known angle to one another; the procedure for a two-surface analysis is given in Section 2.3. The determination of the glide direction presents a little more difficulty. It is clear from Fig. 6.2 that the slip steps on the surface of a crystal will be of maximum height on a crystal face lying normal to the direction of glide, but should produce

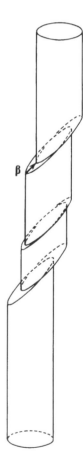

Figure 6.2 Schematic diagram of glide occurring in the direction β in a rod-shaped crystal of circular cross-section

no observable effect on a crystal face lying parallel to the direction of glide.[†]
One method of finding the direction of glide is to observe in which faces of a
crystal the slip lines appear least well marked or in which face they disappear.
The glide direction must be parallel to this plane; since it must also lie in the
slip plane it is then identifiable. Careful measurements of the change of shape of
the crystal can of course reveal the direction in which glide is occurring. This
method is used particularly with metallic crystals which undergo large amounts
of glide without breaking (see Section 6.4).

Only rarely are the slip steps on crystals deep enough to be seen with the naked
eye and the electron microscope shows that these, and those appearing under the
optical microscope as single steps, are in reality composed of a complicated fine
structure of smaller steps (Fig. 6.3). As a measure of the amount of glide in a
crystal one uses a macroscopic average over a volume of the crystal containing
many individual slip steps. If s is the relative translation in the direction of glide
of two planes, both parallel to the glide plane, and separated by a distance h,
measured normal to the glide plane (Fig. 6.4), the crystallographic glide strain α

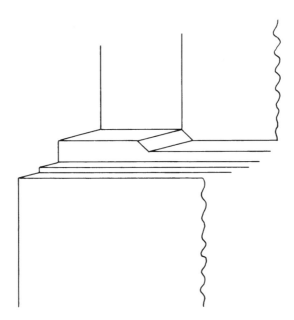

Figure 6.3   Schematic diagram of the fine structure observed in slip lines under
the electron microscope

[†] If the crystal face parallel to the glide direction is perfectly smooth and clean then glide produces
no effect on this face. In practice a crystal face usually contains dirt or corrosion products which
give rise to a surface film. This surface film is torn during glide and some effects due to glide are
then observable on a face parallel to the glide direction.

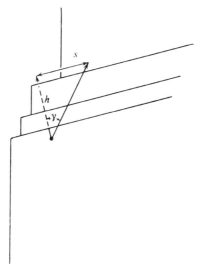

Figure 6.4

is defined as

$$\alpha = \frac{s}{h}$$

so that $\alpha = \tan \gamma$ in Fig. 6.4. When α is very small it can be written in terms of the pure strain tensor ε_{ij}. If x_1 is in the glide direction and x_3 is normal to the glide plane,

$$\alpha = 2\varepsilon_{13} = 2\varepsilon_{31}$$

and is thus equal to the engineering shear strain $\gamma_{13} = \gamma_{31}$. In terms of the tensor e_{ij} (Section 5.3), $\alpha = e_{13}$ provided α is defined for a sufficiently large volume of the crystal for the strain to be considered homogeneous.

6.2 GLIDE ELEMENTS

The glide elements of a crystal are the glide direction and the glide plane. These are also named the slip direction and slip plane respectively. A glide plane and a glide direction in that plane together constitute a glide system.

The deformation taking place during glide is a simple shear, so that the displacements relative to the centre of the crystal are, as Fig. 6.5 shows, inherently centrosymmetric. Because of this, a single glide system defined, say, by the slip plane, unit normal $\mathbf{n}$, and a slip direction normal to $\mathbf{n}$, say $\boldsymbol{\beta}$, produces exactly the same deformation as slip on $-\mathbf{n}$ in the direction $-\boldsymbol{\beta}$. The point group of the crystal governs the multiplicity of the slip planes and directions and because of

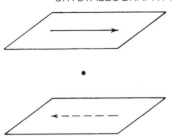

Figure 6.5

the centrosymmetric nature of the process it is the Laue group that must be used to determine the multiplicity of glide systems.

When a slip direction and a slip plane are given to define a glide system, then in all crystals of symmetry higher than the triclinic system there is usually a multiplicity of glide systems because the glide plane and glide direction are repeated by the point group symmetry of the crystal. All those combinations of slip planes and slip directions that arise from the point group symmetry of the crystal if one slip plane and one slip direction are given are called the *family of slip systems*.

A given glide system, say **n**, β, is taken to be capable of operation in either a positive or negative sense so that on the plane of normal **n** slip can proceed in either the direction β or in the direction $-\beta$. The deformation produced by slip in the direction β is, of course, just the reverse of that produced by slip in the direction $-\beta$. However, slip in these two directions will only be *crystallographically equivalent* if one of certain conditions is fulfilled. These conditions are:

(a) An even-fold axis of rotational symmetry (diad, tetrad, hexad) lies parallel to **n**.
(b) A mirror plane lies parallel to the slip plane, i.e. a $\bar{2}$ axis lies parallel to **n**.
(c) β is parallel to an even-fold axis of rotational symmetry.
(d) A mirror plane lies normal to β, i.e. β is parallel to a $\bar{2}$ axis.

It is noteworthy from Table 2.1 that if a centre of symmetry is added to the point group, fulfilment of condition (a) implies that (b) is obeyed and, correspondingly, fulfilment of (c) implies that (d) is fulfilled. If none of the conditions (a) to (d) is fulfilled then slips in the directions β and $-\beta$ on the same slip plane are not crystallographically identical and cannot be necessarily expected to occur under the same shear stress. Some examples are considered later in this section.

Crystals of NaCl slip on {1 1 0} planes in $\langle 1\bar{1}0 \rangle$ directions. The crystal is cubic and the point group is $m3m$. The normals to the glide plane — the {1 1 0} poles — occupy special positions. There are twelve of these. Each slip plane contains two slip directions, which are of opposite sense, e.g. $[1\bar{1}0]$ and $[\bar{1}10]$ in (1 1 0). There are then initially 24 glide systems in the {1 1 0}$\langle 1\bar{1}0 \rangle$ family to consider. Since the process of glide is centrosymmetric, these 24 glide systems yield twelve systems which produce different values of the tensor components e_{ij} defining the deformation (regarding positive and negative values of the same component as being different). In this point group conditions (a) to (d) are all satisfied so that reversal of the glide direction produces crystallographically equivalent effects. We say then that there are twelve physically distinct glide systems (different values of the e_{ij}) but, since reversal of the slip direction produces crystallographically equivalent effects, we regard a slip system as capable of operation in both a positive and a negative sense and hence we speak of there being six different glide systems in the $\{1\,1\,0\}\langle 1\,\bar{1}\,0 \rangle$ family in this case.

The glide elements of a large number of crystals are listed in Table 6.1. With few exceptions the glide direction is parallel to the shortest lattice translation vector of the Bravais lattice. This is so even when the shortest lattice translation vector is considerably larger than the interatomic distance, as, for example, in bismuth. This rule finds a simple explanation in terms of the dislocations present in the crystal (Section 8.1) because glide always occurs by the motion of dislocations. As a rule, the slip planes in simple crystal structures at low temperatures are parallel to the closest packed atomic planes in the crystal, but this rule has many exceptions. In addition, crystals often show a number of crystallographically different slip planes, e.g. the metal magnesium glides on $\{0\,0\,0\,1\}$, $\{1\,0\,\bar{1}\,0\}$, $\{1\,0\,\bar{1}\,1\}$. In many crystals, at moderate temperatures, whilst the slip direction remains fixed the slip plane in a given slip line varies so that what is called *pencil glide* or *wavy glide* occurs with the slip plane being any plane with the slip direction as the zone axis (Fig. 6.6). The slip trace then appears to be irrational on all crystal faces except those parallel to the slip direction.

Crystals can show all variations between very well defined slip planes and wavy glide. For instance, in sodium chloride at temperatures less than 200 °C the slip plane is accurately $\{1\,1\,0\}$ and slip traces on all crystal faces appear to

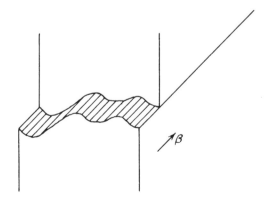

Figure 6.6

Table 6.1 Glide elements of crystals. (At room temperature and at atmospheric pressure except where stated)

Crystal type and examples	Class	Lattice	Direction	Plane	Remarks
F.c.c. metals and solid solutions Al, Cu, α-Cu–Zn	$m3m$	F	$\langle 1\,\bar{1}\,0\rangle$	$\{1\,1\,1\}$	
Diamond C, Si, Ge	$m3m$	F	$\langle 1\,\bar{1}\,0\rangle$	$\{1\,1\,1\}$	At temperatures above one-half the melting temperature
B.c.c. metals Fe, Nb, Ta, W, Na, K	$m3m$	I	$\langle 1\,\bar{1}\,1\rangle$	$\{1\,1\,0\}$	Predominant system wavy glide frequent
			$\langle 1\,\bar{1}\,1\rangle$	$\{\bar{2}\,1\,1\}^{a}$	
			$\langle 1\,\bar{1}\,1\rangle$	$\{\bar{1}\,2\,3\}^{a}$	At high homologous temperatures in alkali metals
Sodium chloride NaCl, LiF, MgO, NaF, AgCl, NH$_4$I, KI, UN, LiCl, LiBr, KCl, NaBr, RbCl, KBr, NaI, AgBr	$m3m$	F	$\langle 1\,\bar{1}\,0\rangle$	$\{1\,1\,0\}$	
TiC, UC	$m3m$	F	$\langle 1\,\bar{1}\,0\rangle$	$\{1\,1\,1\}$	At high temperature
PbS, PbTe	$m3m$	F	$\langle 1\,\bar{1}\,0\rangle$	$\{0\,0\,1\}$	$\langle 0\,0\,1\rangle$ not certain for PbS
			$\langle 0\,0\,1\rangle$	$\{1\,1\,0\}$	
Caesium chloride CsBr, NH$_4$Cl, NH$_4$Br, Tl(Br, I) LiTl, MgTl, AuZn, AuCd	$m3m$	P	$\langle 0\,0\,1\rangle$	$\{1\,1\,0\}$	
β-CuZn	$m3m$	P	$\langle 1\,\bar{1}\,1\rangle$	$\{1\,1\,0\}$	
Fluorite CaF$_2$, UO$_2$	$m3m$	F	$\langle 1\,\bar{1}\,0\rangle$	$\left.\begin{array}{l}\{0\,0\,1\}\\ \{1\,1\,0\}\\ \{1\,1\,1\}\end{array}\right\}$ At higher temperatures	
α-Al$_2$O$_3$	$\bar{3}m$	R	$\langle 1\,1\,\bar{2}\,0\rangle$	$(0\,0\,0\,1)$	Above 1000° C
			$\langle 1\,0\,\bar{1}\,0\rangle$	$\{1\,\bar{2}\,1\,0\}$	Above 1200° C Indices refer to triply-primitive hexagonal cell (p. 112 in Section 3.6)
Graphite	$6/mmm$	P	$\langle 1\,1\,\bar{2}\,0\rangle$	$(0\,0\,0\,1)$	

Table 6.1 (*continued*)

Crystal type and examples	Class	Lattice	Direction	Plane	Remarks
H.c.p. metals Zn, Cd, Mg	6/*mmm*	*P*	$\langle 11\bar{2}0\rangle$	(0001)	Predominant
				$\{10\bar{1}1\}$	
				$\{10\bar{1}0\}$	
			$\langle 11\bar{2}3\rangle$	$\{11\bar{2}2\}^a$	
Ti, Zr	6/*mmm*	*P*	$\langle 11\bar{2}0\rangle$	$\{10\bar{1}0\}$	Predominant
				$\{10\bar{1}1\}$	Zr shows only
				$\{0001\}$	$\{10\bar{1}0\}$ slip
Te	32	*P*	$\langle 11\bar{2}0\rangle$	$\{10\bar{1}0\}$	
Be and AgMg	6/*mmm*	*P*	$\langle 11\bar{2}0\rangle$	$\{0001\}$	Predominant
				$\{10\bar{1}0\}$	
Ga	*mmm*	*A*	$[010]$	(001)	
			$[010]$	(102)	
			$\langle 0\bar{1}1\rangle$	$\{011\}^a$	
Sphalerite α-ZnS, InSb, GaAs	$\bar{4}3m$	*F*	$\langle 1\bar{1}0\rangle$	$\{111\}$	Slip direction not certain for α-ZnS
β-Sn	4/*mmm*	*I*	$\langle 001\rangle$	$\{110\}$	
				$\{100\}$	
			$\langle 10\bar{1}\rangle$	$\left.\begin{array}{l}\{101\}\\\{121\}\end{array}\right\}^a$	Rare
Rutile TiO$_2$	4/*mmm*	*P*	$\langle 10\bar{1}\rangle$	$\{101\}^a$	
			$\langle 001\rangle$	$\{110\}$	
Bi	$\bar{3}m$	*R*	$\langle 10\bar{1}\rangle$	(111)	Primitive cell
Hg	$\bar{3}m$	*R*	$\langle 0\bar{1}1\rangle$	(100)	
			$\langle 100\rangle$	$(100)^a$	Primitive cell
α-U	*mmm*	*C*	$[100]$	(010)	
			$[100]$	(001)	
			$[1\bar{1}0]$	(110)	Rare
Al$_2$O$_3$–MgO spinel	*m3m*	*F*	$\langle 1\bar{1}0\rangle$	$\{111\}$	In non-stoichiometric crystals, with excess Al$_2$O$_3$, $\{110\}$ is the predominant slip plane
				$\{110\}$	

^{a} Indicates systems where reversing the slip direction does not produce crystallographically equivalent motions

be very straight. Above a temperature of about 250 °C slip appears to occur on any plane in the $\langle 1\bar{1}0\rangle$ type zone. For crystals that show wavy glide, it is the well-defined slip plane observed at low temperatures that is listed in Table 6.1.

A number of cases in which slip on a given slip plane in a certain direction is not crystallographically equivalent to slip in the reverse direction are pointed

Table 6.2

Metal	c/a	Predominant slip plane
Cd	1.886	$(0\,0\,0\,1)$
Zn	1.856	$(0\,0\,0\,1)$
Co	1.628	$(0\,0\,0\,1)$
Mg	1.624	$(0\,0\,0\,1)$
Re	1.615	$(0\,0\,0\,1)$
Tl	1.598	—
Zr	1.593	$(1\,0\,\bar{1}\,0)$
Ti	1.587	$(1\,0\,\bar{1}\,0)$
Hf	1.581	$(1\,0\,\bar{1}\,0)$
Y	1.571	$(1\,0\,\bar{1}\,0)$
Be	1.568	$(0\,0\,0\,1)$

out in Table 6.1. A common example is slip in the $\langle 1\,1\,1\rangle$ on the $\{1\,1\,\bar{2}\}$ planes in body-centred cubic metals. There are 24 planes of the type $\{1\,1\,\bar{2}\}$. Each plane contains two directions, e.g. $[1\,1\,1]$ and $[\bar{1}\,\bar{1}\,\bar{1}]$ in $(1\,1\,\bar{2})$, so initially there are 48 slip systems. The centrosymmetric nature of glide reduces this number to 24. None of conditions (a) to (d) of this section is satisfied and hence there are, strictly speaking, two crystallographically different families, each containing twelve members.

Table 6.1 shows that a knowledge of the crystal structure alone is not sufficient to deduce the slip planes and directions of a crystal. Many alkali halides with the NaCl structure glide on $\{1\,1\,0\}\langle 1\,\bar{1}\,0\rangle$, but in PbS and PbTe with this structure this is not so, and titanium carbide also has this structure but slips, as do the face-centred cubic metals. In all metals with the close packed hexagonal structure the most commonly observed direction is $\langle 1\,1\,\bar{2}\,0\rangle$, but the slip plane most commonly observed varies with the metal (Table 6.2) and is not explained simply by variation of the ratio c/a. It is clear that glide planes, and to some extent the glide directions, depend upon the type of interatomic binding in the crystal.

6.3 INDEPENDENT SLIP SYSTEMS

The change of shape produced by glide in a crystal is a simple shear (Section 5.3). Consider a point P in a crystal (Fig. 6.7) distant $\mathbf{r}$ from an origin O. If glide occurs on a plane of unit normal $\mathbf{n}$ in the direction of the unit vector $\boldsymbol{\beta}$, let P move to P' at $\mathbf{r}'$ from the origin. Then

$$PP' = \mathbf{r}' - \mathbf{r} = \alpha(\mathbf{r} \cdot \mathbf{n})\boldsymbol{\beta}$$

or

$$\mathbf{r}' = \mathbf{r} + \alpha(\mathbf{r} \cdot \mathbf{n})\boldsymbol{\beta} \tag{6.1}$$

where α is the amount of crystallographic glide strain.

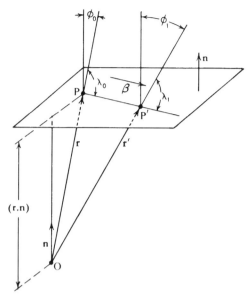

Figure 6.7

Provided α is small we can write down the components of the tensor e_{ij} and of the pure strain tensor ε_{ij}, referred to the axes (x_1, x_2, x_3), describing the deformation. From Eqns (5.9), for example,

$$e_{11} = \frac{\partial u_1}{\partial x_1} = \frac{\partial}{\partial x_1}(\mathbf{r'} - \mathbf{r})_1 = \frac{\partial}{\partial x_1}\alpha(\mathbf{r} \cdot \mathbf{n})\beta_1 \tag{6.2}$$

If we write

$$\mathbf{r} = x_1\mathbf{i} + x_2\mathbf{j} + x_3\mathbf{k}$$
$$\mathbf{n} = n_1\mathbf{i} + n_2\mathbf{j} + n_3\mathbf{k}$$

and

$$\boldsymbol{\beta} = \beta_1\mathbf{i} + \beta_2\mathbf{j} + \beta_3\mathbf{k}$$

then Eqn (6.2) yields

$$e_{11} = \frac{\partial}{\partial x_1}\alpha(\mathbf{r} \cdot \mathbf{n})\beta_1 = \alpha\frac{\partial}{\partial x_1}[(x_1\mathbf{i} + x_2\mathbf{j} + x_3\mathbf{k}) \cdot (n_1\mathbf{i} + n_2\mathbf{j} + n_3\mathbf{k})]\beta_1$$

$$= \alpha\frac{\partial}{\partial x_1}(x_1n_1 + x_2n_2 + x_3n_3)\beta_1$$

$$= \alpha n_1\beta_1$$

Similarly, a component such as e_{23} is given by

$$e_{23} = \frac{\partial u_2}{\partial x_3} = \frac{\partial}{\partial x_3} \alpha (\mathbf{r} \cdot \mathbf{n}) \beta_2$$

or

$$e_{23} = \alpha n_3 \beta_2$$

The tensor e_{ij} defining the deformation thus has components

$$e_{ij} = \begin{pmatrix} \alpha n_1 \beta_1 & \alpha n_2 \beta_1 & \alpha n_3 \beta_1 \\ \alpha n_1 \beta_2 & \alpha n_2 \beta_2 & \alpha n_3 \beta_2 \\ \alpha n_1 \beta_3 & \alpha n_2 \beta_3 & \alpha n_3 \beta_3 \end{pmatrix} \tag{6.3}$$

This tensor contains a pure strain tensor ε_{ij} and also describes the rotation produced by the glide. Since glide corresponds to a simple shear the deformation proceeds without a change of volume. The tensor (6.3) shows this to be the case, for since $\mathbf{n}$ and $\boldsymbol{\beta}$ are orthogonal the diagonal terms of the tensor sum to zero, i.e.

$$\alpha n_1 \beta_1 + \alpha n_2 \beta_2 + \alpha n_3 \beta_3 = 0 = e_{11} + e_{22} + e_{33}$$

Equation (6.3) can be decomposed into the symmetric tensor describing the pure strain produced by the glide and the tensor describing the rotation. From Section 5.2 the pure strain is

$$\varepsilon_{ij} = \begin{pmatrix} \alpha n_1 \beta_1 & \frac{\alpha}{2}(n_1 \beta_2 + n_2 \beta_1) & \frac{\alpha}{2}(n_1 \beta_3 + n_3 \beta_1) \\ \frac{\alpha}{2}(n_1 \beta_2 + n_2 \beta_1) & \alpha n_2 \beta_2 & \frac{\alpha}{2}(n_3 \beta_2 + n_2 \beta_3) \\ \frac{\alpha}{2}(n_1 \beta_3 + n_3 \beta_1) & \frac{\alpha}{2}(n_3 \beta_2 + n_2 \beta_3) & \alpha n_3 \beta_3 \end{pmatrix} \tag{6.4}$$

and the rotation is

$$\omega_{ij} = \begin{pmatrix} 0 & \frac{\alpha}{2}(n_2 \beta_1 - n_1 \beta_2) & \frac{\alpha}{2}(n_3 \beta_1 - n_1 \beta_3) \\ -\frac{\alpha}{2}(n_2 \beta_1 - n_1 \beta_2) & 0 & \frac{\alpha}{2}(n_3 \beta_2 - n_2 \beta_3) \\ -\frac{\alpha}{2}(n_3 \beta_1 - n_1 \beta_3) & -\frac{\alpha}{2}(n_3 \beta_2 - n_2 \beta_3) & 0 \end{pmatrix} \tag{6.5}$$

To see simply what Eqns (6.4) and (6.5) mean, suppose we take axes so that x_1 is parallel to $\mathbf{n}$, the slip plane normal, and $\boldsymbol{\beta}$ parallel to x_2; then Eqns (6.3), (6.4) and (6.5) reduce to

$$e_{ij} = \begin{pmatrix} 0 & 0 & 0 \\ \alpha & 0 & 0 \\ 0 & 0 & 0 \end{pmatrix}$$

$$\varepsilon_{ij} = \begin{pmatrix} 0 & \alpha/2 & 0 \\ \alpha/2 & 0 & 0 \\ 0 & 0 & 0 \end{pmatrix}$$

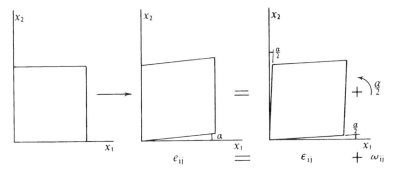

Figure 6.8 A small simple shear deformation e_{ij} is equivalent to a pure strain ε_{ij} plus a rotation ω_{ij}

and

$$\omega_{ij} = \begin{pmatrix} 0 & -\alpha/2 & 0 \\ \alpha/2 & 0 & 0 \\ 0 & 0 & 0 \end{pmatrix}$$

corresponding to the shear shown in Fig. 6.8.

Physically distinct slip systems of the same family may produce similar pure strains. Suppose as an example we consider glide of amount α to occur on the $(0\,1\,1)[0\,\bar{1}\,1]$ system of a cubic crystal. Then if we take axes parallel to the cubic axes of the crystal we have

$$\mathbf{n} = 0\mathbf{i} + \frac{1}{\sqrt{2}}\mathbf{j} + \frac{1}{\sqrt{2}}\mathbf{k}$$

$$\boldsymbol{\beta} = 0\mathbf{i} - \frac{1}{\sqrt{2}}\mathbf{j} + \frac{1}{\sqrt{2}}\mathbf{k}$$

Therefore, from (6.4)

$$\varepsilon_{ij} = \begin{pmatrix} 0 & 0 & 0 \\ 0 & -\alpha/2 & 0 \\ 0 & 0 & \alpha/2 \end{pmatrix} \tag{6.6}$$

and from (6.5)

$$\omega_{ij} = \begin{pmatrix} 0 & 0 & 0 \\ 0 & 0 & -\alpha/2 \\ 0 & \alpha/2 & 0 \end{pmatrix} \tag{6.7}$$

If we consider glide of the same amount α on the system $(0\,\bar{1}\,1)[0\,1\,1]$, which is orthogonal to our first example, then we find that

$$\varepsilon_{ij} = \begin{pmatrix} 0 & 0 & 0 \\ 0 & -\alpha/2 & 0 \\ 0 & 0 & \alpha/2 \end{pmatrix}$$

and

$$\omega_{ij} = \begin{pmatrix} 0 & 0 & 0 \\ 0 & 0 & \alpha/2 \\ 0 & -\alpha/2 & 0 \end{pmatrix}$$

Thus the operation of each of these two physically different systems produces the same pure strain but opposite rigid-body rotations (Fig. 6.9). Clearly, slip (also of amount α) on the system $(0\,\bar{1}\,1)[0\,\bar{1}\,\bar{1}]$, i.e. just reversing the direction of glide in the last case, could produce the opposite pure strain to slip on the system $(0\,1\,1)[0\,\bar{1}\,1]$ but the *same rotation*. Since in the absence of any constraint glide occurs without any deformation of the *lattice* of the crystal, so that the lattice is subject to neither pure strain nor rotation, it is very important to realize that simultaneous slip *on more than one system* can produce a *pure* strain of a crystal without change in the orientation of the lattice and alternatively can rotate the

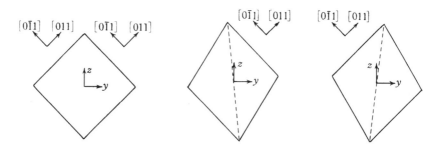

Figure 6.9

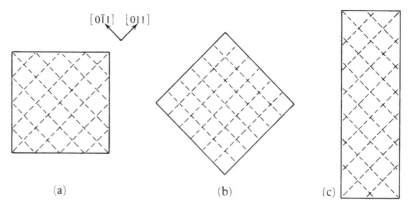

(a)   (b)   (c)

Figure 6.10 Schematic diagram to illustrate that the cubic crystal in (a) slipping by equal amounts on the glide system $(0\,\bar{1}\,1)[0\,1\,1]$ and $(0\,1\,1)[0\,1\,\bar{1}]$ would be rotated to the position in space shown in (b) without pure strain or rotation of the crystal lattice. If (a) slips by equal amounts on $(0\,\bar{1}\,1)[0\,1\,1]$ and on $(0\,1\,1)[0\,\bar{1}\,1]$ no rotation of the crystal in space (nor of the lattice) occurs and a *pure* shear results, leading to (c)

crystal without changing the *orientation of the lattice* (Fig. 6.10). In order for these two operations to be accomplished, slip lines corresponding to the slip on various slip systems must interpenetrate; whether or not this is possible depends on the properties of individual dislocations (Chapter 8).

Physically distinct slip systems may produce the same *pure* strain; we have just seen an example of this. Slip on $(0\,1\,1)[0\,\bar{1}\,1]$ in a cubic crystal and slip on $(0\,\bar{1}\,1)[0\,1\,1]$ produce the same pure strain tensor given by Eqn (6.6). Slip on $(1\,1\,0)[\bar{1}\,1\,0]$ of amount γ in a cubic crystal produces the pure strain

$$\begin{pmatrix} -\gamma/2 & 0 & 0 \\ 0 & \gamma/2 & 0 \\ 0 & 0 & 0 \end{pmatrix} \tag{6.8}$$

This system then produces a pure strain that cannot be produced by slip on either of the systems $(0\,1\,1)[0\,\bar{1}\,1]$ or $(0\,\bar{1}\,1)[0\,1\,1]$. Slip systems that produce different pure strains are said to be *independent*. A general pure strain can be produced in a crystal by glide provided the crystal can slip on five independent systems. The reason that five independent systems are needed is as follows. Plastic deformation of a crystal by glide proceeds without change in volume so that the pure strain tensor describing the deformation always has the property that

$$\varepsilon_{11} + \varepsilon_{22} + \varepsilon_{33} = 0 \tag{6.9}$$

Thus only two of the components ε_{11}, ε_{22}, ε_{33} can be chosen arbitrarily, since the third is always fixed by the condition (6.9). Since the pure strain tensor is symmetric, $\varepsilon_{ij} = \varepsilon_{ji}$, and so there are just five numbers that can be selected arbitrarily to be the strain components of the general pure strain tensor when the volume is conserved. Glide on any one slip system can alter just one component of the pure strain tensor independently of its effect on the others, so five independent glide systems must be available if a general arbitrary pure strain is to be produced by glide. A slip system is said to be independent of others if the pure strain that can be produced by its operation *cannot* be produced by combining suitably chosen amounts of slip on these other systems.

In a cubic crystal we saw that the glide systems $(0\,1\,1)[0\,\bar{1}\,1]$ and $(1\,1\,0)[\bar{1}\,1\,0]$ are independent. The system $(1\,0\,1)[\bar{1}\,0\,1]$ is not independent of these two. This is easily seen by writing down together the pure strain components referred to the same cubic axes for amounts of slip α on $(0\,1\,1)[0\,\bar{1}\,1]$, γ on $(1\,1\,0)[\bar{1}\,1\,0]$ and β on $(1\,0\,1)[\bar{1}\,0\,1]$. These are

$$\begin{pmatrix} 0 & 0 & 0 \\ 0 & -\alpha/2 & 0 \\ 0 & 0 & \alpha/2 \end{pmatrix}, \quad \begin{pmatrix} -\gamma/2 & 0 & 0 \\ 0 & \gamma/2 & 0 \\ 0 & 0 & 0 \end{pmatrix}, \quad \begin{pmatrix} -\beta/2 & 0 & 0 \\ 0 & 0 & 0 \\ 0 & 0 & \beta/2 \end{pmatrix}$$

Obviously, if we adjust the amounts of slip on the first two systems so that $\alpha = \gamma$ and we arrange that $\alpha = \beta$, then the pure strain produced by the last system can be duplicated by appropriate combinations of slip on the other two.

Since five independent systems produce a general pure strain (without volume change) it is clear that a crystal cannot possess more than five independent systems in this sense. The number of independent systems produced by the operation of all members of a given family is listed for some different families of slip systems found in the more symmetric crystals in Table 6.3.

When pencil glide occurs, the slip direction remains fixed but the slip plane can be any plane in the zone whose axis is the glide direction. It is then easily shown (see problem 6.6) that shear in a given slip direction can provide two independent components of the pure strain tensor.

To decide formally whether the members of a given family, or of a set of families, of slip systems will together provide five independent systems we can proceed as follows. Choose a set of orthogonal axes and write down, using Eqn (6.4), the components of the pure strain tensor produced by an arbitrary amount of glide on a given system. Call these components ε_1, ε_2, ε_3, ε_{12}, ε_{13}, ε_{23}. Do the same thing for four other slip systems, referring the strain tensor to the same axes as before. Finally, since the three tensile strains are not independent of one another, form the five-by-five determinant of the quantities $\varepsilon_1 - \varepsilon_3$, $\varepsilon_2 - \varepsilon_3$, ε_{12}, ε_{13}, ε_{23} for each of the systems. If this determinant has a value other than zero the five chosen slip systems are independent of one another, since a determinant will equal zero if any row can be expressed as a linear combination of other rows. If a crystal glides only on a family of slip systems that does not possess five independent members, then there will be directions in which it cannot be extended or compressed, i.e. there will be certain orientations of shear stress that cannot produce glide. These can be easily found from Eqns (6.12) and (6.13) of the next section (see the footnote on p. 201).

If a random polycrystal (i.e. a body consisting of a large number of individual crystals (or grains) with the crystal axes in the various grains distributed randomly in space) is to deform by glide in each of the component grains without holes

Table 6.3 Independent slip systems in crystals

System		Class	Number of independent systems
$\langle 1\bar{1}0 \rangle$	$\{111\}$	$m3m$	5
$\langle 1\bar{1}1 \rangle$	$\{110\}$	$m3m$	5
$\langle 1\bar{1}0 \rangle$	$\{110\}$	$m3m$	2
$\langle 10\bar{1} \rangle$	$\{101\}$	$4/mmm$	4
$\langle 001 \rangle$	$\{110\}$	$m3m$	3
$\langle 1\bar{1}0 \rangle$	$\{001\}$	$m3m$	3
$\langle 11\bar{2}0 \rangle$	(0001)	$6/mmm$	2
$\langle 11\bar{2}0 \rangle$	$\{10\bar{1}0\}$	$6/mmm$	2
$\langle 11\bar{2}0 \rangle$	$\{10\bar{1}1\}$	$6/mmm$	4
$\langle 11\bar{2}3 \rangle$	$\{11\bar{2}2\}$	$6/mmm$	5

appearing at the grain boundaries, then the average grain in the interior of the body must change its shape so as to conform with the changes of shape of the neighbouring grains and with the change of shape of the body as a whole. Provided a grain possesses five independent glide systems, in the sense that we have defined, this grain can undergo a general arbitrary change of shape by glide. This condition is often stated as one that must be obeyed for all grains if the polycrystal is to be capable of large plastic changes of shape by glide, within the grains, without internal cavities forming. It is called the *von Mises condition*. During deformation by slip of a polycrystal the individual grains are constrained by their neighbours and under these conditions the *crystal lattice* often rotates in space.

The part played by lattice rotation can be appreciated by *supposing* that each grain of a polycrystal suffers the same small deformation e_{ij}, so that the deformation of the whole body is homogeneous.[†] We divide the small deformation e_{ij} into a pure strain and a rotation:

$$e_{ij} = \varepsilon_{ij} + \omega_{ij}$$

The pure strain ε_{ij} must be the same in each grain and so must the rotation ω_{ij}. By operating five independent slip systems we can supply the same ε_{ij} to each grain. This will not in general produce the same rotation of each grain. The rotation produced directly by slip does not rotate the lattice, but due to the constraint of neighbouring grains a rigid-body rotation which rotates the lattice may always be added so that the total rotation, given by

$$(\omega_{ij}) = (\omega_{ij})_{\text{slip}} + (\omega_{ij})_{\text{lattice}}$$

is the same for all grains. Each grain will therefore receive the lattice rotation that is needed to bring its total rotation to the common value.

This argument assumes that each grain may rotate freely. Whether this is so or not can only be decided by consideration of the constraints at the intercrystalline boundaries.

6.4 LARGE STRAINS

In many experiments on glide a single crystal of the substance under investigation is pulled in tension or compressed along a given direction (Fig. 6.11). It is easy to find the shear stress on the slip plane and resolved in the slip direction (called the *resolved shear stress*) by altering the axes of reference of the tensor representing the applied stress (see problem 6.12). However, this can be very simply derived directly from Fig. 6.11. If a force **F** is applied to the crystal of cross-sectional

---

[†] In a real polycrystal the deformation of each crystal grain may not be homogeneous; indeed, X-ray examination of individual grains in a polycrystal proves that usually it is not.

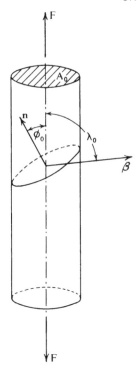

Figure 6.11

area A_0 the tensile stress parallel to **F** is $\sigma = F/A_0$. The force **F** has a component $F \cos \lambda_0$ in the slip direction, where λ_0 is the angle between **F** and the slip direction; this force acts over an area $A_0/\cos \phi_0$ so that the resolved shear stress τ is

$$\tau = \frac{F \cos \lambda_0}{A_0/\cos \phi_0} = \sigma \cos \phi_0 \cos \lambda_0 \qquad (6.10)$$

where ϕ_0 is the angle between **F** and the normal to the glide plane. The stress normal to the slip plane is

$$\sigma_n = \frac{F \cos \phi_0}{A_0/\cos \phi_0} = \sigma \cos^2 \phi_0 \qquad (6.11)$$

It is worth emphasizing that in Fig. 6.11 the directions of **F, n** and β are not necessarily coplanar so that only in a special case is ϕ_0 equal to $(90° - \lambda_0)$.

From Eqn (6.10) the resolved shear stress on any glide system can be evaluated. To a very good approximation it is always found that when a crystal is subjected to an increasing uniaxial tensile or compressive load, as in Fig. 6.11, slip always occurs first on that glide system on which the resolved shear stress is greatest. To a worse approximation it is found that in a given pure crystal at a given temperature glide starts when the resolved shear stress reaches a certain critical value. This last approximation, called the law of *critical resolved shear stress*, is moderately well obeyed in crystals of metals, where there is usually a high density of dislocations present (see Chapter 7), but in non-metals the resolved shear stress at which glide occurs is so dependent on the previous history of the crystal that such a rule is seldom of use.

The particular glide system with the highest resolved shear stress for a particular orientation of **F** with respect to the crystal axes may be found from Eqn (6.10). For common families of slip systems the result of examining all members to find which is subject to the largest resolved shear stress is easily shown on a stereogram. Figure 6.12a is a stereogram of a cubic crystal. If a face-centred cubic metal crystal is considered the point group is $m3m$ and slip occurs on the $\{1\,1\,1\}\langle1\,\bar{1}\,0\rangle$ system. There are twelve physically distinct glide systems. If **F** is the direction of a uniaxial force applied to the crystal the direction of **F** can be plotted on the stereogram, say as $\mathbf{F}_0$ in Fig. 6.12a. The particular glide system having the largest resolved shear stress is indicated by the lettering of the unit triangle on the stereogram, within which $\mathbf{F}_0$ falls. For instance, for the case

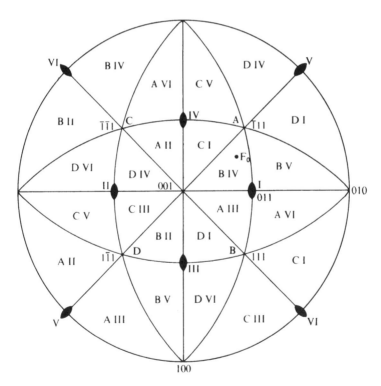

Figure 6.12a   Standard stereogram of an f.c.c. metal crystal which glides on $\{1\,1\,1\}$ in $\langle1\,\bar{1}\,0\rangle$ to show the particular slip system with the maximum resolved shear stress for any orientation of the tensile axis

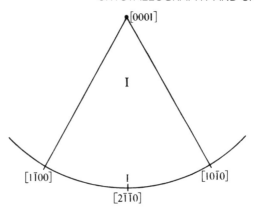

Figure 6.12b A unit triangle of the standard stereogram of a hexagonal crystal of point group 6/*mmm* which glides on (0001) in ⟨1120⟩ showing the slip direction of maximum resolved shear stress

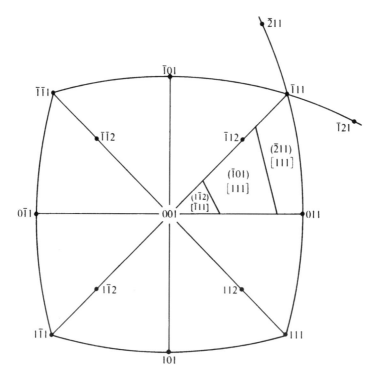

Figure 6.12c A standard stereogram of a b.c.c. metal crystal which glides on {110} in ⟨111⟩ and on {112} in ⟨111⟩ showing the regions of one unit triangle in which the resolved shear stresses on the various slip systems are greatest

shown in Fig. 6.12a the triangle is lettered B IV, meaning that slip will occur first on the octahedral plane B, i.e. (1 1 1), in the direction IV, i.e. $[\bar{1}\,0\,1]$. If $\mathbf{F}_0$ lies on the boundaries of the unit triangle then clearly more than one slip system is equally stressed (see problem 6.8). Similar diagrams for some other common slip systems are given in Figs. 6.12b and 6.12c and are used in the same way as Fig. 6.12a. From Table 6.1 it is worth nothing that in crystals of the NaCl structure slipping on $\{1\,1\,0\}\langle 1\,\bar{1}\,0\rangle$ at least two slip systems are always equally stressed. Face-centred cubic metal crystals have been very intensively studied. Even for crystals with the position of $\mathbf{F}_0$ near to the centre of the unit triangle, small amounts of glide on systems other than the most highly stressed one are observed. Since the slip plane of a glide system is easily observed, these slip systems are usually identified by reference to the glide plane alone. The slip plane of the most highly stressed system is called the primary glide plane and the names given to the others are shown in Fig. 6.13. In cubic crystals slipping on $\{1\,1\,1\}\langle 1\,\bar{1}\,0\rangle$ a simple rule has been given by Diehl to identify the primary glide system when the direction of $\mathbf{F}$ lies in any of the unit triangles shown in Fig. 6.12a. This states: 'Reflect the $\langle 1\,1\,0\rangle$ pole of the triangle in question in the opposite side of the triangle to find the glide direction, and reflect the $\{1\,1\,1\}$ pole of the triangle in the opposite side to find the glide plane normal.'

Returning to Fig. 6.11, if a small amount of glide, dα, occurs on the slip system shown, the crystal will increase in length by dl in the direction of $\mathbf{F}$ where

$$dl = l \cos \phi_0 \cos \lambda_0 \, d\alpha \tag{6.12}$$

where dα is the increment of crystallographic glide strain and l the instantaneous length of the crystal.[†] If e is the (small) elongation parallel to $\mathbf{F}$ produced by dα,

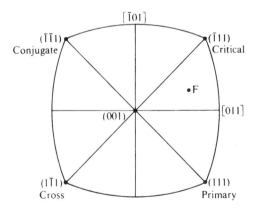

Figure 6.13

[†] The identity of the geometrical factors in Eqns (6.12) and (6.10) is worth noting in passing. This arises because, of course, if a given applied force produces no resolved shear stress on a given glide

then $e = \mathrm{d}l/l$ and so Eqn (6.12) gives

$$e = \mathrm{d}\alpha \cos \phi_0 \cos \lambda_0 \qquad (6.13)$$

The relative translation of the parts of the crystal parallel to β implies that as the crystal gets longer its two ends move with respect to one another in a direction transverse to the applied force $\mathbf{F}$. This effect is not important for very small strains, but when minerals are deformed large amounts in compression (usually under hydrostatic pressure) or metal crystals extended by large amounts in tension it must be taken into account. In compression the ends of the crystal may be prevented from moving sideways by friction and in tension crystals are usually gripped at the ends. Consider the tensile case and suppose that there is no lateral constraint. The crystal will take up the form shown in Fig. 6.14. If however, due to the lateral constraints the point A' is forced to lie above O at O', then if there is glide solely on the system shown, the crystal lattice will be rotated with respect to the direction OO'. From Fig. 6.14 the elongation of the crystal e equals $(l - l_0)/l_0$ so from the triangle OAA'

$$\frac{l}{l_0} = 1 + e = \frac{\sin \lambda_0}{\sin \lambda} \qquad (6.14)$$

and from triangles NAO, NA'O

$$\frac{l}{l_0} = \frac{\cos \phi_0}{\cos \phi} \qquad (6.15)$$

where λ_0 and ϕ_0 are the initial values of λ and ϕ. The crystallographic glide strain α is given by $AA'/l_0 \cos \phi_0$ and $AA' = l \cos \lambda - l_0 \cos \lambda_0$, so

$$\alpha = \frac{(1 + e) \cos \lambda - \cos \lambda_0}{\cos \phi_0} = \frac{\cos \lambda}{\cos \phi} - \frac{\cos \lambda_0}{\cos \phi_0} \qquad (6.16)$$

and the resolved shear stress $\tau[=(F/A) \cos \phi \cos \lambda]$ is

$$\tau = \frac{F \cos \phi}{A} \left(1 - \frac{\sin^2 \lambda_0}{(1 + e)^2}\right)^{1/2}$$

From Fig. 6.14 it can be seen that the area of the glide plane remains constant during deformation (except for negligibly small changes due to the formation of slip steps). Therefore the quantity $A/\cos \phi$ does not alter, and the last equation can be written as

$$\tau = \frac{F \cos \phi_0}{A_0} \left[1 - \left(\frac{l_0}{l}\right)^2 \sin^2 \lambda_0\right]^{1/2} \qquad (6.17)$$

---

system, because $\cos \phi_0$ or $\cos \lambda_0$ is zero, then it does no work if that glide system operates. Certain orientations of an applied force $\mathbf{F}$ will give zero values for τ in Eqn (6.10) or for $\mathrm{d}l$ in Eqn (6.12) if the crystal does not possess five independent slip systems.

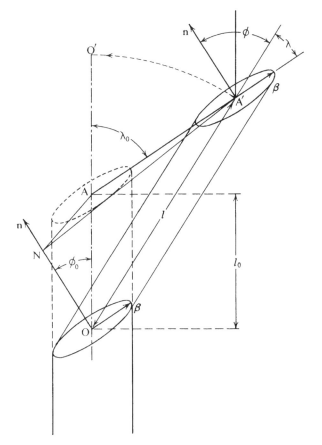

Figure 6.14

As deformation proceeds the value of λ decreases and so for a given force $\mathbf{F}$ the resolved shear stress increases. We can also derive formulae (6.14) to (6.16) directly from Eqn (6.1) and Fig. 6.7. It is useful to do this because the vector formulae make the compression case easy to deal with. From Fig. 6.7 a vector $\mathbf{r}$ is changed by glide of amount α to the vector $\mathbf{r}'$, where

$$\mathbf{r}' = \mathbf{r} + \alpha(\mathbf{r} \cdot \mathbf{n})\boldsymbol{\beta} \qquad (6.1)$$

We take $\mathbf{r}$ parallel to the tensile axis when $\alpha = 0$ and then $\mathbf{r}'$ represents the tensile axis after a crystallographic glide strain α:

$$\frac{l}{l_0} = (1 + e) = \frac{|\mathbf{r}'|}{|\mathbf{r}|}$$

The quantity $\mathbf{n}$ is a unit vector normal to the glide plane, so

$$\cos\phi = \frac{(\mathbf{r'} \cdot \mathbf{n})}{|\mathbf{r'}|} = [(\mathbf{r} \cdot \mathbf{n}) + \alpha(\mathbf{r} \cdot \mathbf{n})\boldsymbol{\beta} \cdot \mathbf{n}]\frac{1}{|\mathbf{r'}|}$$

$$= \frac{l_0 \cos\phi_0}{l}$$

since $\cos\phi_0 = (\mathbf{r} \cdot \mathbf{n})/|\mathbf{r}|$ and $\boldsymbol{\beta} \cdot \mathbf{n} = 0$. Thus we have Eqn (6.15). Also

$$|\mathbf{r'} \times \boldsymbol{\beta}| = |\mathbf{r'}| \sin\lambda$$

since $\boldsymbol{\beta}$ is a unit vector. From Eqn (6.1),

$$\mathbf{r'} \times \boldsymbol{\beta} = \mathbf{r} \times \boldsymbol{\beta}, \quad \text{since } \boldsymbol{\beta} \times \boldsymbol{\beta} = 0$$

and so

$$\frac{l}{l_0} = \frac{\sin\lambda_0}{\sin\lambda}$$

which is Eqn (6.14). Finally, if we form the dot product of both sides of Eqn (6.1) with $\boldsymbol{\beta}$ we obtain

$$l \cos\lambda = l_0 \cos\lambda_0 + \alpha l_0 \cos\phi_0$$

Thus, using (6.15) we have

$$\alpha = \frac{\cos\lambda}{\cos\phi} - \frac{\cos\lambda_0}{\cos\phi_0}$$

which is Eqn (6.16).

When a long thin crystal is used in an experiment such as that illustrated in Fig. 6.11 the change in orientation of the tensile axis with respect to the crystal axes (which is easier to consider than what is the same thing, the change of the crystal axes with respect to the tensile direction) can be measured and plotted on a stereogram. Since the change of orientation is such that the tensile axis rotates towards the slip direction, this can be used to determine the direction of glide. An actual example for an f.c.c. metal crystal is given in Fig. 6.15a. When the orientation of the tensile axis in Fig. 6.15a reaches the great circle joining the $(0\,0\,1)$ and $(\bar{1}\,1\,1)$ poles, which is a mirror plane, the resolved shear stresses on two-slip systems are equal and slip might occur on both systems simultaneously. When two-slip systems operate simultaneously *double slip* or *duplex slip* is said to occur.

In many cases in face-centred cubic metals a crystal continues to slip on the primary system after the orientation has reached the mirror plane. In this case the orientation of the tensile axis 'overshoots' into the neighbouring unit triangle so that slip is not proceeding on the nominally most highly stressed system. An example is shown in Fig. 6.15b. After a certain amount of this additional

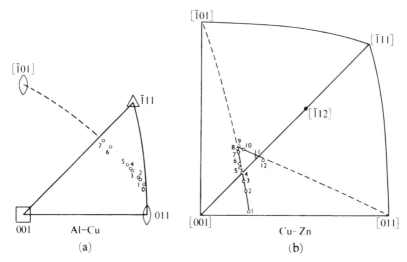

Figure 6.15 Changes in orientation of f.c.c. metal crystals during glide. The numbers indicate successive determinations of the crystal orientation. (From Seeger [1])

primary slip the conjugate system suddenly operates and further slip occurs on this system, sometimes followed by 'overshooting' on the conjugate system as in Fig. 6.15b. When the tensile axis reaches a $\{1\,1\,2\}$ pole — $(\bar{1}\,1\,2)$ in the example shown in Fig. 6.15b — no further orientation change should occur, provided each slip system contributes equally to the deformation.

There is also a change of orientation of the lattice of a crystal during a compression test. To deal with this case it is simplest to use vector formulae. We first use Eqn (6.1) to find how a given plane of normal $\mathbf{m}$ is altered by glide. Let $\mathbf{r}_1$, $\mathbf{r}_2$ be any two non-parallel unit vectors in the given plane. Write $\mathbf{m} = \mathbf{r}_1 \times \mathbf{r}_2$. Now transform each of $\mathbf{r}_1$ and $\mathbf{r}_2$ according to Eqn (6.1) to give $\mathbf{r'}_1$ and $\mathbf{r'}_2$. Then the new vector $\mathbf{m}'$ representing $\mathbf{m}$ after a crystallographic glide strain of amount α is

$$\mathbf{m}' = \mathbf{r'}_1 \times \mathbf{r'}_2$$

Substituting for $\mathbf{r'}_1$ and $\mathbf{r'}_2$ from Eqn (6.1) we have

$$\mathbf{m}' = [\mathbf{r}_1 \times \mathbf{r}_2] + \alpha(\mathbf{r}_1 \cdot \mathbf{n})\boldsymbol{\beta} \times \mathbf{r}_2 - \alpha(\mathbf{r}_2 \cdot \mathbf{n})\boldsymbol{\beta} \times \mathbf{r}_1$$

$$= \mathbf{m} + \alpha\boldsymbol{\beta} \times [(\mathbf{r}_1 \cdot \mathbf{n})\mathbf{r}_2 - (\mathbf{r}_2 \cdot \mathbf{n})\mathbf{r}_1]$$

Remembering the formula for the vector triple product (see Appendix 2), this is seen to be

$$\mathbf{m}' = \mathbf{m} + \alpha\boldsymbol{\beta} \times \mathbf{n} \times [\mathbf{r}_2 \times \mathbf{r}_1]$$

i.e.

$$\mathbf{m}' = \mathbf{m} - \alpha\boldsymbol{\beta} \times [\mathbf{n} \times \mathbf{m}] \tag{6.18}$$

and again using the formula for the vector triple product

$$\mathbf{m'} = \mathbf{m} - \alpha((\boldsymbol{\beta} \cdot \mathbf{m})\mathbf{n} - (\boldsymbol{\beta} \cdot \mathbf{n})\mathbf{m})$$

or

$$\mathbf{m'} = \mathbf{m} - \alpha(\boldsymbol{\beta} \cdot \mathbf{m})\mathbf{n} \qquad (6.19)$$

since $(\boldsymbol{\beta} \cdot \mathbf{n}) = 0$. Equations (6.18) and (6.19) show that any plane moves to its final position in such a way that its normal always lies in the plane containing the initial position of the normal and $\mathbf{n}$, the normal to the glide plane.

If we now consider the case of compression of a thin crystal between flat plates, as in Fig. 6.16, we take the constraint to be that the plane initially normal to the stress axis must be maintained in this orientation with respect to the stress axis as the crystal deforms. To maintain positive values of α during compression we suppose slip to occur in the direction $-\boldsymbol{\beta}$ (Fig. 6.16). Equation (6.19) then reads

$$\mathbf{m'} = \mathbf{m} + \alpha(\boldsymbol{\beta} \cdot \mathbf{m})\mathbf{n} \qquad (6.20)$$

From Eqn (6.20) the normal to the glide plane then approaches the normal to the compression plates by rotating about an axis which is parallel to the line of intersection of the glide plane and the compression plates (Fig. 6.16b). The initial thickness of the crystal is h_0, which will be reduced during compression to a height h such that

$$\frac{h_0}{h} = \frac{|\mathbf{m'}|}{|\mathbf{m}|} \qquad (6.21)$$

because the volume of the crystal must be conserved. With the constraint which we have assumed, the change of orientation of the crystal lattice will be such that the normal to the slip plane rotates towards the axis of compression. If the initial and final angles, after a shear strain α, between the direction of compression and the normal to the slip plane are ϕ_0, ϕ and λ and λ_0 are the corresponding values

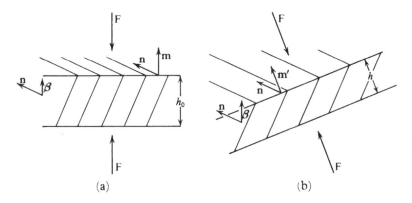

(a)   (b)

Figure 6.16

of the angle between the compression axis and the slip direction, then we have, from Eqn (6.20),

$$\mathbf{m}' \cdot \boldsymbol{\beta} = \mathbf{m} \cdot \boldsymbol{\beta}$$

since

$$\mathbf{n} \cdot \boldsymbol{\beta} = 0$$

Therefore

$$\frac{h_0}{h} = \frac{\cos \lambda_0}{\cos \lambda} \qquad (6.22)$$

and also, from (6.20),

$$\mathbf{m}' \times \mathbf{n} = \mathbf{m} \times \mathbf{n}$$

Thus

$$\frac{h_0}{h} = \frac{\sin \phi_0}{\sin \phi} \qquad (6.23)$$

Finding the modulus of each side of Eqn (6.20) we have

$$|\mathbf{m}'|^2 = |\mathbf{m}|^2 + 2\alpha |\mathbf{m}|^2 \cos \phi_0 \cos \lambda_0 + \alpha^2 |\mathbf{m}|^2 \cos^2 \lambda_0$$

and using Eqn (6.21),

$$\left(\frac{h_0}{h} \right)^2 = 1 + 2\alpha \cos \phi_0 \cos \lambda_0 + \alpha^2 \cos^2 \lambda_0$$

Forming $\mathbf{m}' \cdot \mathbf{n} = \mathbf{m} \cdot \mathbf{n} + \alpha(\boldsymbol{\beta} \cdot \mathbf{m})$ and using Eqns (6.21) and (6.22),

$$\alpha = \frac{\cos \phi}{\cos \lambda} - \frac{\cos \phi_0}{\cos \lambda_0} \qquad (6.24)$$

The resolved shear stress τ is given by

$$\tau = \frac{F}{A} \cos \phi \cos \lambda = \frac{F}{A} \left(\frac{h}{h_0} \right) \cos \lambda_0 \left[1 - \left(\frac{h}{h_0} \right)^2 \sin^2 \phi_0 \right]^{1/2}$$

Since $A_0 h_0 = Ah$ this may be written as

$$\tau = \frac{F}{A_0} \left(\frac{h}{h_0} \right)^2 \cos \lambda_0 \left[1 - \left(\frac{h}{h_0} \right)^2 \sin^2 \phi_0 \right]^{1/2} \qquad (6.25)$$

In compression the normal to the glide plane approaches the direction of compression, so that, for example, in Fig. 6.12a, $\mathbf{F}_0$ would approach the (1 1 1) pole until it meets the boundary between [0 0 1] and [0 1 1]; thereafter systems B IV and A III would be equally stressed and the stable orientation is [0 1 1].

Formulae for the change in orientation of the crystal lattice with respect to the direction of an applied force have been given for *duplex* slip in both compression and tension by Bowen and Christian [2]. During double glide there is a change in orientation of the lattice of the crystal with respect to the direction of a tensile force until the applied force is symmetrically related to the two slip systems. Thus, in tension of f.c.c. metals, $\langle 1\,1\,2 \rangle$ directions represent stable orientations lying midway between two $\langle 1\,1\,0 \rangle$ directions.

6.5 TEXTURE

Because large strains can cause a reorientation of the lattice of a crystal it follows that, in a polycrystalline material, continued plastic flow, such as that which occurs during the working of metals, tends to develop a *texture* or a *preferred orientation* of the lattice within the grains as well as a preferred change of shape of the grains. Working involves very large plastic strains, often hundreds of per cent. The development of a texture significantly changes the crystallographic properties of the material. A fine-grained material in which the grains have random lattice orientations will have isotropic properties (provided that other crystal defects such as inclusions and boundaries are uniformly distributed). However, a specimen with a preferred orientation will have anisotropic properties, which may be desirable or undesirable depending upon the intended use of the material. The final texture that develops may resemble a single orientation or may comprise crystals distributed between two or more preferred orientations. During plastic flow the process of reorientation is gradual. In theory the orientation change proceeds until a texture is reached that is stable against indefinitely continued flow of a given type. The theoretical end distribution of orientations and the manner in which it develops are a characteristic of the material, its crystal structure, its microstructure and the nature of the forces arising from the deformation process. Whether a material actually reaches its stable orientation depends upon many variables, e.g. the rate of working and the temperature at which it is carried out. Textural changes may correspond with marked changes in the ductility of a material. Thus there is an interplay between the role of deformation in creating texture and the role of texture in facilitating deformation. Whilst the basic principles underlying texture development for single crystals have been presented in Section 6.4, the prediction of actual textures during forming processes is necessarily very complicated (see the specialist books on the subject listed at the end of this chapter). The various processes of texture development have been categorized according to their final state and to the kind of working involved, e.g. wire and fibre textures, compression textures, sheet and rolling textures, torsion textures and deep drawing textures. Some examples of ideal end textures are presented in Table 6.4. In addition to deformation textures, anisotropic properties of a polycrystalline material can also arise as a result of recrystallization following working (see the footnote in Section 10.1) or as a

Table 6.4 Some common end or final textures

Class	Description	Example material	Texture	Comments	Ref.
f.c.c.	Rolling texture {plane}/⟨axis⟩	Aluminium	{110}⟨112⟩ then {112}⟨111⟩	Known as 'pure metal' texture at high temperatures and large deformations	3
		Copper	{110}⟨112⟩ and {112}⟨111⟩ {123}⟨412⟩ + {146}⟨211⟩	Orientations between these extremes or spread around these ideal end-textures	3
		Brass(70/30)	{110}⟨112⟩ and {110}⟨001⟩	Known as 'alloy' texture — changes to 'pure metal' texture at high temperatures	3
	Tension (fibre) texture ⟨axis⟩	Al, Cu, Au, Ag, Ni	⟨111⟩ and ⟨100⟩	Relative proportions dependent upon temperature, strain rate, % reduction and stacking fault energy and twinning	4
		Some single crystals	⟨211⟩	Except for starting orientation ⟨111⟩	5
			⟨111⟩	For starting orientation ⟨111⟩	4
	Compression texture ⟨axis⟩	Single and polycrystalline Al	Range from ⟨110⟩ to ⟨311⟩	Tendency to rotate away from ⟨111⟩ and ⟨100⟩	5
		Brass (70/30)	Range from ⟨110⟩ to ⟨311⟩ Some ⟨111⟩	Tendency to rotate away from ⟨100⟩	4
	Deposition texture ⟨growth direction⟩	Gold on cleaved NaCl by thermal evaporation	⟨111⟩	Random orientation in the plane of growth	6

(continued overleaf)

Table 6.4 (continued)

Class	Description	Example material	Texture	Comments	Ref.
b.c.c.	Rolling texture {plane}/⟨axis⟩	Iron single and polycrystalline	Between {001}⟨110⟩ and {112}⟨110⟩ {111}⟨112⟩ {111}⟨112⟩ {100}⟨110⟩	Often both components together Occasionally	3
		Aluminium single crystals		Up to 90% reduction After 90% reduction	5
	Tension (fibre) texture ⟨axis⟩	Mo, Nb, Ta, V, W, Si–iron, beta brass	⟨110⟩		4
	Compression texture ⟨axis⟩	Iron	⟨111⟩, some ⟨100⟩		4
	Deposition texture ⟨growth direction⟩	Mo on (001) Si by magnetron sputtering	{111} ⟨110⟩	High growth rates Low growth rates and low substrate temperatures Some in-plane texture in both cases	7,8
h.c.p.	Rolling texture {plane}/⟨axis⟩	Magnesium	{0001}⟨11$\bar{2}$0⟩		4
		Zinc, cadmium	20°–25° off {0001}⟨11$\bar{2}$0⟩		4
		Zirconium, titanium	30°–40° off {0001}⟨10$\bar{1}$0⟩	Deviation is in the transverse sense i.e. about the rolling direction	4
	Tension (fibre) texture ⟨axis⟩	Magnesium	⟨10$\bar{1}$0⟩ ⟨$\bar{2}$110⟩	Low temperatures High temperatures	4
		Zinc	⟨0001⟩ 70° from ⟨0001⟩ spiral texture	Small reductions Large reductions	4
	Compression texture ⟨axis⟩	Zirconium, titanium	⟨10$\bar{1}$0⟩		4
		Magnesium	⟨0001⟩		4
	Deposition texture ⟨growth direction⟩	Ti on oxidized Si (001) by magnetron sputtering	⟨0001⟩	Random in-plane orientation	9

result of material fabrication processes that do not involve plastic deformation. For example, the vapour deposition of layers on substrates often results in the layer having a polycrystalline texture with a preferred orientation in the growth direction and random orientations perpendicular to the growth direction. Some examples of deposition textures are included in Table 6.4.

There is usually a statistical distribution of orientations of the various crystals in a sample. This distribution is measured and represented on a pole figure such as in Fig 6.17. A pole figure is similar to a stereographic projection. On it is presented the statistical distribution of the orientations of the plane normals from a particular set of planes. A pole figure is usually obtained using diffraction from X-rays, neutrons or electrons[†]. In the typical case of X-ray diffraction in

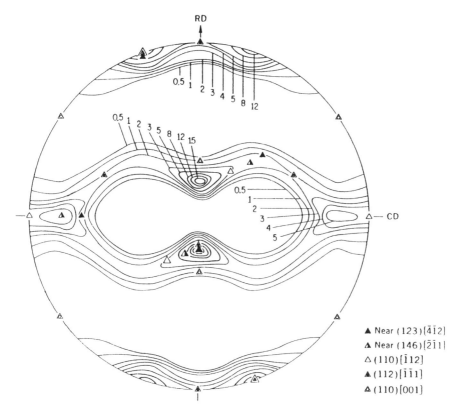

Figure 6.17 The {1 1 1} pole figure of electrolytic copper rolled to 96.6% reduction at room temperature where RD is the rolling direction and CD is the cross direction. The numbers represent relative intensities of the diffraction signal. (From Hu and Goodman [10])

[†] For X-ray and neutron diffraction the experiment obtains an average texture over volumes of the order of cubic millimetres. Using electron diffraction, textures from different regions of the sample

the laboratory the detector position is fixed to collect the diffraction from one particular set of planes. The sample is rotated about two axes so as to scan over all possible orientations. The corresponding pole figure is usually presented as a contour map, where the contours are labelled according to the relative strength of the diffraction signal. For example, Fig. 6.17 is a pole figure showing the relative intensities of the reflections from $\{1\,1\,1\}$ planes in a cold-rolled copper sample. The relative populations of planes in the various orientations are proportional to the numbers against the contour lines. The pole figure is interpreted in terms of an ideal end orientation. In uniaxial textures, such as those obtained in tension, compression and wire- and rod-forming processes, it is often sufficient to specify which crystallographic direction or directions lie parallel to the principal stress axis. For sheet textures, ideal orientations are given by a plane or planes lying parallel to the plane of the sheet and a direction or directions in the plane lying parallel to a significant direction in the sheet, e.g. the rolling direction (RD). For example, to identify $\{1\,1\,0\}\langle0\,0\,1\rangle$ and $\{1\,1\,0\}\langle1\,1\,2\rangle$ textures one might follow the procedures illustrated in Fig. 6.18. Firstly, identify the orientations of the $\{1\,1\,1\}$ poles relative to the $\{1\,1\,0\}$ plane and the $\langle0\,0\,1\rangle$ and $\langle1\,1\,2\rangle$ directions. Figure 6.18a shows a stereographic projection with the $(1\,1\,0)$ plane normal perpendicular to the plane of the paper. All of the $\{1\,1\,1\}$ poles are shown along with those $\langle0\,0\,1\rangle$ and $\langle1\,1\,2\rangle$ directions that are in the $(1\,1\,0)$ plane and thus could lie parallel to the rolling direction. Figures 6.18b and c show the orientations of the $\{1\,1\,1\}$ poles when the rolling direction corresponds to $\langle0\,0\,1\rangle$ and $\langle1\,1\,2\rangle$ respectively. Note that there are two possible $\langle1\,1\,2\rangle$ orientations, which results in there being twelve $\{1\,1\,1\}$ pole positions. Figure 6.18d illustrates the case where the end texture contains both $\{1\,1\,0\}\langle0\,0\,1\rangle$ and $\{1\,1\,0\}\langle1\,1\,2\rangle$ orientations. A pole figure from an ideal end texture will have the appearance of a pole plot from one or a few single crystals, depending upon how many orientations comprise the end texture. A real textured material will show a broader distribution of intensities about the ideal orientation. How close the actual texture is to the ideal texture is a qualitative judgement. Actual textures are described as varying from 'weak' (a small proportion of grains having rotated to the endpoint) to 'strong' (a large volume fraction of grains having rotated to or close to the endpoint).

A more involved but more representative measure of texture is in the *orientation distribution function* (ODF). Using data from pole figures taken for several reflections, the orientations of three primary axes of the crystallites are plotted with respect to three standard axes. ODFs can be displayed as three-dimensional plots or as a series of two-dimensional sections. Figure 6.19 shows an ODF of heavily rolled α-brass, where ϕ_1, Φ and ϕ_2 are the angles between principal directions in the crystallites and significant directions in the forming process and

---

can be discriminated, giving rise to the terms *microtexture* and *mesotexture* used in the study of textural variations within a sample.

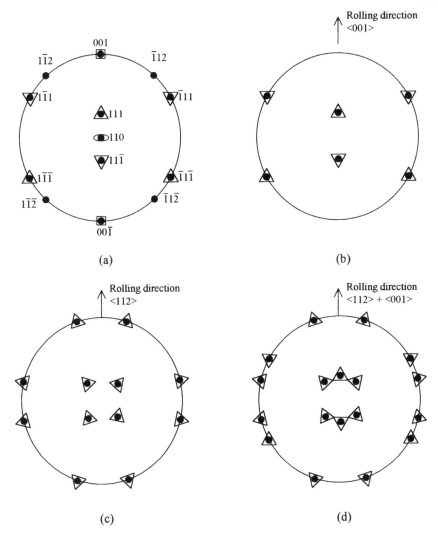

Figure 6.18 (a) Standard stereographic projection in the (1 1 0) orientation showing the poles for the {0 0 1}, {1 1 2} and {1 1 1} plane normals. Orientations of the {1 1 1} plane normals for (b) the (1 1 0)⟨0 0 1⟩ rolling texture, (c) the (1 1 0)⟨1 1 2⟩ rolling texture and (d) a combination of (b) and (c)

are known as the Euler angles. Typical Euler angles for rolled sheet are illustrated in Fig. 6.20, where ϕ_1 is the angle between the crystallite [0 1 0] axis and the rolling direction (RD), Φ is the angle between the crystallite [0 0 1] direction and the sheet normal (ND) and ϕ_2 is the angle between the crystallite [1 0 0] direction and the direction orthogonal to both ND and RD. Thus we see in Fig. 6.19 that

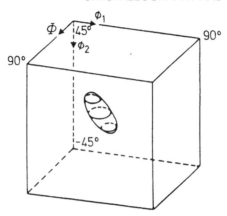

Figure 6.19 Definition of the orientation of a crystallite in rolled sheet by means of three Euler angles, ϕ_1, Φ and ϕ_2. RD is the rolling direction and ND is the sheet normal. (From Cahn, Haasen and Kramer [11])

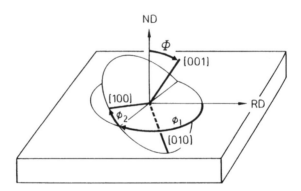

Figure 6.20 The orientation distribution function (ODF) of rolled α-brass calculated using the data from pole figures measured for (1 1 1), (2 0 0), (2 2 0) and (3 1 1). (From Bunge [12])

the spread in ϕ_1 is about $10°$, ϕ is about $10°$ and ϕ_2 is about $20°$, indicating a pronounced texture with only a small spread in orientation. Using the complicated mathematical analysis of the ODF (see the specialist books listed at the end of this chapter), a near-quantitative description of texture is possible.

PROBLEMS

6.1 α-ZnS has the point group $\bar{4}3m$. (a) Does it possess a centre of symmetry? Assume that the slip planes are $\{1\,1\,1\}$ and the slip directions $\langle 1\,\bar{1}\,0 \rangle$. (b) Is slip in the $[1\,\bar{1}\,0]$ direction crystallographically equivalent to that in $[\bar{1}\,1\,0]$?

6.2 During the process of glide the displacements imposed upon the surfaces of a crystal are centrosymmetric. Use this notion and Fig. 6.5 to deduce the rules (a) to (d) given in Section 6.2.

6.3 Enumerate the number of physically distinct glide systems for a cubic crystal (e.g. an f.c.c. metal) of point group $m3m$ which slips on $\{1\,1\,1\}$ in $\langle 1\,\bar{1}\,0 \rangle$ directions. Along which directions should such a crystal be stressed in uniaxial tension if the resolved shear stress is to be the same on (a) eight systems, (b) six systems and (c) four systems?

6.4 From Tables 6.1 and 3.2 find the ratio of shortest lattice repeat distance in the direction of glide to the nearest neighbour separation for bismuth.

6.5 Find the number of independent slip systems possessed by CsBr which is cubic ($Pm3m$) and slips on $\{1\,1\,0\}$ in $\langle 0\,0\,1 \rangle$. How many independent systems does such a crystal possess if pencil glide occurs with the same glide directions?

6.6 Consider an amount of glide α to occur on the $(0\,1\,1)$ plane in the $[0\,1\,\bar{1}]$ direction of a cubic crystal. Take axes parallel to the cubic axes. (a) Write down the components of the pure strain tensor. (b) Write down the components of the pure strain tensor if glide of amount α' occurs on $(1\,0\,0)$ in the same direction, $[0\,1\,\bar{1}]$, as before. (c) How many components of the pure strain tensor can be altered independently of one another by glide on these two systems? (d) If α' was always equal to a constant fraction of α so that $\alpha' = k\alpha$, what would be the answer to (c)? (e) Use the above results to prove generally that if pencil glide occurs in a crystal then slip with a given slip direction can provide two independent components of the pure strain tensor.

6.7 Can you think of three slip directions which operating together would provide a contradiction of the statement 'When pencil glide occurs a crystal need only possess in general three non-coplanar slip directions in order for a general strain to be possible'?

6.8 Copper is cubic ($Fm3m$) and slips on $\{1\,1\,1\}\langle 1\,\bar{1}\,0 \rangle$. Sketch a stereogram of a crystal showing the components of the slip systems. Which slip system(s) are expected to operate if the tensile axis is (a) $[1\,2\,3]$, (b) $[1\,1\,3]$ and (c) $[1\,2\,2]$?

6.9 Rock-salt is cubic ($Fm3m$) and slips on $\{1\,1\,0\}\ \langle 1\,\bar{1}\,0 \rangle$. Sketch a stereogram showing the components of the slip systems. Which slip system(s) are expected to operate if the tensile axis is (a) $[1\,2\,3]$, (b) $[2\,1\,3]$ and (c) $[0\,1\,3]$?

6.10 Magnesium oxide is cubic with the NaCl structure and slips on $\{1\,1\,0\}$ in $\langle 1\,\bar{1}\,0 \rangle$.
   (a) List the directions along which a tensile or compressive force applied to the crystal cannot produce glide.
   (b) Can the crystal be twisted plastically about $[1\,0\,0]$?

(c) List the same directions as in (a) for graphite (Table 6.1 lists the glide element).

6.11 The analysis of strain in Chapter 5 requires that the strain be homogeneous or infinitesimal. Do these restrictions invalidate the analysis of Section 6.3 (which shows that five independent glide systems are necessary for a general deformation) because when glide takes place large strains occur in some volumes of the crystal?

6.12 A tetragonal crystal is subject to a tensile stress of magnitude σ along $[1\,0\,0]$ and to a shear stress τ on $(1\,0\,0)$ in $[0\,1\,0]$. (a) Write down the components of the stress tensor referred to the crystal axes. (b) Find the shear stress on $(1\,1\,0)$ in the $[1\,\bar{1}\,0]$ direction by altering the axes of reference of the stress tensor. (c) Find, by the same procedure, the tensile stress normal to $(1\,1\,0)$.

6.13 Prove formula (6.12).

6.14 A single crystal of zinc which slips on $(0\,0\,0\,1)$ in $\langle 1\,1\,\bar{2}\,0 \rangle$ directions is stressed in tension and is oriented so that the tensile axis makes an angle of $45°$ with $[1\,1\,\bar{2}\,0]$ and is at $60°$ to the normal to $(0\,0\,0\,1)$. (a) Along which direction will slip first occur? (b) If slip starts when the tensile stress on the crystal is 2.1×10^6 Pa, calculate the critical resolved shear stress necessary to produce slip in zinc.

6.15 Copper slips on $\{1\,1\,1\}$ in $\langle 1\,\bar{1}\,0 \rangle$. A cylindrical single crystal 10 cm long is stressed in tension. The tensile axis is $[\bar{1}\,2\,3]$.
   (a) Which is the slip system with the highest resolved shear stress? The specimen is strained in tension until the tensile axis reaches a symmetry position where two slip systems have equal resolved shear stresses.
   (b) Which are the two systems?
   (c) What is the orientation of the tensile axis at this stage?
   (d) What is the new length of the specimen?

6.16 Discuss the following statement. 'A general small homogeneous change of shape of a body can be described by a pure strain and a rotation. If the volume of the body is conserved the pure strain tensor has five independent components and the pure rotation tensor has three. Therefore eight quantities must be specified to describe the change of shape of a single crystal within a polycrystalline body. It follows from this that eight slip systems must in general operate in all grains of a polycrystal which is subject to a general change of shape.'

SUGGESTIONS FOR FURTHER READING

Bowen, D. K. and Christian J. W., 'The calculation of shear stress and shear strain for double glide in tension and compression', *Phil. Mag.*, **12**, 369 (1965).

Bunge H. J., *Texture Analysis in Materials Science, Mathematical Methods*, Butterworths, London (1982).

Cahn, R. W., 'Measurement and control of texture', in Vol. 15, *Materials Science and Technology* (series eds. R. W. Cahn, P. Haasen and E. J. Kramer), *Processing of Metals and Alloys* (eds. R. W. Cahn), Wiley–VCH, Chichester and Weinheim (1991), Ch. 10.

Chin, G. Y. Thurston, R. W. and Nesbitt, E. A., 'Finite deformations due to crystallographic slip', *Trans. Amgnst., Mining Metall. and Petrol. Engrs*, **236**, 69 (1966).

Dieter, G. E., *Mechanical Metallurgy* (SI metric edition adapted by D. Bacon), McGraw-Hill, London (1988).

Groves, G. W. and Kelly, A., 'Independent slip systems in crystals', *Phil. Mag.*, **8**, 877 (1963).

McGregor Tegart, W. J., *Elements of Mechanical Metallurgy*, Macmillan (1966).

Schmid, E. and Boas, W., *Plasticity of Crystals*, F. A. Hughs and Co. Ltd, London (1950).

REFERENCES

1. A. Seeger, *Handbuch der Physik*, Vol. VII/2, Springer-Verlag, Berlin.
2. D. K. Bowen and J. W. Christian, 'The calculation of shear stress and shear strain for double glide in tension and compression', *Phil. Mag.*, **12**, 369 (1965).
3. I. L. Dillamore and W. T. Roberts, *Metall. Rev.*, **10**, 271 (1965).
4. C. S. Barrett and T. B. Massalski, *Structure of Metals: Crystallographic Methods, Principles and Data*, Pergamon, Oxford, (1980), pp. 541–583.
5. H. Hu, R. S. Cline and S. R. Goodman, *Trans. Am. Soc. for Metals*, 295 (1965).
6. K. E. Harris and A. H. King, in *Mechanisms of Thin Film Evolution* (eds. S. M. Yalisove, C. V. Thompson and D. J. Eaglesham), MRS Vol. 317, MRS. Pennsylvania (1994) p. 425.
7. M. Vill, S. G. Malhotra, Z. Rek, S. M. Yalisove and J. C. Bilello, in *Mechanisms of Thin Film Evolution* (eds. S. M. Yalisove, C. V. Thompson and D. J. Eaglesham), MRS Vol. 317, MRS, Pennsylvania (1994), p. 413.
8. O. P. Karpenko, M. Vill, S. G. Malhotra, J. C. Bilello and S. M. Yalisove, in *Mechanisms of Thin Film Evolution* MRS Vol. 317, (eds. S. M. Yalisove, C. V. Thompson and D. J. Eaglesham), MRS Vol. 317, MRS, Pennsylvania (1994), p. 467.
9. R. Ahuja and H. L. Fraser, *Mechanisms of Thin Film Evolution* (eds. S. M. Yalisove, C. V. Thompson and D. J. Eaglesham) MRS Vol. 317, MRS, Pennsylvania (1994), p. 479.
10. H. Hu and S. R. Goodman, *Trans. Met. Soc. AIME*, **227**, 627 (1963).
11. R. W. Cahn, P. Haasen and E. J. Kramer (eds.), *Materials Science and Technology*, Vol. 15, *Processing of Metals and alloys* (ed. R. W. Cahn), VCH, Weinheim (1991).
12. H. Bunge, *J. Int. Mater. Rev.*, **32**, 265 (1987).

7

Dislocations

7.1 INTRODUCTION

In Chapter 6 the geometry of glide was studied on a macroscopic scale. The geometrical results on this scale are adequately described by the model of blocks of crystal sliding over one another rigidly. However, when glide occurs in a real crystal, all the atoms above the slip plane do not move simultaneously over those below. At any given time some of the atoms have moved into their new positions while others have not yet done so, and the displacement of the upper block of crystal relative to the lower varies from one region of the slip plane to another. Lines in the slip plane separating regions where slip has occurred from those where it has not are called *dislocations*. This is the simplest way in which to introduce the idea of dislocations; we shall see later that they can be formed in other ways.

For ease of illustration we will consider slip in a crystal with the primitive cubic lattice and one atom associated with each lattice point. Figure 7.1 shows a block of such a crystal cut with faces parallel to {1 0 0} planes. A (0 0 1) slip plane, PQRT, is shown, and slip in the [0 1 0] direction has occurred over only the region PQSS'. Evidently the crystal lattice is distorted, especially close to the dislocation which lies at SS'. The distortion decreases as the amount of slip over PQSS' is reduced. In Fig. 7.1 the displacement over PQSS' is equal to $[0\,1\,0]$,[†] which is the smallest displacement that can be made without producing a fault in the crystal structure at the plane PQSS'. The displacement is a constant vector quantity over the whole region PQSS' (except within a few atoms of the dislocation line SS'), and it is this quantity that characterizes the dislocation. As a dislocation moves through the crystal, the displacement associated with it does not change. The characteristic quantity is called the *Burgers vector* of the dislocation. As the

---

[†] A specific lattice vector is written in terms of its components along the crystal axes. Thus the vector $[p\,q\,r]$ is the vector from a lattice point to the point reached by travelling along the crystallographic x axis a distance of p lattice point spacings, then a distance of q spacings parallel to the y axis and then a distance of r spacings parallel to the z axis. In cubic crystals the lattice parameter a is often included in the notation, although it is superfluous. Thus the vector $[0\,1\,0]$ is often written $a[0\,1\,0]$, showing that its components along the x, y and z axes are of length zero, a and zero respectively.

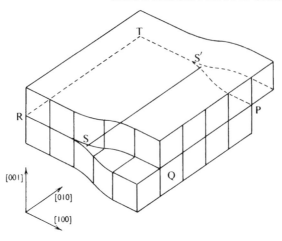

Figure 7.1   A screw dislocation in a primitive cubic lattice

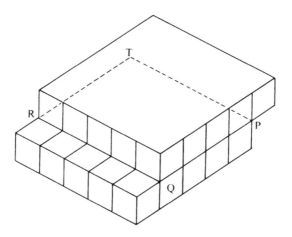

Figure 7.2

dislocation SS' in Fig. 7.1 moves from right to left, the slipped region of the crystal increases, and when the dislocation reaches the surface at RT, the whole crystal has been slipped by [0 1 0], as shown in Fig. 7.2. A dislocation may turn into a new orientation; this does not change its Burgers vector. Figure 7.3 shows the result of rotating SS' in the slip plane through a right angle. The dislocation which results, EE', has the same [0 1 0] displacement vector associated with it, the crystal now having slipped by this amount over QREE' and not having slipped over PTEE'. A length of dislocation which lies normal to its Burgers vector, such as EE', is called an *edge dislocation*; one like SS' which lies parallel to its Burgers vector is called a *screw dislocation*. In general, a dislocation may lie at any angle

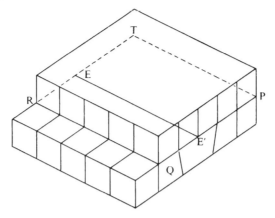

Figure 7.3   Edge dislocation in a primitive cubic lattice

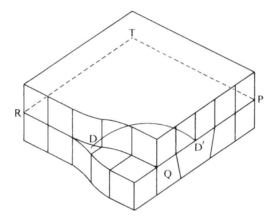

Figure 7.4

to its Burgers vector or may be curved like the dislocation DD' in Fig. 7.4. At D' the dislocation is in edge orientation and at D in screw orientation. Elsewhere the dislocation has both edge and screw components and is referred to as *mixed*.

The reason for the names 'edge' and 'screw' becomes clear when we look at the positions of the atoms around the dislocation. The positions close to the dislocation cannot be calculated exactly without a detailed knowledge of the laws of force between the atoms, although in some cases they can be observed at the surface of a crystal with the field-ion and other microscopes. For the present it is enough to recognize that there is a small amount of arbitrariness in the location of the atoms shown in Figs 7.5 and 7.6. Figure 7.5 shows the arrangement of atoms in a plane normal to the edge dislocation EE' of Fig. 7.3. There is an

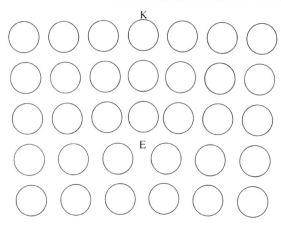

Figure 7.5   Atom positions around an edge dislocation (according to Eqn 7.21)

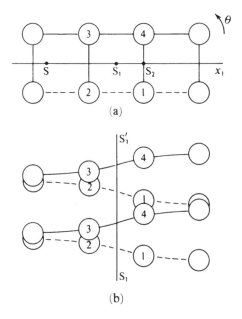

Figure 7.6.   Screw dislocation in a simple cubic crystal (a) looking along the dislocation and (b) looking normal to the dislocation which lies along $S_1 S'_1$

extra incomplete plane of atoms KE above the slip plane. The edge of the extra plane coincides with the dislocation and gives it its name. A similar view of a plane normal to a screw dislocation does not show the distortion of the lattice, because the displacements of the atoms are parallel to the dislocation in this case (Fig. 7.6a). A view of the atom positions in the atom planes immediately above and below the slip plane is given by Fig. 7.6b. The term 'screw' arises because the successive planes of atoms normal to the dislocation are converted by the presence of the dislocation into a single screw surface, or spiral ramp. This can be visualized from Fig. 7.1 by imagining oneself to be walking around the dislocation on these planes. Successive intersections of such a tour with the plane of Fig. 7.6 occur at 1, 2, 3, 4, etc.

Crystals normally contain a great number of dislocations, even before they are plastically deformed. The total length of dislocation line in 1 cm^3 of crystal (the *dislocation density*) may typically be 10^6 cm, although a wide variation is possible.

So far we have considered only dislocations occurring as a result of slip and lying in a slip plane. This is an unnecessary restriction, and dislocations can, in fact, be produced in other ways than by glide, as we shall see in Section 7.2. A general method is therefore needed for describing the Burgers vector — that characteristic discontinuity of displacement which defines a dislocation in a crystal. The following method is often used. Firstly, a closed circuit is made around the suspected dislocation, called a Burgers circuit. This is a circuit made by jumping from one lattice point to a neighbouring lattice point until the starting point is reached again. The circuit must be made in 'good' material, i.e. in regions of the crystal in which, even though they may be strained, each jump is *recognizably* associated with a jump in a perfect crystal of the same structure. With this vital condition fulfilled, a corresponding set of jumps can then be traced in the perfect crystal; if it fails to close then the circuit in the real crystal does indeed surround a dislocation. A little thought will show that if the circuit fails to close, then it must do so by a lattice vector. The lattice vector that is needed to complete the circuit in the ideal crystal is defined as the Burgers vector of the dislocation. A dislocation that can be dealt with by this procedure must have a Burgers vector equal to a translation vector of the Bravais lattice; such a dislocation is called a perfect dislocation. This method of defining the Burgers vector is illustrated in Fig. 7.7. The circuit of seventeen successive jumps in the real crystal has a corresponding set of jumps in the perfect crystal which fails to close. The vector [0 1 0] is needed to complete the circuit in the ideal crystal; this is the Burgers vector. Some sign convention is needed to define the *sense* of the Burgers vector. The FS/RH convention works as follows. First a sense is arbitrarily attached to the dislocation line. This permits a choice of Burgers circuit with a right-hand screw relationship to the line (the RH of the notation). In Fig. 7.7 the line is taken to run into the paper. The Burgers vector is then taken to run from the finish to the start of the path in the perfect crystal (the FS of the notation), e.g. from F to S in Fig. 7.7.[†] If the sense of the line is reversed, the Burgers vector is also reversed, so that some consistency in choosing a line sense is desirable. As an

[†] The opposite convention, SF/RH, is used by some authors. For an account of where one convention has been used, and where the other, see de Wit [1].

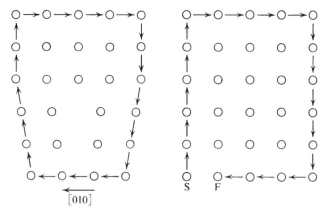

Figure 7.7 A Burgers circuit around an edge dislocation

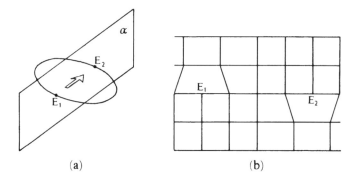

Figure 7.8

example, consider a closed loop of dislocation lying in a slip plane (Fig. 7.8a). Suppose that within the loop the top half of the crystal has slipped by [0 1 0] with respect to the bottom half, as shown by the arrow. Then the positions of the atoms in the (1 0 0) plane (the plane α in Fig. 7.8a) are as shown in Fig. 7.8b. Since the edge dislocations at E_1 and E_2 would annihilate one another if they glided together (their extra planes would join), they are commonly referred to as edge dislocations of opposite sign. The Burgers vector at E_2 is in the opposite direction to that at E_1 if the direction assigned to the line at E_2 is the same as that assigned to the line at E_1, e.g. into the page. If the whole loop is being considered, however, it is more logical to make the sense of the line continuous around the loop. Then if the direction of the line at E_1 is into the page, the direction of the line at E_2 must be out of the page, and when the Burgers circuits round E_1 and E_2 are each taken in the RH sense, the same sense of Burgers vector is obtained at E_1 as at E_2. This second way of stating the Burgers vector emphasizes the fact that the displacement discontinuity, which the Burgers vector measures, remains constant along a given dislocation line. This follows because if we move the Burgers circuit along the dislocation line it must remain in good material.

The principle of the conservation of the Burgers vector is a general one, and from it, it follows that a dislocation line cannot end inside a crystal. It must end at a surface, form a closed loop or meet other dislocations. A point at which dislocation lines meet is called a *node*. The conservation of the Burgers vector in this case can be stated in the following way. If the directions of all the dislocation lines are taken to run out from the node, then the sum of the Burgers vectors of all the dislocations is zero. This is the dislocation analogue of Kirchhoff's law of the conservation of electric current in a network of conductors.

7.2 DISLOCATION MOTION

We have seen that the motion of a dislocation on a slip plane changes the area over which slip has occurred. In this section the effect of dislocation motion in general will be discussed.

Consider two neighbouring atoms, P and Q, lying at lattice points in a plane which is threaded by a dislocation DD' (Fig. 7.9a). Suppose the dislocation passes between P and Q so as to move inside a Burgers circuit, part of which is the step from P to Q. The corresponding circuit in the ideal crystal then acquires a closure failure equal to the Burgers vector, **b**. This must be attributed to a displacement of P relative to Q of **b** in the real crystal (it cannot be attributed to the relative displacement of the members of any other neighbouring pair of atoms in the Burgers circuit since it is possible to draw Burgers circuits through these that do not contain the dislocation either before or after its movement).

The determination of the sense of the relative displacement requires some care. Define a positive direction **QP** by the cross product $\mathbf{l} \times \mathbf{d}$ (Fig. 7.9b); i.e. **l**, **d** and **QP** are related as a right-handed set of axes x, y, z. Here **l** is the vector

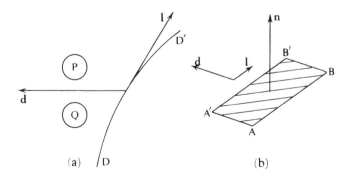

(a)   (b)

Figure 7.9

parallel to the dislocation line which is used in defining the Burgers vector **b** by the FS/RH convention; **d** is a vector in the direction of motion of the dislocation. With this definition, the displacement of the atom on the positive side, P, relative to that on the negative, Q, is $+\mathbf{b}$ when the dislocation passes between P and Q, as shown. We shall use this result to study the relative displacement of material on either side of the surface that a dislocation sweeps out during its motion.

Suppose a segment of a dislocation line AB moves to A'B', sweeping out a plane surface whose unit normal is **n**. The sense of **n** is defined in accordance with the previous paragraph as the sense of $\mathbf{l} \times \mathbf{d}$ (Fig. 7.9b). The atoms above ABB'A' are displaced by **b** relative to those below. If $\mathbf{b} \cdot \mathbf{n}$ is positive the atoms above move away from those below, creating a void, whose volume is $\mathbf{b} \cdot \mathbf{n}$ per unit area swept out. Motion for which $\mathbf{b} \cdot \mathbf{n} < 0$ requires the disposal of extra material of volume $\mathbf{b} \cdot \mathbf{n}$ per unit area. When $\mathbf{b} \cdot \mathbf{n} \neq 0$, the dislocation motion is normally accompanied by diffusion which brings in atoms to fill the void when $\mathbf{b} \cdot \mathbf{n} > 0$ and disperses the extra atoms when $\mathbf{b} \cdot \mathbf{n} < 0$. Dislocation motion of this type is called climb. The origin of this name can be seen from Fig. 7.5. If the atoms on the edge of the extra plane are removed, the edge dislocation in Fig. 7.5 will climb up out of its slip plane. Climb downwards could be accomplished by adding atoms to the extra plane.

Dislocation glide is motion that does not involve the addition or removal of material, i.e. motion for which $\mathbf{b} \cdot \mathbf{n} = 0$. The glide surface is defined accordingly as that surface which contains the dislocation line and whose normal is everywhere perpendicular to the Burgers vector. For a straight edge dislocation this defines a single plane which is the slip plane. For a screw dislocation any plane passing through the dislocation fulfils the condition $\mathbf{b} \cdot \mathbf{n} = 0$. For a closed loop of dislocation a cylinder is defined as the glide surface. The surface of the cylinder is generated by lines parallel to the Burgers vector passing through all points on the dislocation (Fig. 7.10a). The glide of the loop is not confined to the surface of its glide cylinder, because if it has any screw parts they can glide on any plane; however, the area of the loop in projection on a plane normal to the axis of the cylinder remains constant (Fig. 7.10b). This area is zero for a loop on a slip plane (Fig. 7.8a). When the area is not zero the loop is called prismatic. A special case is a pure edge dislocation loop. It could be produced by prismatic punching, as shown in Fig. 7.11, in which case its structure corresponds to a penny-shaped disc of extra atoms in the plane of the loop. With a Burgers vector of opposite sense its structure corresponds to a disc of missing atoms. In the latter case the loop could be formed by collecting together a disc of vacant lattice sites.

When a dislocation glides, there is no problem of acquiring or disposing of extra atoms. This is in sharp distinction to climb, and it leads us to suppose that a dislocation will be able to glide much more quickly than it can climb. The ease of gliding may be quite different for different crystallographic surfaces, and may

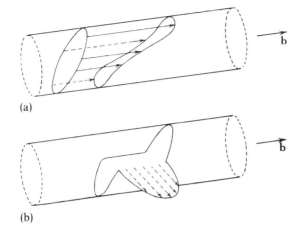

(a)

(b)

Figure 7.10 The dislocation loop shown in (a) can glide so that its area projected normal to the Burgers vector does not change. It moves in the surface of a cylinder. (b) This shows that a screw segment can leave the surface of this cylinder without changing the projected area

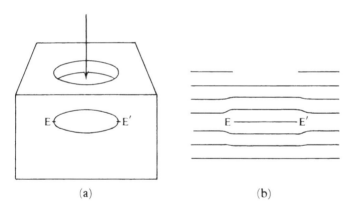

(a)   (b)

Figure 7.11 (a) Production of a prismatic dislocation loop by punching. (b) Section normal to the plane of the loop and passing through E and E'

be so much easier for one crystallographic plane that slip is effectively confined to that plane.

7.3 THE FORCE ON A DISLOCATION

In Section 7.1 we found that when a dislocation moved on a slip plane, the area of the slip plane across which slip had occurred was increased or decreased, according to the direction of motion. As a result, work may be done by the forces applied to the crystal. For example, suppose forces acting in the slip direction are applied to the top and bottom surfaces of the crystal in Fig. 7.1, in the directions SS' on the top surface and S'S on the bottom. When the dislocation, of unit length say, moves to the left and sweeps out an area dx of slip plane, applied forces of magnitude $\sigma \, dx$ move through the slip distance b, if the applied forces are of strength σ per unit area of the top and bottom surfaces. The work done is then

$$dw = \sigma b \, dx \qquad (7.1)$$

We can use this result to define a force[†] acting on a unit length of the dislocation in the direction Ox:

$$F = -\frac{dE}{dx} \qquad (7.2)$$

where E is the energy of the system, comprising the crystal and the device applying the forces to it. If we examine the work done by the forces acting on the surfaces at the slip plane, which are displaced relative to one another when the dislocation moves a distance dx, we see that it has the same magnitude $\sigma b dx$ as the work done by the externally applied forces, but is opposite in sign. This work done at the slip plane is dissipated, like the work done at sliding surfaces in a fraction problem. The dissipated energy is supplied by the device applying the forces to the crystal, which therefore loses energy:

$$-dE = dw = \sigma b \, dx \qquad (7.3)$$

Substituting Eqn (7.2) in Eqn (7.3) gives

$$F = \sigma b \qquad (7.4)$$

This definition of the force acting on a dislocation can be extended to the general case of a dislocation moving in any direction in a crystal which is stressed internally or by external forces. To do this we use the result of the preceding section, that when a segment of dislocation line sweeps out an area of internal surface, the material on the positive side of that surface is displaced by **b** relative to the material on the negative side. The crystal is conceived to be cut at this surface. The force acting on one of the surfaces of the cut is calculated from the stress in the crystal, and the work done by this force on the crystal during the displacement is then calculated, the other surface being held fixed.

The dislocation will be driven in such a direction that this work is negative. In the absence of external forces the energy of the crystal will then decrease (its

---

[†] Defining a 'force' exerted on a dislocation line is a way of describing the tendency of the configuration which constitutes the dislocation to move through the crystal. There is no physical force on a dislocation line in the sense in which there is a force on a rod when a string attached to it is pulled, for example.

self-stress will be reduced), and the magnitude of the force on the dislocation is defined by the decrease in the energy of the crystal when the dislocation moves a unit distance. If the dislocation is being driven by externally applied forces, this energy loss is made up by the work done on the crystal by the applied forces, and the energy of the crystal itself does not change. The decrease in energy of a larger system which includes the device applying the force can then be considered to define the force acting on the dislocation, and identical equations are obtained.

Using the sign convention of the preceding section, the force $\mathbf{F}$ acting on a unit area of the surface bounding the negative side of the cut is given by (see Section 5.4)

$$F_i = \sigma_{ij} n_j \qquad (7.5)$$

or, in vector notation,

$$\mathbf{F} = \sigma \cdot \mathbf{n} \qquad (7.6)$$

This force is exerted by the material on the positive side of the swept-out surface. The stress is σ and $\mathbf{n}$ is a unit vector normal to the swept-out surface. The sense of $\mathbf{n}$ is given by $\mathbf{l} \times \mathbf{d}$, where $\mathbf{l}$ is a unit vector parallel to the dislocation line and $\mathbf{d}$ is a unit vector in the direction of motion (Fig. 7.9). Without loss of generality we can take $\mathbf{d}$ perpendicular to $\mathbf{l}$ since any motion of the dislocation parallel to itself sweeps out no area. Therefore

$$\mathbf{n} = \mathbf{l} \times \mathbf{d} \qquad (7.7)$$

The movement of the dislocation displaces the surface bounding the negative side of the cut by $-\mathbf{b}$, if the opposite surface remains at rest, so that the work done on the crystal is

$$w = -(\sigma \cdot \mathbf{n}) \cdot \mathbf{b} \qquad (7.8)$$

which, because σ is a symmetrical tensor, can be written as

$$w = -(\sigma \cdot \mathbf{b}) \cdot \mathbf{n} \qquad (7.9)$$

The decrease in energy is given by the negative of this so

$$-w = (\sigma \cdot \mathbf{b}) \cdot (\mathbf{l} \times \mathbf{d})$$

or

$$-w = (\sigma \cdot \mathbf{b}) \times \mathbf{l} \cdot \mathbf{d} \qquad (7.10)$$

using the theorems on the scalar triple product of vectors (see Appendix 2).

We now have the work done by the crystal when a unit length of dislocation moves a unit distance in the form

$$-w = \mathbf{F} \cdot \mathbf{d} \qquad (7.11)$$

where $\mathbf{d}$ is a unit vector in the direction of motion. $\mathbf{F}$ is therefore the force per unit length acting on the dislocation, so

$$\mathbf{F} = (\sigma \cdot \mathbf{b}) \times \mathbf{l} \qquad (7.12)$$

The expression $\sigma \cdot \mathbf{b}$ represents the force acting on a plane normal to the Burgers vector, of area b. Since $\mathbf{l}$ is a unit vector parallel to the dislocation we see that the force is always normal to the dislocation line.

As an example of the application of this formula, consider the edge dislocation of Fig. 7.12 in a region of general stress σ. The stress may be the result of forces applied to the exterior of the crystal, or it may be a local self-stress[†]. Suppose that the dislocation runs along the x_2 axis and that its Burgers vector is in the direction $-x_1$. The traction on the surface whose outward normal is $\mathbf{b}$ is the resultant of components due to σ_{11}, σ_{12} and σ_{13} (Fig. 7.12). The component of $(\sigma \cdot \mathbf{b})$ due to σ_{12} is parallel to $\mathbf{l}$ and so makes no contribution to $(\sigma \cdot \mathbf{b}) \times \mathbf{l}$. The component σ_{11} produces a traction in the direction of $\mathbf{b}$ and a component of force on the dislocation of magnitude $\sigma_{11}b$, acting downwards. If σ_{11} is negative, the force on the dislocation is directed upwards. This confirms the intuition that a compression will tend to squeeze the extra plane out of the crystal and that a tensile stress will tend to draw it downwards. The component of the force normal to the glide plane is called the climb force. The glide force acts in the glide plane and in this case it is given by the component σ_{13}, which produces a force per unit length of magnitude $\sigma_{13}b$ directed along Ox_1. Sometimes it is only important to know the glide force, i.e. the component of force in the slip plane. As can be seen from Eqn (7.8), the glide force on any dislocation is simply b times the component of the traction on the slip plane which acts in the slip direction.

When an edge dislocation climbs under the influence of a stress, void or extra material is created, and it is intuitively clear that the climb will be brought to a

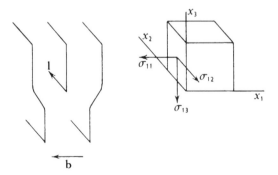

Figure 7.12

[†] A self-stress, sometimes called an internal stress, is a stress that exists in a crystal that is free from externally applied forces.

halt if the extra material is allowed to accumulate or if the void is not filled in. This leads to the idea of a 'chemical' force on a dislocation, due, for instance, to the absence of atoms on some of the sites in the surrounding lattice. To understand this force we must anticipate somewhat. It is shown in Chapter 9 that a crystal in *equilibrium* contains a certain number of empty atomic sites, called vacancies. To define the chemical force in this instance we write down the equilibrium concentration of vacancies in the presence of an edge dislocation under stress. When the concentration has this equilibrium value the dislocation has no tendency to create or annihilate vacancies by climbing, and so we define a chemical force equal and opposite to the climb force exerted by the stress.

Suppose that when a vacancy is created by adding a lattice atom to the edge of the extra plane of an edge dislocation, a length αb of dislocation climbs by βb. Here α and β are constants of order unity which depend upon the structure of the crystal. Then when a vacancy is created, a stress that exerts a climb force F per unit length does work W given by*

$$W = F\alpha\beta b^2 \qquad (7.13)$$

Consequently, the equilibrium concentration of vacancies is increased from its value C_0 in the absence of the climbing force[†] to

$$C = C_0 \exp\left(\frac{F\alpha\beta b^2}{kT}\right) \qquad (7.14)$$

We define a chemical (or osmotic) force equal and opposite to F as being exerted by the concentration C of vacancies

$$-F = -\frac{kT}{\alpha\beta b^2} \ln \frac{C}{C_0} \qquad (7.15)$$

In the absence of stress a concentration C which is not equal to its equilibrium value C_0 gives a chemical force that is not balanced by a force due to stress. The magnitude of this force is given by Eqn (7.15). The dislocation will then climb upwards in Fig. 7.12 if $C/C_0 > 1$, i.e. in the presence of a 'supersaturation' of vacancies.

It is interesting to compare the magnitude of the 'chemical stress' F/b with the magnitude of a typical mechanical stress. Setting $\alpha = \beta = 1$, $\mu b^3 \sim 5$ eV,[‡] where μ is the rigidity modulus, $kT \approx 1/40$ eV at room temperature we obtain

$$\frac{F}{b} \sim \frac{\mu}{200} \ln \frac{C}{C_0} \qquad (7.16)$$

---

*Here we are neglecting work that will be done by the hydrostatic component of the stress if the volume of the crystal changes when an atom is removed from an interior site. For the effect of this see Weertman [2].

[†] See Chapter 9, Section 9.1.

[‡] The eV (electron volt) is a unit of energy equal to 1.59×10^{-12} ergs or 1.59×10^{-19} J.

Many crystals cannot support so large an applied stress as $\mu/200$, so quite trifling supersaturations $(C/C_0 \sim 3)$ produce a force larger than that arising from any applied stress. Larger supersaturations, such as may be produced by quenching from close to the melting point, can cause small prismatic dislocation loops to grow. This has been observed, for example, in aluminium.

7.4 THE DISTORTION IN A DISLOCATED CRYSTAL

It is important to know exactly how the structure of a crystal is distorted by a dislocation. This information is needed in tackling such problems as how dislocations interact with each other and with other lattice defects. An obvious way to approach the problem is to investigate an elastic continuum containing a dislocation. If a satisfactory solution is found for the dislocated continuum, the atoms in a crystal can then be assigned the displacements that are suffered by the corresponding points in the continuum. As a first approximation, isotropic elasticity can be assumed.

For a screw dislocation in an isotropic continuum the problem is simple enough for an intuitive approach to give the correct solution. We expect the displacement at any point to be parallel to the dislocation. Further, because of the radial symmetry of the situation, a total displacement b must be spread out evenly over a circular path around the dislocation. This suggests a displacement u_z given by

$$u_z = \frac{b\theta}{2\pi} \tag{7.17}$$

If a path at a radial distance r from the dislocation is imagined to be straightened out, as in Fig. 7.13, the shear strain is seen to be

$$\gamma_{z\theta} = \frac{b}{2\pi r} \tag{7.18}$$

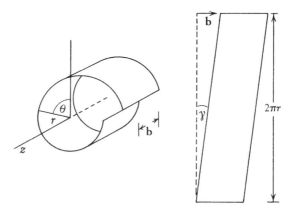

Figure 7.13 Strain due to a screw dislocation lying along the z axis

The non-zero stress component is the corresponding shear stress, given by Hooke's law as

$$\sigma_{z\theta} = \sigma_{\theta z} = \frac{\mu b}{2\pi r} \tag{7.19}$$

It remains only to check that the stresses given by Eqn (7.19) are such that every part of the body containing the dislocation is in equilibrium. Because the stress $\sigma_{z\theta}$ does not change with z or θ the opposite faces of all small elements, like the one shown in Fig. 7.14, are acted on by equal and opposite forces, so that the element is in equilibrium. Further, since σ_{zr}, $\sigma_{\theta r}$ and σ_{rr} are everywhere, zero, the surface of a cylinder around the dislocation is free of force. A difficulty with the shear strain given by Eqn. (7.18) is that it becomes infinite as r approaches zero. Since Hooke's law applies only to small strains, it is usual to cut off the solution at a radius where the strains become large, say $r \sim 5b$. The material within a cylindrical surface of this radius is called the dislocation core. The core surface is in effect a boundary of the elastic body. The forces that would be exerted from point to point on this boundary by the core material are not known, but it is known that no net force can be exerted in any direction, nor any couple. This condition is satisfied by Eqn (7.19), which gives a cylindrical surface that is entirely free of force. Equations (7.17) and (7.19) therefore give satisfactory displacements and stresses at distances greater than about $5b$ from a screw dislocation in an infinitely long isotropic cylinder. For a finite cylinder with free ends, the solution needs to be modified in order to remove the effect of the couple on the end surfaces that is produced by the stress $\sigma_{z\theta}$.

The problem of finding the displacements around other dislocations can be approached in the same way.[†] A displacement function is sought that gives the required discontinuity of displacement in a circuit around the dislocation and in addition satisfies the equations of equilibrium. Expressed in Cartesian coordinates,

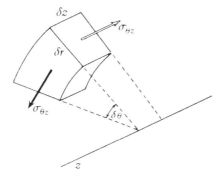

Figure 7.14   Stress due to a screw dislocation along the z axis

---

[†] The remainder of this section is somewhat advanced and can be omitted at a first reading.

with the dislocation lying along the x_3 axis, these are the three equations

$$\frac{\partial \sigma_{i1}}{\partial x_1} + \frac{\partial \sigma_{i2}}{\partial x_2} = 0 \qquad (i = 1, 2, 3,) \tag{7.20}$$

The stress σ_{ij} can be expressed in terms of the derivatives of the components of displacement u_1, u_2 and u_3 through Hooke's law (Section 5.5). Because the stress and strain given by any displacement having the required discontinuity becomes infinite at the dislocation, it is further necessary to ensure that a cylinder of material containing the dislocation is in equilibrium or, in other words, that the force exerted on the core surface vanishes.

The solution for an edge dislocation in an isotropic medium is more complicated than that for a screw. If the dislocation lies along the x_3 axis with its Burgers vector parallel to x_1, the displacements are

$$u_1 = \frac{b}{2\pi} \left[\tan^{-1} \frac{x_2}{x_1} + \left(\frac{\lambda + \mu}{\lambda + 2\mu} \right) \frac{x_1 x_2}{x_1^2 + x_2^2} \right] \tag{7.21a}$$

$$u_2 = \frac{b}{2\pi} \left[-\frac{\mu}{2(\lambda + 2\mu)} \log \left(x_1^2 + x_2^2 \right) + \left(\frac{\lambda + \mu}{\lambda + 2\mu} \right) \frac{x_2^2}{x_1^2 + x_2^2} \right] \tag{7.21b}$$

Here, λ is the Lamé constant (see section 5.5). The first term in Eqns (7.21a) has the same form as the solution for a screw dislocation, and provides the necessary cyclic discontinuity in u_1 equal to the Burgers vector b. The first term in u_2, the displacement normal to the slip plane, is needed in order that the equations of equilibrium (7.20) may be satisfied. The second terms in Eqns (7.21) are needed to make the core force vanish. The pattern produced when these displacements are imposed on a square array of points in the $x_1 x_2$ plane is shown in Fig. 7.5. The elastic constants used in Eqns (7.21) to obtain Fig. 7.5 were those of tungsten, which happens to be elastically isotropic, to a good approximation.

The stresses which follow from the displacements given by Eqns (7.21) are

$$\sigma_{11} = -\frac{\mu b}{2\pi(1 - \nu)} \frac{x_2(3x_1^2 + x_2^2)}{(x_1^2 + x_2^2)^2}$$

$$\sigma_{22} = \frac{\mu b}{2\pi(1 - \nu)} \frac{x_2(x_1^2 - x_2^2)}{(x_1^2 + x_2^2)^2} \tag{7.22}$$

$$\sigma_{33} = \nu(\sigma_{11} + \sigma_{22})$$

$$\sigma_{12} = \frac{\mu b}{2\pi(1 - \nu)} \frac{x_1(x_1^2 - x_2^2)}{x_1^2 + x_2^2)^2}$$

where ν is Poisson's ratio.[†]

[†] The stresses and strains due to a mixed dislocation in an isotropic medium whose line makes an angle θ with its Burgers vector can be obtained by simply adding the stresses and strains due to a

Tungsten, together with aluminium, are exceptional crystals in being, elastically, almost isotropic. For most crystals, getting accurate values for the positions of atoms outside the core of a dislocation requires the use of the true elastic constants of the crystal, rather than the averaged values, which describe the isotropic behaviour of a polycrystal of the material. This means that the correct form of Hooke's law must be used in converting the equations of equilibrium (7.20) into equations in the displacements. In the general case, this introduces all nine terms of the type $\partial^2 u_i / \partial x_\alpha \partial x_\beta$ into Eqns (7.20) ($i = 1, 2, 3$; $\alpha, \beta = 1, 2$). The cases that have been solved are ones in which this number of terms is reduced because of symmetry. This requires not only that the crystal structure be of high symmetry but also that the dislocation lies in some special direction in the crystal, e.g. along a twofold axis of symmetry. Full solutions have been published for an edge dislocation along $\langle 0\,0\,1 \rangle$ in a cubic crystal, a screw normal to a mirror plane and a dislocation lying in the basal plane of a hexagonal crystal [3,4].

The simplest case is that of a screw dislocation lying normal to a mirror plane. Then, if the x_3 axis is chosen to be parallel to the dislocation, the elastic constants c'_{14}, c'_{15}, c'_{24}, c'_{25}, c'_{34}, c'_{35}, c'_{46} and c'_{56} vanish (see Table 5.2; primed elastic constants are constants referred to axes that are not chosen with a conventional relationship to the crystal axes). The displacement is

$$u_3 = \frac{b}{2\pi} \tan^{-1} \frac{x_2 - hx_1}{\sigma x_1} \qquad (7.23)$$

where

$$\alpha = \frac{(c'_{44}c'_{55} - c'^2_{45})^{1/2}}{c'_{55}}, \qquad h = \frac{c'_{45}}{c'_{55}}$$

For a screw dislocation along $\langle 1\,1\,0 \rangle$ in an f.c.c. crystal, Eqn (7.23) reduces to

$$u_3 = \frac{b}{2\pi} \tan^{-1} \frac{A^{1/2}x_2}{x_1} \qquad (7.24)$$

where the x_1 and x_2 axes are shown in Fig. 7.15 and A is the anisotropy factor given by Eqn (5.54):

$$A = \frac{2c_{44}}{c_{11} - c_{12}}$$

The anisotropy factor destroys the radial symmetry of the displacement which is otherwise of the same form as in an isotropic medium. When $A > 1$, as it is for f.c.c. metals, the shear strain is greatest on the $\{1\,1\,0\}$ plane; when $A < 1$, as it is for some alkali halides and some b.c.c metals, the shear strain is greatest on the

---

screw dislocation of the Burgers vector $b\cos\theta$ to those due to an edge dislocation of the Burgers vector $b\sin\theta$. This is because the terms in the stress fields of the screw and edge components are independent of one another.

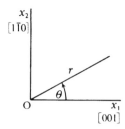

Figure 7.15

$\{1\,0\,0\}$ plane. Because the shear modulus varies with orientation, a shear stress σ_{zr} is introduced:

$$\sigma_{zr} = \frac{c_{44}b(1-A)\sin\theta\cos\theta}{2\pi A^{1/2}r(\cos^2\theta + A\sin^2\theta)} \tag{7.25}$$

in addition to the shear stress $\sigma_{z\theta}$ which has the same form as in the isotropic case, namely

$$\sigma_{z\theta} = \frac{c_{44}b}{2\pi A^{1/2}r} \tag{7.26}$$

7.5 ATOM POSITIONS CLOSE TO A DISLOCATION

Although the solutions for dislocations in an elastic continuum can be used to assign displacements to the atoms around a dislocation in a crystal, these displacements are unlikely to be correct for atoms close to the dislocation, because the elastic continuum solutions are not valid in this region of high strain. To determine the displacements in this core region of the dislocation, a detailed knowledge of the forces acting between atoms would be needed. In the absence of this knowledge, our discussion will necessarily be rather qualitative.

As a starting point, consider a simple cubic lattice and take the displacements for a screw dislocation given by Eqn (7.17):

$$u_z = \frac{b\theta}{2\pi}$$

If the position of the dislocation line is taken to be as in Fig. 7.6 and θ is taken as the angle measured anticlockwise from sx_1, which can be considered to be the trace of a plane on which the screw glides, the displacements of the atoms numbered 1 to 4, for example, are

$$\frac{b}{8}, \frac{3b}{8}, \frac{5b}{8}, \frac{7b}{8}$$

respectively. The displacement across the slip plane sx_1 is

$$\Delta u = b - \frac{b\theta}{\pi} \tag{7.27}$$

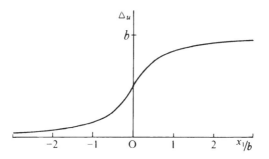

Figure 7.16 Plot of Eqn (7.28)

No displacement has occurred across the slip plane to the far left of the dislocation ($\theta = \pi$) while the relative displacement is b to the far right ($\theta = 0$).
Equation (7.27) can be written as

$$\Delta u = b - \frac{b}{\pi} \tan^{-1} \frac{b}{2x_1}$$

$$= b - \frac{b}{\pi} \left(\frac{\pi}{2} - \tan^{-1} \frac{2x_1}{b} \right)$$

or

$$\Delta u = \frac{b}{2} + \frac{b}{\pi} \tan^{-1} \frac{2x_1}{b} \tag{7.28}$$

This function is plotted in Fig. 7.16. The distance on the slip plane over which
the displacement lies between $\frac{1}{4}$ and $\frac{3}{4}$ of its extreme values is defined as the
width of the dislocation on that plane. The width of the dislocation in Fig. 7.16
is b. Intuitively, we suspect that a dislocation in a real crystal may be wider than
this.

Suppose now that we double the width of the dislocation by making Δu change
with x_1 according to the equation

$$\Delta u = \frac{b}{2} + \frac{b}{\pi} \tan^{-1} \frac{x_1}{b} \tag{7.29}$$

This is equivalent to giving the atoms the displacements that they would have,
according to the symmetrical solution (7.17), if they were arranged on a tetragonal
lattice with a spacing across the slip plane twice that within the plane. The
displacements of the atoms 1 to 4 become approximately

$$\frac{b}{6}, \frac{b}{3}, \frac{2b}{3}, \frac{5b}{6}$$

The widened dislocation is shown in Fig. 7.17. The shear strain in the bonds 1–4
and 2–3 is twice that in the bonds 1–2 and 3–4. A similar situation would be

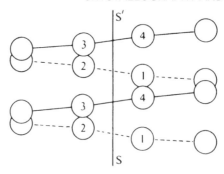

Figure 7.17 A screw dislocation of twice the width of the one shown in Fig 7.6

expected in a real crystal if the bonds between the slip planes were weaker than those between atoms within a slip plane.

A model which gives special consideration to the bonds across the slip plane is the Peierls dislocation. In this model the crystal is supposed to be divided into two blocks at the slip plane, each of which is an elastic continuum. The atomic structure is taken into account only at the slip plane, where the two blocks join. A periodic force law is allowed to act between the rows of atoms immediately above the slip plane and those immediately below. For simplicity, a sine law is chosen

$$\sigma = \frac{b\mu}{2\pi a} \sin \frac{2\pi \Delta u}{b} \tag{7.30}$$

The spacing between the slip planes is a, so that $\Delta u/a$ is the shear in the bond between two rows of atoms across the slip plane, while b is the atomic spacing within a slip plane. Eqn (7.30) reduces to Hooke's law at small strains (small values of Δu). A dislocation is made by straining the elastic blocks and then joining them at the slip plane. The restoring force acting between sheared rows of atoms according to Equation (7.30) tends to narrow the dislocation. The forces due to the elastic strains within the blocks tend to widen it. A balance is reached with a certain distribution of displacements. Either an edge or screw dislocation can be made, although the model is more attractive for an edge dislocation, where there are large shear strains on a unique slip plane.

The displacements across the slip plane of a Peierls screw are identical to those predicted by the elastic continuum solution (7.17) and (7.28), i.e.

$$\Delta u = \frac{b}{2} + \frac{b}{\pi} \tan^{-1} \frac{2x_1}{a} \tag{7.31}$$

The displacements for a Peierls edge differ from those for an elastic edge dislocation and give

$$\Delta u = \frac{b}{2} + \frac{b}{\pi} \tan^{-1} \frac{2x_1(1 - v)}{a} \tag{7.32}$$

The Peierls edge is wider than the screw by the factor $1/(1 - v)$, where v is Poisson's ratio. As v increases towards its greatest possible value of 0.5, the resistance of the material to elastic shear decreases relative to its resistance to dilatation. An increase in the width of an edge dislocation as v increases therefore accords with the qualitative idea that if the atoms are hard, but shear over one another easily, then squeezing in the extra plane of atoms will spread the dislocation out over its slip plane. On the other hand, if there are strong directed bonds across the slip plane, their resistance to shear will lead to a narrow dislocation. Such directed bonds occur in covalent crystals like silicon.

An important property which is controlled by the atom positions in the dislocation core is the force needed to start a straight dislocation moving in an otherwise perfect lattice. As the dislocation moves, the displacements around it will change, and in general the energy stored in the core will change slightly. Consequently, a force is exerted on the dislocation, given by

$$F_x = -\frac{dE'}{dx} \tag{7.33}$$

Where E is the energy due to the dislocation.

The maximum value of this force is called the Peierls force, because it was first investigated by Peierls using a model of an edge dislocation. The method can be briefly illustrated with the help of Fig. 7.6, taking the dislocation S to be a Peierls screw. For any position of S on the plane Sx_1 (Fig. 7.6a), Eqn (7.31) can be used to assign displacements to the atoms. The energy stored in the bonds across the slip planes 2–3, 1–4, etc., can then be summed as a function of the position of the dislocation. The energy sum E' turns out to be a minimum for the position S_1 and a maximum for position S_2 (Fig. 7.6a), where the highly strained 1–4 bond makes a large contribution. At an intermediate position, dE'/dx_1 can be shown to have a maximum value of the form

$$\frac{dE'}{dx_1} = F_p \approx \mu \exp\left(-\frac{\pi a}{b}\right) \tag{7.34}$$

This result has no quantitative significance because of the unrealistic nature of the model, and indeed it gives far too large a value for the force needed to move a dislocation in, say, an f.c.c. metal. The decrease in F_p with the parameter a, which on this model is equivalent to the width of the dislocation, is physically reasonable. As a dislocation becomes narrower, the change in displacement as it moves is distributed among fewer atoms. When the width is very small, a few core atoms have to make large jumps, for which there may be a substantial energy barrier, in order to move the dislocation by one atomic spacing.

7.6 THE INTERACTION OF DISLOCATIONS WITH ONE ANOTHER

When a crystal is plastically deformed, some of its dislocations are lost as they glide to the surfaces of the crystal. However, dislocations must also be replenished during plastic flow, because a strain of 10% typically increases their density by 10^8 cm^{-2} or 10^9 cm^{-2} (dislocation density is defined as the length of dislocation line in 1 cm^3 of crystal). As the density of dislocations increases, their interaction with one another becomes important. For example, the stress needed to plastically deform the crystal increases and becomes controlled by the forces that dislocations exert on one another, rather than by the stress needed to move an isolated dislocation.

A simple mechanism by which dislocations can multiply was suggested by Frank and Read. Its operation requires only that a dislocation in a slip plane be pinned at two points. In Fig. 7.18 the dislocation is pinned at the points x', y' where it changes its plane. Figures 7.18 a and b show that this configuration could occur if a screw segment xy glided off its original slip plane, α, to the position $x'y'$. This manoeuvre is called *cross-slip*. When the segment $x'y'$ changes back to a plane parallel to α, it is easy to see that it can generate any number of expanding loops, using x' and y' as pivots. This variant of the Frank–Read

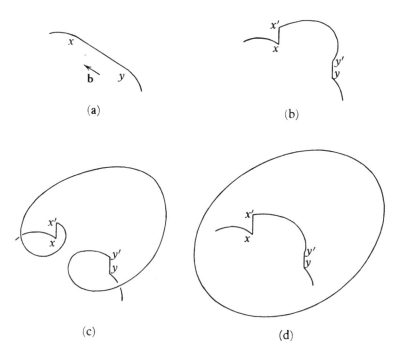

Figure 7.18 Dislocation multiplication

mechanism is due to Koehler. It allows slip to spread out sideways from a single active slip plane and so it can account for the commonly observed *slip band*, which is a set of very closely spaced parallel planes on which slip has occurred (see Fig. 6.3).

When slip occurs in bands, or even when it occurs more diffusely, dislocations on nearby parallel slip planes must on occasion come close to one another. Then the stress due to the one dislocation will exert a force on the other dislocation, according to Eqn (7.12), where we substitute for σ the stress due to one dislocation at the position of the other. Two special cases, parallel screws and parallel edges, will be briefly reviewed, assuming isotropic elasticity.

Parallel screw dislocations a distance r apart experience a simple central force

$$F = \frac{\mu b^2}{2\pi r} \tag{7.35}$$

The force per unit length, F, is repulsive when the screws are identical, so that their stress fields reinforce one another, and attractive when they are of opposite sign so that their stress fields cancel as they come together. There is no position of equilibrium.

Parallel edge dislocations, on the other hand, have two positions of equilibrium in respect of the component of force which lies in the slip plane. Since an edge dislocation would have to climb to move out of its slip plane, this is the only component that can cause an edge dislocation to move at low temperature. Its magnitude per unit length can be derived from Eqns (7.22) and it is convenient to write it in terms of polar coordinates as follows:

$$F_g = \frac{\mu b^2}{2\pi(1-v)r} \cos\theta(\cos^2\theta - \sin^2\theta) \tag{7.36}$$

The angle θ is defined in Fig. 7.19, which illustrates Eqn (7.36) for edge of *opposite* sign. There is an unstable equilibrium at $\theta = \pi/2$ and a stable one at $\theta = \pi/4$, $3\pi/4$. The force per unit length, F_m, which would be needed to push the parallel edges past one another on slip planes a distance y apart is

$$F_m = \frac{\mu b^2}{8\pi(1-v)y} \tag{7.37}$$

Since yield stresses are typically in the range 10^{-3}–10^{-4} μ, edge dislocations of opposite sign are likely to become stuck against one another if they lie on slip planes which are a few hundred Å or less apart. Close pairs of dislocations of opposite sign are in fact commonly observed in deformed crystals; they are called *dislocation dipoles*.

The forces acting between *like* edges are equal in magnitude but opposite in sign to those acting between unlike edges. The stable equilibrium therefore occurs at $\theta = \pi/2$. Two like edge dislocations are less likely to trap one another

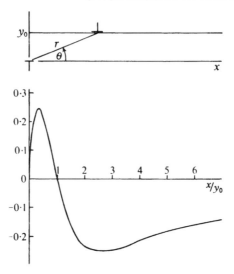

Figure 7.19 Force between parallel edge dislocations on slip planes a distance y_0 apart. The force is given in units of $\mu b^2 / 2\pi(1 - \nu)y$

during glide than two unlike ones, because the applied stress moves them both in the same direction. However, the equilibrium configuration at $\theta = \pi/2$ becomes prominent when climb can occur. The component of force normal to the slip plane, the climb force, is given by

$$F_c = \frac{\mu b^2}{2\pi(1 - \nu)r} \sin\theta(1 + 2\cos^2\theta) \tag{7.38}$$

This force is one of attraction between unlike edges and repulsion between like edges, for all θ. If the temperature is high enough to allow climb, unlike edges from different slip planes annihilate one another, while any excess edges of the same sign which are left over arrange themselves in walls perpendicular to the slip plane. This is known as *polygonization*.

The elastic interaction of two dislocations in an anisotropic medium can differ quite significantly from the corresponding interaction in an isotropic medium. In copper, for example, two parallel screws lying along $\langle 1\,1\,0 \rangle$ in the same $\{1\,1\,1\}$ slip plane exert forces on one another which are directed out of the slip plane, because of the shear stress σ_{rz} given by Eqn (7.25). Substitution of the appropriate constants into Eqn (7.25) shows that the component of force normal to the slip plane is no less than 60% of the component of attraction or repulsion in the slip plane, which would be the sole component if the crystal of copper were elastically isotropic.

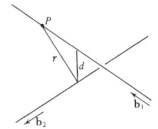

Figure 7.20 Orthogonal screw dislocations, a perpendicular distance d apart

Dislocations on intersecting slip planes also exert forces on one another due to their stress fields. For example, the stress field of a screw in an *isotropic* medium exerts a force on an orthogonal screw, given by

$$F = \frac{\mu b_1 b_2}{2\pi} \frac{d}{r^2} \qquad (7.39)$$

where F is the force per unit length at the point P on the orthogonal screw, defined in Fig. 7.20. The direction of F is normal to both dislocations. Its sign can be checked by comparing the shear stresses due to each screw at the midpoint of the shortest line joining them (the line of length d in Fig. 7.20). The screws attract if the two stresses cancel one another and repel if they reinforce one another.

Dislocations on intersecting slip planes interact in a second way, when they cut through one another. Because the material on one side of the surface swept out by a moving dislocation is displaced by the Burgers vector relative to the material on the other side, each dislocation acquires an extra length of line, equal in magnitude and direction to the Burgers vector of the other dislocation. If the extra length of line happens to lie on the plane on which the remainder of the dislocation is already gliding it is called a *kink*. A kink can be eliminated immediately by glide. When the extra length of line does not lie on this plane, it is called a *jog*. When two orthogonal screws intersect, each acquires a jog which is a segment of edge dislocation. A jog of this type is shown in Fig. 7.21, where

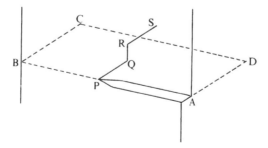

Figure 7.21

the sign of the screw is indicated by means of the step which it produces on the surface of the crystal. If the jogged dislocation PQRS moves as a whole towards AD, the jog QR is forced to climb, and careful inspection of Fig. 7.21 will show that a void is produced of volume QR times b times the distance moved. This void will take the form of a trail of vacant lattice sites. If PQRS moves in the opposite direction, towards BC, a trail of interstitial defects will be produced.

A third way in which two intersecting dislocations may interact with one another is that in certain circumstances they can come together and combine to form a new dislocation. This process is discussed in detail in the next chapter.

PROBLEMS

7.1 A screw dislocation of the Burgers vector **b** passes straight through a crystal, meeting the front and back surfaces at 90°. Keeping this orientation, it moves so that its ends trace out circular paths and the dislocation line sweeps out a cylindrical surface. Describe the deformation produced by n revolutions of the dislocation.

7.2 Show how a pure edge dislocation loop, like that shown in Fig. 7.11, can develop screw segments by gliding into a new shape. Hence show that it is geometrically possible for an edge dislocation loop to become a source for glide on any plane containing its Burgers vector.

7.3 An edge dislocation in a crystal of NaCl lies along [0 0 1] and its Burgers vector is $\frac{1}{2}$[1 1 0]. If the dislocation is to be moved through a distance of 10^{-4} cm in the [1 0 0] direction, calculate the volume of material that must be added to or subtracted from the edge of its extra plane, per cm of dislocation.

7.4 An edge dislocation in a crystal of NaCl lies along [0 0 1] and its Burgers vector is $\frac{1}{2}$[1 1 0]. Calculate the glide force acting upon the dislocation under the action of a tensile stress σ (a) when the tensile axis is [1 0 0] and (b) when the tensile axis is [1 1 1].

7.5 A screw dislocation in a crystal of AgCl has the Burgers vector $\frac{1}{2}$[1 1 0]. If the dislocation could glide with equal ease on any plane, on which plane would it glide when under the action of a tensile stress (a) when the tensile axis is [1 0 0] and (b) when the tensile axis is [1 1 1]?

7.6 A screw dislocation lies along the axis of an isotropic cylinder of radius R. If the stress field were given by Eqn (7.19), there would be forces acting on the end faces of the cylinder. Obtain an expression for the couple which would then be acting on the cylinder. (In a free cylinder the stress field would not be exactly given by (7.19) but would contain an additional term, unimportant near the dislocation line, corresponding to a twisting of the cylinder, which neutralizes the couple implied by (7.19) alone.)

7.7 From either Eqns (7.21) or Eqns (7.22) and (5.63), obtain the strain at a point directly above or below an edge dislocation ($x_1 = 0$). Calculate the dilatations at points $5b$ above and $5b$ below an edge dislocation in aluminium.

7.8 Prepare a diagram of a plane which is normal to an edge dislocation in an isotropic medium, showing the lines in the plane on which one of the stress components σ_{11}, σ_{22} and σ_{12} vanishes (the dislocation lies along the x_3 axis with its Burgers vector parallel to Ox_1). Show the signs of the stresses in the various regions of the diagram.

7.9 Consider a screw dislocation in an f.c.c. crystal lying along $[1\,1\,0]$. Show (from Eqn 7.24) that at equal distances from the dislocation, the shear strain on the $(1\,\overline{1}\,0)$ plane that passes through the dislocation is A times the shear strain on the $(0\,0\,1)$ plane that passes through the dislocation, where $A = 2c_{44}/(c_{11} - c_{12})$.

7.10 Obtain an expression for the width on the $(0\,0\,1)$ plane of a screw dislocation lying along $[1\,1\,0]$ in a sodium chloride crystal. Assume that the elastic continuum solution for the displacements (Eqn 7.24) applies to all the atoms around the dislocation.

7.11 A cubic crystal is being stretched in a $[1\,0\,0]$ direction. Its deformation is being produced by edge dislocations lying parallel to $[0\,0\,1]$ and gliding on $(1\,1\,0)$ and $(1\,\overline{1}\,0)$ planes. If the density of these dislocations is ρ, the magnitude of the Burgers vector of each one is b and their average velocity is $\overline{V}$, determine the rate of increase of tensile strain of the crystal ($=(1/l)(dl/dt)$, l = length of crystal).

7.12 Analyse the interaction between two edge dislocations whose lines are parallel to one another but whose Burgers vectors are orthogonal. Determine all positions of equilibrium with respect to glide forces and divide these into cases of stable, unstable and neutral equilibrium.

7.13 Find the total force experienced by an infinitely long screw dislocation of the Burgers vector $\mathbf{b}_1$ due to the presence of an infinitely long orthogonal screw dislocation of the Burgers vector $\mathbf{b}_2$

7.14 In a crystal of sodium chloride, two orthogonal screw dislocations, of Burgers vectors $\frac{1}{2}[1\,1\,0]$ and $\frac{1}{2}[1\,\overline{1}\,0]$ respectively, are being driven towards one another by a tensile stress applied along $[1\,0\,0]$. Demonstrate that each dislocation acquires a jog which produces interstitial defects if the dislocations continue to move under the action of the applied stress.

7.15 What is the minimum number of dislocations each with Burgers vector equal to the smallest lattice translation vector that can meet at a node in the following crystals: (a) b.c.c. iron, (b) diamond, (c) zinc and (d) caesium chloride? Under what conditions will the node be glissile?

SUGGESTIONS FOR FURTHER READING

1. Cottrell, A. H., *Dislocations and Plastic Flow in Crystals*, Oxford (1953).
2. Eshelby, J. D., Read W. T., and Shockley, W., 'Anisotropic elasticity with applications to dislocation theory', *Acta. Met.*, **1**, 251 (1953).
3. Hull D., and Bacon, D. J., *Introduction to Dislocations*, Pergamon (1984).
4. Read Jr, W. T., *Dislocations in Crystals*, McGraw-Hill, New York (1953).
5. Steeds, J. W., *Introduction to Anisotropic Theory of Dislocations*, Clarendon Press, Oxford (1973).
6. Weertman, J. and Weertman, J. R., *Elementary Dislocation Theory*, Oxford University Press, Oxford (1992).

REFERENCES

1. R. de Wit, *Acta Met.*, **13**, 1210 (1965).
2. J. Weertman, *Phil, Mag.*, **11**, 1217 (1965).
3. J. D. Eshelby, W. T. Read and W. Shockley, *Acta Met.*, **1**, 251 (1953).
4. Y. T. Chou and J. D. Eshelby, *J. Mech. Phys. Solids*, **10**, 27 (1962).

8

Dislocations in Crystals

8.1 THE STRAIN ENERGY OF A DISLOCATION

A considerable amount of strain energy is stored in the elastically distorted region around a dislocation line. The magnitude of the energy stored in an elastically strained region is always of the form $\frac{1}{2} \times$ (elastic modulus) $\times$ (strain)^2, per unit volume (cf. Eqn 5.40). Since the strain at a given point is proportional to the magnitude of the Burgers vector of the dislocation $\mathbf{b}$, the total strain energy will be proportional to $\mathbf{b}^2$. To find the absolute magnitude of the energy stored, it is necessary to make a more detailed calculation. This can be done in two ways.

In the first method, the energy stored in a small element of volume is found and an integration is performed. For a screw dislocation the appropriate element is a cylindrical shell of radius r, thickness dr, because the shear strain γ within such a shell is constant (Eqn 7.18):

$$\gamma = \frac{b}{2\pi r}$$

An elastically isotropic medium is assumed. The strain energy per unit length of shell is then

$$dE = (\tfrac{1}{2}\mu\gamma^2)2\pi r \, dr \qquad (8.1)$$

and the energy per unit length of dislocation in the region from the core, radius r_0, out to a radius R is[†]

$$E = \int_{r_0}^{R} \frac{\mu b^2}{4\pi r} \, dr$$

or

$$E = \frac{\mu b^2}{4\pi} \ln \left(\frac{R}{r_0} \right) \qquad (8.2)$$

The strain energy due to an edge dislocation is more easily calculated by a second method. Suppose that the dislocation is formed in initially unstrained

---

[†] To be exact, Eqn (8.2) should be corrected to allow for the relaxation of stress at the free end surfaces of the cylinder. The uncertainty in the value of r_0 makes this correction unimportant.

material by making a cut from the surface of the material up to the site of the dislocation line, and then applying forces to the faces of the cut so as to give them a relative displacement of b. Then the work done by these forces equals the elastic energy stored around the completed dislocation. As the faces of the cut are displaced, the Burgers vector increases smoothly from zero to $\mathbf{b}$. The work done in increasing the Burgers vector of unit length of dislocation from β to $\beta + d\beta$ is

$$dW = F(\beta)\,d\beta \tag{8.3}$$

where $F(\beta)$ is the force acting on one of the faces of the cut in the direction of β. This force is given by the stress field of the dislocation at the cut surface. For an edge dislocation which is made by cutting along the slip plane in a cylinder of radius R (Fig. 8.1),

$$F = \int_{r_0}^{R} \sigma_{12}\,dx_1 \tag{8.4}$$

The fact that only the stress on a single plane is involved gives this method an advantage over volume integration, when the stress field is complicated. From Eqns (7.22), with $x_2 = 0$ and the Burgers vector $= \beta$,

$$\sigma_{12} = \frac{\mu\beta}{2\pi(1 - \nu)x_1} \tag{8.5}$$

and therefore

$$F = \frac{\mu\beta}{2\pi(1 - \nu)} \ln\left(\frac{R}{r_0}\right) \tag{8.6}$$

The strain energy per unit length is then

$$E = W = \int_0^b F\,d\beta \tag{8.7}$$

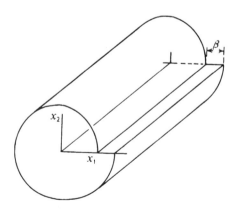

Figure 8.1

$$E = \frac{\mu b^2}{4\pi(1 - v)} \ln \left(\frac{R}{r_0}\right) \tag{8.8}$$

In an isotropic medium, an edge dislocation has more elastic strain energy than a screw, in the ratio of $1:1 - v$. For aluminium, $v = 0.34$ and the edge dislocation has 1.5 times as much energy as the screw; for tungsten, $v = 0.17$ and the ratio is only 1.2.

The strain due to a mixed dislocation, with a Burgers vector at an angle θ to the dislocation line, is equal to the sum of the strains due to parallel edge and screw dislocations of Burgers vectors $b \sin \theta$ and $b \cos \theta$ respectively. Therefore the energy of a mixed dislocation is given by

$$E = \frac{Kb^2}{4\pi} \ln \left(\frac{R}{r_0}\right) \tag{8.9}$$

where

$$K = \frac{\mu}{1 - v} \sin^2 \theta + \mu \cos^2 \theta$$

To get a rough idea of the absolute magnitude of the strain energy, it is sufficient to consider Eqn (8.2):

$$E = \frac{\mu b^2}{4\pi} \ln \left(\frac{R}{r_0}\right)$$

The difficulty in applying this equation is that the strain energy increases without limit as R increases. In a real crystal, either the surfaces of the crystal or more probably other dislocations will limit R. Fortunately, the magnitude of E is not very sensitive to the particular value of R which is assumed. For aluminium, which is approximately isotropic elastically, with $\mu = 28$ GPa, $b = 2.9 \times 10^{-10}$ m, $r_0 = 5b$, the energy per unit length is 1.24×10^{-9} J m^{-1} when R is set at 1 μm and 2.95×10^{-9} J m^{-1} when R is set at 0.01 m. The value of 2×10^{-9} J m^{-1} corresponds to an energy per atomic length of dislocation of the 3.6 eV. This energy is large enough to insure that a crystal containing a dislocation is never in thermodynamic equilibrium. (Even at high temperatures the entropy introduced by a dislocation subtracts only a relatively small amount from its energy, so that the free energy of a dislocated crystal of ordinary size is always greater than that of a perfect crystal.)

To obtain the total energy of a dislocation, the energy stored within the dislocation core must be added to the energy given by Eqn (8.9). The core energy can be roughly estimated in various ways, which agree in giving its order of magnitude as $\mu b^3/10$ per atomic length. For aluminium this amounts to 0.7 eV per atomic length, considerably less than the energy stored *outside* the core. In a real crystal, the importance of the core energy lies less in its magnitude than in the *variation* of this magnitude as the dislocation glides through the crystal. If the core energy increases sharply as the dislocation moves from its position of minimum energy, then its motion is strongly resisted (Section 7.5).

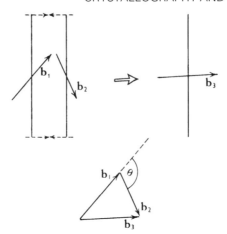

Figure 8.2

It is always geometrically possible for two dislocations to combine to form a single dislocation, as shown in Fig. 8.2. When this happens, the Burgers vector of the resultant dislocation, $\mathbf{b}_3$, is simply the sum of the Burgers vectors of the combining dislocations

$$\mathbf{b}_3 = \mathbf{b}_1 + \mathbf{b}_2 \qquad (8.10)$$

To determine whether such a reaction is energetically favourable, we must compare the energy of a crystal containing a dislocation of the Burgers vector $\mathbf{b}_3$ with that of a crystal containing two widely separated dislocations of the Burgers vectors $\mathbf{b}_1$ and $\mathbf{b}_2$. The former can be written as $K_3 b_3^2$ and the latter $K_1 b_1^2 + K_2 b_2^2$. Now although Eqn (8.9) shows that the constants K_1, K_2 and K_3 depend on the angles between the dislocation lines and their Burgers vectors, the difference between $K_3 b_3^2$ and $K_1 b_1^2 + K_2 b_2^2$ is usually so large that the sign of $K_3 b_3^2 - (K_1 b_1^2 + K_2 b_2^2)$ is not affected by setting $K_1 = K_2 = K_3$. Consequently, the sign of $b_3^2 - (b_1^2 + b_2^2)$ is a useful test for determining whether the reaction of two dislocations to form a third leads to a reduction in energy. A reduction will occur if

$$b_3^2 < b_1^2 + b_2^2 \qquad (8.11)$$

This rapid test for deciding whether two dislocations will react is known as Frank's rule. It can be expressed in terms of the angle θ between $\mathbf{b}_1$ and $\mathbf{b}_2$ (Fig. 8.2). If $\theta < 90°$, the dislocations will not react; if $\theta > 90°$ they will attract and react with one another. The same rule can be applied in reverse to decide whether a dislocation of the Burgers vector $\mathbf{b}_3$ is stable. If there are available two lattice vectors summing to $\mathbf{b}_3$ and making an angle $\theta < 90°$ with each other, then the dislocation will dissociate into two dislocations having these vectors as Burgers vectors. The rule is obviously most reliable when θ is far from $90°$.

From Frank's stability rule it follows that most dislocations other than those with the shortest lattice vector as the Burgers vector are unstable. For example, a dislocation of the Burgers vector $\frac{1}{2}\langle 2\,1\,1\rangle$ in an f.c.c. crystal would decompose into two $\frac{1}{2}\langle 0\,0\,1\rangle$ dislocations:

$$\tfrac{1}{2}[2\,1\,1] \rightarrow \tfrac{1}{2}[1\,1\,0] + \tfrac{1}{2}[1\,0\,1] \tag{8.12}$$

However, it sometimes happens that a dislocation whose Burgers vector is greater than the least lattice vector cannot decompose spontaneously. An example is the $[1\,0\,0]$ dislocation in a b.c.c. crystal. Although the $\frac{1}{2}\langle 1\,1\,1\rangle$ dislocation has a smaller energy, Frank's rule predicts that the decomposition

$$[1\,0\,0] \rightarrow \tfrac{1}{2}[1\,\overline{1}\,\overline{1}] + \tfrac{1}{2}[1\,1\,1] \tag{8.13}$$

will not occur, nor is any other geometrically possible decomposition favourable energetically.

Table 8.1 lists the Burgers vectors that are definitely stable in various lattices. Doubtful cases in which the vector is the sum of two orthogonal vectors ($\theta = 90°$) are omitted.

The fact that the dislocation with the smallest Burgers vector has the least elastic energy fits in well with the observation that the slip direction is almost always that of the shortest lattice translation vector in the crystal (Table 6.1). It is natural to enquire whether the observed slip planes are those on which a dislocation of minimum Burgers vector has the smallest elastic energy, when the elastic anisotropy of the crystal is taken into account. The elastic energy per unit length of a dislocation is an anisotropic medium and is given by

$$E = \frac{K'b^2}{4\pi} \ln\left(\frac{R}{r_0}\right) \tag{8.14}$$

where K' depends on the elastic constants. Foreman has computed the energy factor K' in a number of cases [1]. In copper, for example, an edge dislocation of the $\frac{1}{2}\langle 1\,1\,0\rangle$ Burgers vector has a lower elastic energy when it lies in a $\{1\,1\,0\}$ plane ($K' = 64.5$ G Pa) than when it lies in a $\{1\,1\,1\}$ plane ($K' = 74.5$ G Pa). Since the observed slip plane is $\{1\,1\,1\}$ it appears that the elastic energy alone does not determine which slip plane operates. In order to understand the choice of slip plane it is evidently necessary to take account of the atomic structure of the crystal (see also Section 6.1).

The fact that a dislocation line has a certain amount of energy *per unit length* associated with it causes it to try to shorten itself or, in other words, gives it a line tension. It is not easy to define this line tension precisely because the energy is not concentrated at the line itself, but is contained in its far-reaching strain field. Let us consider the tendency of a circular dislocation loop to shrink. To prevent it from shrinking, a radial force F_r must be applied, given by the condition that

Table 8.1 Definitely stable dislocations in some Bravais lattices

Lattice	Burgers vector	Number of equivalent vectors	Square of Burgers vector
Simple cubic	$\langle 1\,0\,0\rangle$	6	a^2
B.c.c.	$\frac{1}{2}\langle 1\,1\,1\rangle$	8	$3a^2/4$
	$\langle 1\,0\,0\rangle$	6	a^2
F.c.c.	$\frac{1}{2}\langle 1\,1\,0\rangle$	12	$a^2/2$
Hexagonal	$\frac{1}{3}\langle 1\,1\,\bar{2}\,0\rangle$	6	a^2
	$\langle 0\,0\,0\,1\rangle$	2	c^2
Rhombohedral $\alpha < 90°$	$\langle 1\,0\,0\rangle$	6	a^2
	$\langle 1\,\bar{1}\,0\rangle$	6	$4a^2\sin^2\alpha/2$
	$\langle 1\,\bar{1}\,1\rangle$	6	$a^2(1 + 4\sin^2\alpha/2)$
Rhombohedral $\alpha > 90°$	$\langle 1\,0\,0\rangle$	6	a^2
	$\langle 1\,1\,0\rangle$	6	$4a^2\cos^2\alpha/2$
	$\langle 1\,1\,1\rangle$	2	$9a^2(1 - \frac{4}{3}\sin^2\alpha/2)$
Simple tetragonal	$\langle 1\,0\,0\rangle$	4	a^2
	$\langle 0\,0\,1\rangle$	2	c^2
Body-centred tetragonal $c/a < \sqrt{2}$	$\frac{1}{2}\langle 1\,1\,1\rangle$	8	$a^2/2 + c^2/4$
	$\langle 1\,0\,0\rangle$	4	a^2
	$\langle 0\,0\,1\rangle$	2	c^2
Body-centred tetragonal $c/a > \sqrt{2}$	$\frac{1}{2}\langle 1\,1\,1\rangle$	8	$a^2/2 + c^2/4$
	$\langle 1\,0\,0\rangle$	4	a^2

the energy change accompanying an infinitesimal contraction of the loop must vanish. Therefore

$$F_r \times 2\pi r\,\mathrm{d}r = E \times 2\pi\,\mathrm{d}r \qquad (8.15)$$

or

$$F_r = \frac{E}{r}$$

where E is the energy per unit length of the dislocation loop. Since the strains due to the opposite sides of the loop cancel one another at distance that are large compared to r, we may determine E approximately by setting $R = r$ in Eqn (8.2):

$$E \approx \frac{\mu b^2}{4\pi} \ln\left(\frac{r}{r_0}\right) \qquad (8.16)$$

If we consider a small segment of the loop, shown in Fig. 8.3, we see that to keep this segment in equilibrium its ends must experience a tension T given by

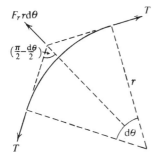

Figure 8.3

setting the sum of radial forces equal to zero:

$$2T \sin \frac{d\theta}{2} = F_r r \, d\theta \tag{8.17}$$

From Eqns (8.15) and (8.17),

$$T = E \tag{8.18}$$

We may extend this result with the aid of Eqn (8.16) to say that the line tension of a dislocation bowed to a radius of curvature ρ is approximately

$$T \approx \frac{\mu b^2}{4\pi} \ln \left(\frac{\rho}{r_0} \right) \tag{8.19}$$

With $\rho \approx 500 r_0$, Eqn (8.19) gives the value of $0.5 \; \mu b^2$, which is often used as the value of the line tension of a dislocation in rough calculations. Figure 8.4 illustrates a calculation of the stress needed to extrude a dislocation between obstacles a distance l apart. The radial force on the dislocation is supplied by a shear stress τ on the slip plane, parallel to the Burgers vector. Equation (7.4) gives

$$F_r = \tau b \tag{8.20}$$

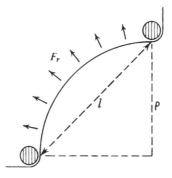

Figure 8.4

From Eqns (8.15) and (8.18), if the radius of curvature of the dislocation is ρ,

$$\tau = \frac{T}{b\rho} \qquad (8.21)$$

or, setting $T = 0.5\ \mu b^2$,

$$\tau = \frac{0.5\mu b}{\rho} \qquad (8.22)$$

Since the minimum value of ρ is 0.5 l, occurring when the dislocation has been bowed out to a semicircle, the stress required to force the dislocation between the obstacles is given by

$$\tau \approx \frac{\mu b}{l} \qquad (8.23)$$

Equation (8.23) also gives the stress needed to operate a Frank–Read source whose pinning points are a distance l apart.

An interesting improvement to Eqn (8.19) can easily be made by allowing for the fact that the strain energy of a dislocation varies with the angle θ between the line and its Burgers vector. (For an isotropic medium, this variation is given by Eqn (8.9).) As a consequence, if the equilibrium of an isolated segment of dislocation is considered, not only must a force E be applied to its ends to stop it shrinking, but also a couple Γ per unit length must be applied to it to stop it turning into an orientation of lower energy.

The value of Γ is found by considering an infinitesimal rotation $d\theta$ of the segment from its equilibrium position. Setting the energy change equal to zero:

$$\Gamma\,d\theta - dE = 0$$

or

$$\Gamma = \frac{dE}{d\theta} \qquad (8.24)$$

As the dislocation curves through the infinitesimal angle $d\theta$, the couple per unit length acting upon it changes by $(d^2E/d\theta^2)\,d\theta$. This results in unbalanced radial forces acting on the ends of a curved segment, as shown by Fig. 8.5. The total radial force per unit length needed to keep the dislocation bowed out is now given by

$$F_r\,ds = E\,d\theta + \frac{d^2E}{d\theta^2}\,d\theta$$

or

$$F_r = \frac{E}{\rho} + \frac{1}{\rho}\frac{d^2E}{d\theta^2} \qquad (8.25)$$

If the line tension T is *defined* by the equation

$$F_r = \frac{T}{\rho} \qquad (8.26)$$

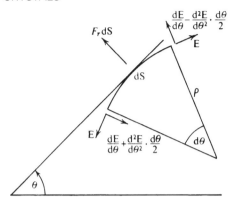

Figure 8.5

then, from Eqn (8.25), T is given by

$$T = E + \frac{d^2E}{d\theta^2} \tag{8.27}$$

Substitution of Eqn (8.9) into Eqn (8.27) shows that a screw dislocation is much stiffer than an edge, even though the edge has the greater elastic energy. Physically, this is because in order to bend a screw dislocation, it must be given some edge character, of relatively high energy. In aluminium, for example, the screw dislocation is four times as stiff as the edge.

8.2 STACKING FAULTS AND PARTIAL DISLOCATIONS

So far, it has been assumed that the smallest Burgers vector that a crystal dislocation may possess is the shortest lattice vector. Although such a dislocation could reduce the elastic strain energy by dissociating into two dislocations with smaller Burgers vectors, this would create a fault in the crystal structure which, in general, would have high energy. In certain special cases, however, the fault has a rather small energy. When this is so, the dislocation does indeed dissociate into two dislocations, called *partial dislocations* or, briefly, partials, which leave a planar fault between them as they move apart. The mutual repulsion of the partials due to their stress fields decreases as the partials separate, but the attractive force exerted by the surface tension of the fault remains constant. An equilibrium separation is therefore reached at which the forces of repulsion and attraction balance.

It is established that $\langle 1\,1\,0 \rangle$ dislocations in some f.c.c. metals dissociate into partials on the $\{1\,1\,1\}$ slip plane. The fault that lies between the partials is a fault in the stacking sequence of the $\{1\,1\,1\}$ planes, which does not change the relative positions of atoms that are nearest neighbours of one another. The production of

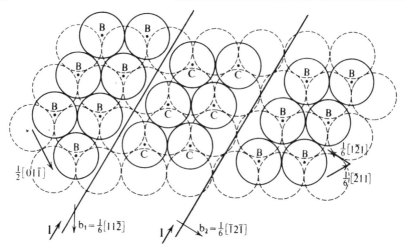

Figure 8.6 Dislocation on a (1 1 1) plane of an f.c.c. metal which has split into two Shockley partials

this stacking fault can best be visualized with the aid of sheets of closely packed balls, stacked one on top of another (Section 3.2). By sliding an upper block of sheets through $a/6\langle 2\,1\,1\rangle$ over those underneath, the so-called intrinsic stacking fault is produced (Fig. 8.6). For example, if the second ABC block in the sequence ABC ABC is slid over the first, the sequence becomes ABCBCA .... . The sheets at the fault are now in the sequence BCBC which, if continued, produces a close packed hexagonal structure. Such a fault occurs when the $\frac{1}{2}\langle 1\,1\,0\rangle$ dislocation dissociates into $\frac{1}{6}\langle 2\,1\,1\rangle$ dislocations, which are called Shockley partials:

$$\tfrac{1}{2}[0\,1\,\bar{1}] \rightarrow \tfrac{1}{6}[1\,1\,\bar{2}] + \tfrac{1}{6}[\bar{1}\,2\,\bar{1}] \tag{8.28}$$

Figure 8.6 shows the pattern of atoms at the (1 1 1) slip plane around an $\frac{1}{2}\langle 1\,1\,0\rangle$ 60° dislocation that has dissociated into Shockley partials. (For simplicity in drawing, the partials are shown to be very narrow; in a real crystal they would be wider.)

The separation of the Shockley partials is determined by the energy per unit area, γ, of the stacking fault between them. The fault draws the partials together with a force γ per unit length, so that the equilibrium separation r is given by

$$\frac{G}{r} = \gamma \tag{8.29}$$

where the elastic repulsion factor G can be calculated. Assuming isotropic elasticity, the screw component of the first partial, $b_1 \cos\theta_1$, gives a shear stress on the slip plane

$$\sigma_{23} = \frac{\mu b_1 \cos\theta_1}{2\pi r} \tag{8.30}$$

where Ox_2 is normal to the slip plane and Ox_3 is parallel to the dislocations. From Eqns (7.22), the edge component gives the stress

$$\sigma_{12} = \frac{\mu b_1 \sin \theta_1}{2\pi(1 - \nu)r} \tag{8.31}$$

The component of the traction on the slip plane in the direction of the Burgers vector of the second partial is given by[†]

$$\tau = \sigma_{23} \cos \theta_2 + \sigma_{12} \sin \theta_2. \tag{8.32}$$

Since, by Section 7.3, $G = \tau b_2$, therefore

$$G = \frac{\mu b_1 b_2}{2\pi} \left(\cos \theta_1 \cos \theta_2 + \frac{\sin \theta_1 \sin \theta_2}{1 - \nu} \right) \tag{8.33}$$

Substituting typical values for the constants in Eqn (8.33) shows that a separation of the order of 100 Å occurs only when the stacking fault energy is as low as 10^{-4} J m^{-2}. Most f.c.c. metals and alloys have a higher stacking fault energy than this, so the separation cannot usually be resolved, even by the electron microscope.

The remaining sections of this chapter are devoted to the study of dislocations in specific crystal structures.

8.3 DISLOCATIONS IN F.C.C. METALS

The $\langle 1\,1\,0 \rangle$ slip direction of f.c.c. metals is easy to understand, because $\frac{1}{2}\langle 1\,1\,0 \rangle$ is the shortest lattice vector. The $\{1\,1\,1\}$ slip plane is the plane on which a 'perfect' dislocation with a $\frac{1}{2}\langle 1\,1\,0 \rangle$ Burgers vector may split into two Shockley partial dislocations, as described in Section 8.2. The separation of the two partials is determined by the energy of the stacking fault between them. The lower the stacking fault energy, the wider the split and the more rigorously should slip be confined to a $\{1\,1\,1\}$ plane. The stacking fault energy is therefore a property of great physical importance; unfortunately, it is also very difficult to measure accurately. Various values are listed in Table 8.2.

The production of a stacking fault during slip is effectively illustrated by a rigid ball model, such as that shown in Fig. 8.6. In this model, the slip path that requires the least 'riding up' of the balls follows a zig-zag route along the valleys in the $\{1\,1\,1\}$ surface, in two alternating $\langle 1\,1\,2 \rangle$ directions. The displacements along $\langle 1\,1\,2 \rangle$ alternately create and remove a stacking fault. From inspection of this model one might wrongly conclude that only one type of stacking fault could occur on $\{1\,1\,1\}$. In particular, the stacking fault that would be obtained by

[†] Care must be taken to measure θ_1 and θ_2, the angles between the partial dislocation lines and their Burgers vectors, in the same sense.

Table 8.2 The stacking fault energies of f.c.c. metals

Metal	Stacking fault energy (mJ m^{-2})	Ref.
Ag	16	2
Al	135	3
Au	32	4
Cu	41	5
Ni	240	3

interchanging the positions of the two partials in Fig. 8.6 appears at first to be impossible, since it seems to involve A–A stacking, with a large ride-up of the balls when the leading dislocation advances. However, this conclusion can be modified if the necessary displacements are spread over two successive {1 1 1} planes. For example, at the position of the dislocation on the left of Fig. 8.6 starting with a perfect crystal, a displacement $\frac{1}{6}[\bar{2}\,1\,1]$ of all the balls above the bottom {1 1 1} layer to the right of this position, followed by a displacement $\frac{1}{6}[1\,\bar{2}\,1]$ of all the balls above the next {1 1 1} layer, produces a net $\frac{1}{6}[\bar{1}\,\bar{1}\,2]$ displacement without violating nearest neighbour relationships. Formally, the partial dislocation then consists of two Shockley partials, $\frac{1}{6}[2\,\bar{1}\,\bar{1}]$ and $\frac{1}{6}[\bar{1}\,2\,\bar{1}]$, on adjacent planes, but except at its core this dislocation cannot be distinguished from a single $\frac{1}{6}[1\,1\,\bar{2}]$ Shockley.

Therefore, by spreading the displacements over two planes, the $\frac{1}{6}[1\,1\,\bar{2}]$ Shockley can be made the leading partial for glide from left to right in Fig. 8.6. When an interchange of the order of the partials is accomplished in this way, they are separated by a 'double' stacking fault, which is called an extrinsic fault. The energy of this fault is expected to be slightly higher than that of the intrinsic (single) fault. Consequently, the form of splitting shown in Fig. 8.6 is believed to be the usual one.

The distinction between the structures of intrinsic and extrinsic stacking faults can be neatly displayed by a notation due to Frank. When a close packed plane is stacked upon the plane beneath according to a pair in the sequence ABC ... , the stacking of that plane is depicted by a triangle Δ. When a plane is stacked so that its relation to the plane beneath it is a pair in the reverse sequence BAC ... , its stacking is then depicted by an inverted triangle ∇ (pronounced *nabla*). These symbols are derived from the stacking of balls — they represent the orientation of the triangles of balls on to which the balls in the next layer are stacked. The f.c.c. structure is then shown as $\Delta\Delta\Delta\Delta$ and the c.p.h. structure as $\Delta\nabla\Delta\nabla\ldots$. The intrinsic stacking fault is $\Delta\Delta\nabla\Delta\Delta\Delta$ while the extrinsic stacking fault is $\Delta\Delta\nabla\nabla\Delta\Delta$, emphasizing that the extrinsic fault is equivalent to two intrinsic faults on adjacent planes.

It is possible to produce stacking faults by means other than slip on the close packed plane. Figure 8.7a shows that an intrinsic stacking fault is produced by

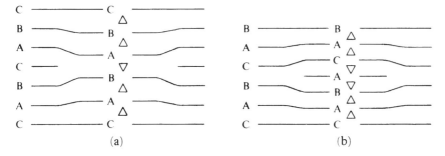

Figure 8.7 (a) Faulted vacancy loop in an f.c.c. metal, showing traces of the (1 1 1) planes in a plane normal to the loop. (b) Extra atom loop

removing part of a {1 1 1} layer of atoms, provided that the gap is then closed directly, without any shear parallel to the {1 1 1} plane. Physically, this may happen when excess vacancies collect together on a {1 1 1} plane. Squeezing in a piece of an extra {1 1 1} layer creates an extrinsic fault, as shown in Fig. 8.7b. The faults shown in Fig. 8.7 are both bounded by partial dislocations whose Burgers vectors are normal to the {1 1 1} and of the magnitude of the spacing of these planes, i.e. $\frac{1}{3}\langle 1\,1\,1\rangle$. In distinction from Shockley partials, these partials could glide only on surfaces that are normal to the usual {1 1 1} slip plane; if they did so they would leave a high-energy stacking fault behind them. They are called *Frank partials* or, because of their immobility, Frank sessiles.

Although a stacking fault must be bounded by a partial dislocation, or end at a surface, it is now clear that the Burgers vector of the partial is not uniquely determined by the type of fault. In a given case, any lattice vector may be added to the Burgers vector of the partial without affecting the fault. For example, an intrinsic fault on (1 1 1) bounded by a $\frac{1}{6}[1\,1\,\bar{2}]$ Shockley might equally well be bounded by two other Shockleys, $\frac{1}{6}[\bar{2}\,1\,1]$ or $\frac{1}{6}[1\,\bar{2}\,1]$. Thus

$$\frac{1}{6}[\bar{2}\,1\,1] = \frac{1}{6}[1\,1\,\bar{2}] + \frac{1}{2}[\bar{1}\,0\,1] \tag{8.34}$$

$$\frac{1}{6}[1\,\bar{2}\,1] = \frac{1}{6}[1\,1\,\bar{2}] + \frac{1}{2}[0\,\bar{1}\,1] \tag{8.35}$$

or, by the Frank partial with the Burgers vector $\frac{1}{3}[\bar{1}\,\bar{1}\,\bar{1}]$,

$$\frac{1}{3}[\bar{1}\,\bar{1}\,\bar{1}] = \frac{1}{6}[1\,1\,\bar{2}] + \frac{1}{2}[\bar{1}\,\bar{1}\,0] \tag{8.36}$$

Before going on to consider reactions between dislocations in f.c.c. metal crystals, it will be helpful to introduce a device known as Thompson's tetrahedron. This gives a picture of various Burgers vectors in relation to the four {1 1 1} planes, which can be chosen to form the four faces of a tetrahedron. Figure 8.8a shows this tetrahedron and its relation to the crystal structure and Fig. 8.8b shows the faces of the tetrahedron when they are opened out from the vertex D. The

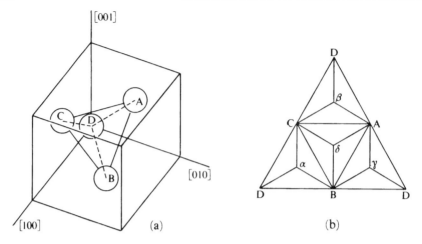

Figure 8.8 Thompson's tetrahedron. (After Thompson [6]. Reproduced by permission of the Institute of Physics and Physical Society)

lettering of the various points is conventional and follows Thompson's original paper.[†] An edge of the tetrahedron, such as CB, represents the Burgers vector of a perfect dislocation, $\frac{1}{2}\langle 1\,1\,0\rangle$. Vectors given by Greek–Roman letter combinations such as δC represent Shockley partials, except when the Greek and Roman letters are the same, such as γC, which indicates a Frank partial.

The reaction between Frank, Shockley and perfect dislocations given by Eqn (8.36) becomes, in Thompson's notation,

$$\delta D = \delta B + BD$$

The relationship between the three Burgers vectors is shown in Fig. 8.9. This reaction has some interesting applications. The stacking fault within the 'vacancy loop of Fig. 8.7a can be removed if the Frank dislocation decomposes according to (8.36) and if then its Shockley component shrinks to nothing by gliding in the plane of the loop. This removes the stacking fault at the expense of an increase in elastic energy due to the increase in the Burgers vector of the loop from $\frac{1}{3}[\bar{1}\,\bar{1}\,\bar{1}]$ to $\frac{1}{2}[\bar{1}\,\bar{1}\,0]$. In comparing the energies of faulted and unfaulted loops, allowance must be made for the shortening of the unfaulted loop when it glides into a $(\bar{1}\,\bar{1}\,0)$ orientation normal to its Burgers vector. The unfaulted loop is favoured when the stacking fault energy is high and also when the loop is large, because the energy gained by destroying the fault is proportional to the *area* of the unfaulted region, whereas the extra elastic energy is approximately proportional to its *perimeter*.

[†] An alternative convention would be to letter the tetrahedron as the mirror image of Fig. 8.8a.

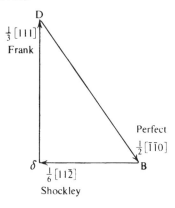

Figure 8.9

Another application is to the possible dissociation of a perfect dislocation into a Shockley and a Frank:

$$BD \rightarrow B\delta + \delta D$$

$$\tfrac{1}{2}[\bar{1}\,\bar{1}\,0] \rightarrow \tfrac{1}{6}[\bar{1}\,\bar{1}\,2] + \tfrac{1}{3}[\bar{1}\,\bar{1}\,\bar{1}] \tag{8.37}$$

Since the Burgers vectors of these two partials are orthogonal and since a stacking fault would have to be created between them, this dissociation does not ordinarily occur. It may occur in special cases, as when the Shockley components from the two members of a perfect dislocation dipole can annihilate one another, leaving behind a Frank dipole. This reaction is simply the reverse of the unfaulting of a prismatic loop described above.

In order to define the *sense* of the Burgers vector of a dislocation, when it is written in Thompson's notation it is necessary to introduce a convention that will assign a *sense* to the slip displacement signified by the symbol δC, for instance. The following convention can be used.

The sense of the slip displacement signified by δC is defined as follows. The observer looks at the slip plane ABC from outside the tetrahedron, i.e. with the apex D pointing away from him (Fig. 8.8a). Then the atoms above the slip plane ABC closest to the observer are displaced by δC. Referring to the crystal structure, it is clear that the displacements δA, δB and δC produce an intrinsic stacking fault, whereas $A\delta$, $B\delta$ and $C\delta$ produce an extrinsic fault. The dissociation of the perfect dislocation CB into Shockley partials on ABC can be written as

$$CB \rightarrow C\delta + \delta B \tag{8.38}$$

The Burgers vectors of the partials in the Thompson notation are then as shown in Fig. 8.10, according to the FS/RH convention. (The line vectors have been chosen to run down the page, so that the observer looks in the negative line direction,

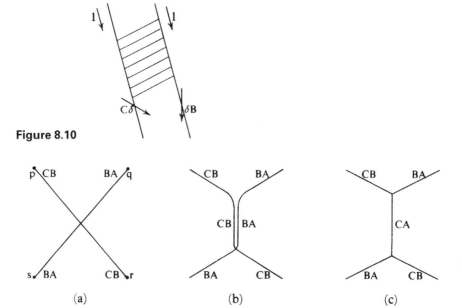

Figure 8.10

Figure 8.11 Reaction of two dislocations on a common slip plane. The dislocations are assumed to be pinned at the points *p, q, r* and *s*

in order to match the figure with the logical form of Eqn (8.38).) Reversing the order of the partials would produce an extrinsic fault between them.

The order of the Shockley partials is an important factor in dislocation reactions. The only reaction between perfect dislocations that definitely leads to a reduction in energy is of the type

$$CB + BA \rightarrow CA \qquad (8.39)$$

If the dislocations of Burgers vectors BA and CB meet on their common slip plane, they will react over a certain length, forming two nodes (Fig. 8.11). Figure 8.12 shows that the nodes will have different structures, because of the sequence of the Shockley partials (it is assumed that only intrinsic faults occur). Alternately, expanded and contracted nodes have been seen, in stainless steel, for example.

The same reaction (8.39) can also occur when the dislocations are on different slip planes. For example, suppose that dislocations CB and BA lie on the planes opposite vertices A and C respectively, and let their lines lie parallel to the intersection of these planes, BD, as they glide towards one another. This event is likely to occur during the later stages of the tensile deformation of a crystal because it corresponds to the intersection of primary and conjugate slip (Section 6.4). If the two dislocations are dissociated into Shockley partials, as shown in Fig. 8.13,

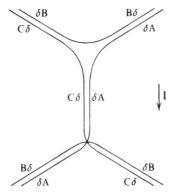

Figure 8.12 Same reaction as Fig. 8.11 but showing splitting into partials

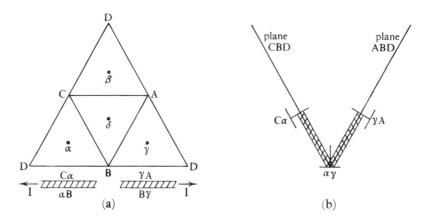

Figure 8.13 Cottrell–Lomer lock

and if they come together in such a way that just before they meet their stacking faults make an acute angle (70°) with one another, the leading partials αB and Bγ will react to form a new partial along BD:

$$\alpha B + B\gamma \rightarrow \alpha\gamma$$

$$\tfrac{1}{6}[1\,2\,\bar{1}] + \tfrac{1}{6}[\bar{2}\,\bar{1}\,1] \rightarrow \tfrac{1}{6}[\bar{1}\,1\,0] \qquad (8.40)$$

The edge dislocation $\alpha\gamma$ is called a stair-rod, from the way in which it joins two stacking faults at a bend rather like a stair-rod holding down a stair-carpet. The whole configuration is called a *Cottrell–Lomer lock*. Because it cannot glide, it may be an important obstacle to other dislocations.

The stair-rod dislocation of the type $\alpha\gamma$ can also be produced by the dissociation of a Frank partial:

$$\alpha A \rightarrow \alpha\gamma + \gamma A$$

$$\tfrac{1}{3}[\bar{1}\,1\,1] \rightarrow \tfrac{1}{6}[\bar{1}\,1\,0] + \tfrac{1}{6}[\bar{1}\,1\,2] \qquad (8.41)$$

This reaction makes it possible to understand the small tetrahedra of stacking faults that have been produced experimentally by quenching gold from near its melting point to room temperature, and then ageing at $100\,^{\circ}$ C.[†] Imagine that the excess vacancies which are trapped by the quench first form a Frank loop similar to that shown in Fig. 8.7a. Let the loop be a triangle, with sides along the $\langle 1\,1\,0 \rangle$ directions AB, BC, CA. Each side can then dissociate on a different $\{1\,1\,1\}$, as shown in Fig. 8.14. The Shockley dislocations produced by the dissociations attract one another and react at the intersections of the $\{1\,1\,1\}$ planes to form stair-rod dislocations. The final product is a tetrahedron whose faces are intrinsic stacking faults and whose edges are stair-rod dislocations. Each edge of the tetrahedron is exactly like the corner of a Cottrell–Lomer lock. The formation of the tetrahedron decreases the elastic energy at the expense of additional stacking fault energy.

The $\tfrac{1}{6}\langle 1\,1\,0 \rangle$ dislocation is only one of four possible stable stair-rods. Other dislocations at an acute-angle bend can be derived by adding lattice vectors to $\tfrac{1}{6}[1\,1\,0]$. The only resultant which is stable, according to Frank's rule, is $\tfrac{1}{3}[\bar{1}\,\bar{1}\,0]$, formed by adding $\tfrac{1}{2}[\bar{1}\,\bar{1}\,0]$. One of the interesting occurrences that would produce an *obtuse-angle* stacking fault bend is the reaction of the leading partials of the dislocations BD on the plane opposite the vertex A and AC on the plane opposite

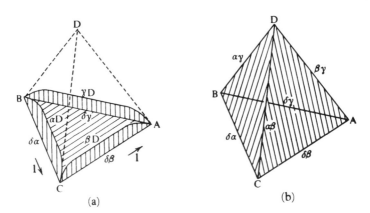

(a)　　　　　　　　　　　(b)

Figure 8.14 Formation of a stacking fault tetrahedron from a Frank vacancy loop

[†] Ageing means holding at a constant temperature for a period of time, usually in order to allow some change in the structure of the material to occur.

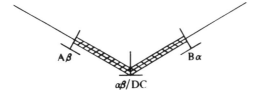

Figure 8.15 Hirth lock

Table 8.3 Dislocations in f.c.c. metals

Name of dislocation	Burgers vector **b**		Relative b^2
	Indices	Thompson notation	
Perfect	$\frac{1}{2}\langle 1\,1\,0\rangle$	AB	1
Shockley	$\frac{1}{6}\langle 2\,1\,1\rangle$	$A\delta$	$\frac{1}{3}$
Frank	$\frac{1}{3}\langle 1\,1\,1\rangle$	$A\alpha$	$\frac{2}{3}$
Thompson stair-rod	$\frac{1}{6}\langle 1\,1\,0\rangle$	$\alpha\delta$	$\frac{1}{9}$
Acute stair-rod	$\frac{1}{3}\langle 1\,1\,0\rangle$	—	$\frac{4}{9}$
Obtuse stair-rod	$\frac{1}{3}\langle 1\,0\,0\rangle$	$\delta\alpha/CB$	$\frac{2}{9}$
Obtuse stair-rod	$\frac{1}{6}\langle 3\,1\,0\rangle$	—	$\frac{5}{9}$

the vertex B:

$$BD \rightarrow B\alpha + \alpha D$$

$$AC \rightarrow A\beta + \beta C \tag{8.42}$$

$$\tfrac{1}{6}[\bar{2}\,1\,\bar{1}] + \tfrac{1}{6}[2\,\bar{1}\,1] \rightarrow \tfrac{1}{3}[0\,\bar{1}\,0] \tag{8.43}$$

The resulting configuration, shown in Fig. 8.15, is called the *Hirth lock*. In Thompson's notation, Eqn (8.41) may be written as

$$\alpha D + \beta C \rightarrow \alpha\beta/DC \tag{8.44}$$

The $\frac{1}{3}\langle 0\,1\,0\rangle$ vector denoted by $\alpha\beta/DC$ is in the direction of the line joining the point bisecting $\alpha\beta$ to the point bisecting DC and has twice the magnitude of this line.

Some of the properties of dislocations in f.c.c. metals are summarized in Table 8.3.

8.4 DISLOCATIONS IN THE ROCK-SALT STRUCTURE

The dislocations that have been identified in crystals of the rock-salt (NaCl) structure have a $\frac{1}{2}\langle 1\,1\,0\rangle$ Burgers vector, as would be expected (Table 8.1). The slip plane is usually {1 1 0}, but not always (Table 6.1).

When slip on {1 1 0} is examined, it is impossible to find a stacking fault that looks as though it might have a low energy and could serve as the halfway stage of an indirect $\frac{1}{2}\langle 1\,1\,0\rangle$ displacement. Therefore the $\frac{1}{2}\langle 1\,1\,0\rangle$ dislocation is not expected to split into partials on $\{1\,\bar{1}\,0\}$. The fact that only perfect dislocations need be considered greatly simplifies the analysis of dislocation reactions. The relationship between the $\langle 1\,1\,0\rangle$ Burgers vectors and the $\{1\,\bar{1}\,0\}$ slip planes is shown in Fig. 8.16, which may be compared with Thompson's tetrahedron. The figure emphasizes that each Burgers vector lies in only one slip plane (e.g. AC lies in OAC only) and each slip plane contains only one Burgers vector (OAC contains only AC). This one-to-one correspondence contrasts with the slip systems of the f.c.c. metals, and still further simplifies the study of dislocation interactions. In fact, there is only one dislocation reaction to consider — that between dislocations on planes at 60° to one another:

$$AB + BC \rightarrow AC$$
$$\tfrac{1}{2}[\bar{1}\,0\,1] + \tfrac{1}{2}[0\,1\,\bar{1}] \rightarrow \tfrac{1}{2}[\bar{1}\,1\,0] \qquad (8.45)$$

If the reacting dislocations AB and BC are confined to their slip planes, the resultant AC is laid down along the line of intersection of the slip planes, OB. Since OB is $\langle 1\,1\,1\rangle$, the glide plane of the resultant is $\{2\,1\,1\}$, and for this reason it is very immobile at low temperatures. In LiF at room temperature, the dislocations on one plane will block slip on another which intersects it at 60° so efficiently that, after 1% shear strain in single slip, a resolved shear stress of about fifteen

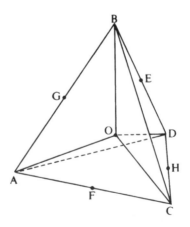

Figure 8.16 Relation between slip planes and directions in {1 1 0}⟨1 $\bar{1}$ 0⟩ or {1 1 0} ⟨1 $\bar{1}$ 1⟩ slip. The point O lies at the centre of the tetrahedron ABCD whose edges are parallel to ⟨1 1 0⟩. The planes OAB, OAC, etc., are {1 1 0} and the directions OA, OB, OC, OD are ⟨1 1 1⟩

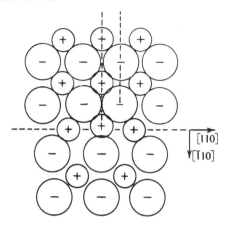

Figure 8.17 Edge dislocation in NaCl. The slip plane and the two sheets of ions constituting the extra plane are shown dotted

times that which was needed to activate the first system must be put upon the intersecting system in order to bring it into operation [7].

Although less complex in their interactions than dislocations in f.c.c. metals, rock-salt dislocations do have a property of special interest — they may carry an electric charge. The edge dislocation on {1 1 0} has an 'extra plane' consisting of a sheet of Na^+ ions and a sheet of Cl^- ions (Fig. 8.17). This sheet of NaCl 'molecules' has been isolated in Fig. 8.18. Its edge defines the dislocation line, which has a number of jogs in it. At A an extra anion defines two jogs. The section BC has a net charge of $+e$; therefore the jogs at B and C are each assigned a charge of $+e/2$. The 'whole' jog at D is uncharged. This is the type of jog which is produced by intersection with an orthogonal screw dislocation; charged jogs of types B and C may be called half-jogs or, because they are produced when a single ion jumps on to or off the edge of the extra plane, diffusional jogs. Evidently the line as a whole can be charged, up to a maximum of $e/2$ per atom length. Physically, a line may become charged by acting as a source or sink for vacancies (Section 9.1). For example, because the energy needed to make a single cation vacancy is less than that needed to make an anion vacancy in NaCl, either a surface or an edge dislocation raised to a high temperature will generate excess cation vacancies until their emission is throttled by the electric field which is built up. The dislocation or surface acquires a positive charge which is balanced by nearby vacant cation sites. On the other hand, a dislocation in an impure crystal may acquire a negative charge in order to compensate, partly, for the extra positive charge carried by divalent impurity cations, such as Ca^{2+}. At low temperatures the effect of impurities should predominate over the equilibrium properties that the dislocation would have in a pure crystal.

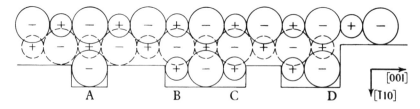

Figure 8.18

8.5 DISLOCATIONS IN HEXAGONAL METALS

By analogy with f.c.c. metals, the slip plane of hexagonal metals should be the closest packed basal plane $(0\,0\,0\,1)$, the Burgers vector of a slip dislocation should be the **a** vector of the hexagonal lattice $\frac{1}{3}[1\,1\,\bar{2}\,0]$ and, furthermore, this dislocation may be expected to split into two $\frac{1}{3}\langle1\,0\,\bar{1}\,0\rangle$ partials, the analogues of Shockley partials. However, the analogy is certainly incomplete, because other slip planes, and another slip direction, sometimes operate (Table 6.2). With this reservation in mind, we will first examine dislocations on the basal plane, produced by slip and by the aggregation of point defects.

Since the three $\langle1\,1\,\bar{2}\,0\rangle$ slip vectors are coplanar, reaction between slip dislocations can only form a network of dislocations in the basal plane. Written in the notation of Fig. 8.19, the reaction is

$$AB + BC \rightarrow AC$$

$$\tfrac{1}{3}[\bar{1}\,2\,\bar{1}\,0] + \tfrac{1}{3}[\bar{1}\,\bar{1}\,2\,0] \rightarrow \tfrac{1}{3}[\bar{2}\,1\,1\,0] \tag{8.46}$$

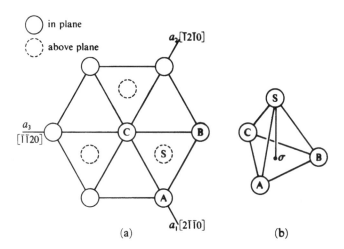

Figure 8.19 Atoms and lattice vectors in a hexagonal metal

The splitting of a perfect dislocation can also be shown on Fig. 8.19:

$$AB \rightarrow A\sigma + \sigma B$$

$$\tfrac{1}{3}[\bar{1}\,2\,\bar{1}\,0] \rightarrow \tfrac{1}{3}[\bar{1}\,1\,0\,0] + \tfrac{1}{3}[0\,1\,\bar{1}\,0] \qquad (8.47)$$

Between the partials $A\sigma$ and σB, which are analogous to Shockley partials, there is a small layer of f.c.c. stacking. The sequence ABABABA becomes ABACBCB or, in Frank's notation, $\triangle\triangledown\triangle\triangledown\triangle\triangledown$ is changed to $\triangle\triangledown\triangledown\triangledown\triangle\triangledown$. The order of the partials on successive basal planes must alternate; i.e. if it is $A\sigma$–σB on one plane it must be σB–$A\sigma$ on the next. This is in contrast to the splitting of perfect dislocations in an f.c.c. metal, where the order of the Shockley partials must always be the same if the same type of stacking fault is to be produced.

In metals of small c/a ratio, the $\tfrac{1}{3}\langle\bar{2}\,1\,1\,0\rangle$ dislocation slips quite easily on a prism plane, $\{1\,0\,\bar{1}\,0\}$, indicating that it is not widely split on the basal plane. Zirconium and titanium, for example, slip much more easily on prism planes than on the basal plane. It is interesting to note that the reaction $AB + BC \rightarrow AC$ between prism plane dislocations cannot produce a strong barrier, because the product, formed along the intersection of two of the prism planes, is always oriented so that it can glide on the third.

At this time, no reliable value of a stacking fault energy can be given, although that of Co is probably low, say $20\ \mathrm{mJ\ m}^{-2}$ at room temperature. (Cobalt transforms to an f.c.c. structure at $420\,^{\circ}\mathrm{C}$ — see Section 11.2.) The stacking faults that have been seen, in zinc for example, were probably the result of point defect condensation. It will now be demonstrated that such a fault is different from the fault produced by the splitting of a $\tfrac{1}{3}\langle1\,1\,\bar{2}\,0\rangle$ dislocation.

When a disc of vacancies collects on a basal plane, a prismatic dislocation loop forms at its perimeter. If the disc is one vacancy thick, and if A–A stacking is to be avoided, the loop may be faulted in one of two ways. Either a single basal layer or the whole crystal on one side of the loop must be sheared over by a vector of the $A\sigma$ type. These alternatives are shown in Fig. 8.20. In case (a), the single layer can be shifted from an A into a C position by passing $A\sigma$ partials of opposite sign on either side of it. The Burgers vector of the prismatic dislocation remains $\tfrac{1}{2}[0\,0\,0\,1]$, σS in Fig. 8.19, at the cost of a stacking fault which is triple, in terms of the number of next-nearest neighbour stacking violations. In case (b), a single $A\sigma$ partial is passed, so that the loop encloses a single stacking fault, but its Burgers vector increases to AS, as follows:

$$A\sigma + \sigma S = AS$$

$$\tfrac{1}{3}[\bar{1}\,1\,0\,0] + \tfrac{1}{2}[0\,0\,0\,1] \rightarrow \tfrac{1}{6}[\bar{2}\,2\,0\,3] \qquad (8.48)$$

Similar reasoning applies to faulted loops produced by discs of extra atoms. The single stacking fault has been seen in zinc; it should have a slightly smaller energy than the double fault produced by the splitting of a slip dislocation.

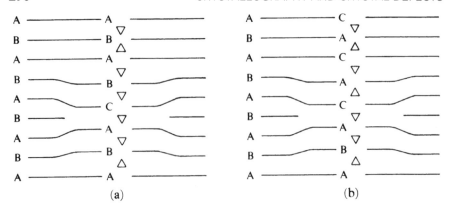

Figure 8.20 Two possible structures for a vacancy loop in a hexagonal metal

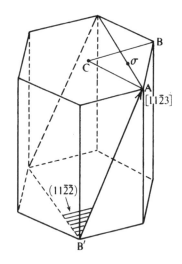

Figure 8.21 A {1 1 $\overline{2}\,\overline{2}$}⟨1 1 $\overline{2}$ 3⟩ slip system

A more surprising slip system is the {1 1 $\overline{2}\,\overline{2}$}⟨1 1 $\overline{2}$ 3⟩, called second-order pyra-midal glide. The remarkable feature is the large size of the slip vector $\frac{1}{3}$⟨1 1 $\overline{2}$ 3⟩, which is the sum of the **c** vector and an **a** vector. Even more puzzling is the fact that this system operates, second only to basal slip, in zinc and cadmium, which have the largest c/a ratios. A slip plane and its slip vectors are shown in the hexagonal cell in Fig. 8.21. Each slip plane contains one slip vector and each slip vector lies in one slip plane. Figure 8.22 shows the relationship between the six slip planes and six Burgers vectors, by means of a surface analogous to that used to illustrate the rock-salt slip systems (Fig. 8.16). Dislocation reactions in this

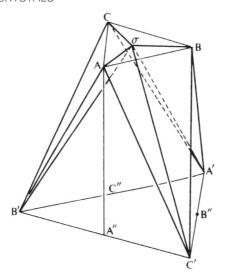

Figure 8.22   Relation between slip planes and directions in $\{1\,1\,\bar{2}\,\bar{2}\}\langle1\,1\,\bar{2}\,3\rangle$ slip

system can be dealt with easily by decomposing each $\frac{1}{3}\langle1\,1\,\bar{2}\,3\rangle$ vector into its **c** and **a** components. The **c** components either add or cancel; provided $c > a$, there is an increase in energy when they add and a decrease when they cancel, whatever the **a** components may be. There are three reactions, then, corresponding to the three possible combinations of the **a** components when the **c** components cancel. The reaction is least favourable energetically when the **a** components are alike:

$$C'A + AB' \rightarrow C'B'$$

$$\tfrac{1}{3}[\bar{1}\,\bar{1}\,2\,3] + \tfrac{1}{3}[\bar{1}\,\bar{1}\,2\,\bar{3}] \rightarrow \tfrac{2}{3}[\bar{1}\,\bar{1}\,2\,0] \qquad (8.49)$$

The product C'B' is formed along the line $A\sigma$ in the basal plane and can immediately split into two perfect $\frac{1}{3}[\bar{1}\,\bar{1}\,2\,0]$ dislocations. The most favourable reaction is

$$C'A + BC' \rightarrow BA$$

$$\tfrac{1}{3}[\bar{1}\,\bar{1}\,2\,3] + \tfrac{1}{3}[\bar{2}\,\bar{1}\,\bar{1}\,\bar{3}] \rightarrow \tfrac{1}{3}[1\,\bar{2}\,1\,0] \qquad (8.50)$$

Figure 8.22 shows that the product BA is formed along the line C'σ, so that it may well be immobile and an obstacle to other dislocations. The result of the third reaction is less easy to see from Fig. 8.22. The reaction is

$$C'A + BA' \rightarrow C'C''$$

$$\tfrac{1}{3}[\bar{1}\,\bar{1}\,2\,3] + \tfrac{1}{3}[\bar{2}\,1\,1\,\bar{3}] \rightarrow \tfrac{1}{3}[\bar{3}\,0\,3\,0] \qquad (8.51)$$

The product C'C" should split into two **a** dislocations, AC + BC, but since its line does not lie in the basal plane, these dislocations may be immobile.

It is conceivable that the $\langle 1\,1\,\overline{2}\,3\rangle$ dislocation might itself split into partials of the type AS (Fig. 8.22):

$$\tfrac{1}{3}[1\,1\,\overline{2}\,3] \rightarrow \tfrac{1}{6}[0\,2\,\overline{2}\,3] + \tfrac{1}{6}[2\,0\,\overline{2}\,3] \tag{8.52}$$

However, there is no reason to assign a low energy to the fault, which would have to form on $\{1\,1\,\overline{2}\,2\}$. Furthermore, $\tfrac{1}{3}[1\,1\,\overline{2}\,3]$ screw dislocations in zinc and cadmium cross-slip frequently, which is evidence against splitting.

There are two reactions between a basal plane dislocation and a second-order pyramidal dislocation. The first is simply a rearrangement of reaction (8.50):

$$C'A + AB \rightarrow C'B$$

$$\tfrac{1}{3}[\overline{1}\,\overline{1}\,2\,3] + \tfrac{1}{3}[\overline{1}\,2\,\overline{1}\,0] \rightarrow \tfrac{1}{3}[\overline{2}\,1\,1\,3] \tag{8.53}$$

The product lies along Aσ, out of its pyramidal slip plane C'σB. Therefore it may be a barrier. The second reaction is simply the annihilation of the **a** component of the pyramidal:

$$C'A + A''C' \rightarrow A''A$$

$$\tfrac{1}{3}[\overline{1}\,\overline{1}\,2\,3] + \tfrac{1}{3}[1\,1\,\overline{2}\,0] \rightarrow [0\,0\,0\,1] \tag{8.54}$$

The extra plane of the product A''A consists of two basal layers of atoms. These may be thought of as tending to wedge open the basal plane, which is also a cleavage plane. In this respect the reaction is like that which produces the $\langle 1\,0\,0\rangle$ dislocation in b.c.c. metals (Section 8.6).

8.6 DISLOCATIONS IN B.C.C. CRYSTALS

It is established that a slip dislocation in a b.c.c. metal has the expected Burgers vector, $\tfrac{1}{2}\langle 1\,1\,1\rangle$. The plane on which this dislocation prefers to glide is far less certain (Table 6.1). We shall first assume that slip occurs somewhat more easily on $\{1\,1\,0\}$ than on other planes in the $\langle 1\,\overline{1}\,1\rangle$ zone.

The relationship between the $\tfrac{1}{2}\langle 1\,1\,1\rangle$ Burgers vectors and the $\{1\,1\,0\}$ slip planes can be shown on the same diagram that was used for rock-salt structures (Fig. 8.16). Figure 8.16 shows that any two different Burgers vectors meet either at 70.53° or at 109.47°. In the latter case the dislocations will react:

$$OA + OC \rightarrow EF$$

$$\tfrac{1}{2}[1\,\overline{1}\,1] + \tfrac{1}{2}[1\,1\,\overline{1}] \rightarrow [1\,0\,0] \tag{8.55}$$

The character of the product dislocation depends on the slip planes of the reactants, because these determine the direction in which its line lies. If the reacting dislocations are on the same slip plane, then the glide plane of the product will of course be this plane also. If they are on planes like AOB and AOC which intersect at 60° along a $\langle 1\,1\,1\rangle$, then the product, formed along this line, will still be able to glide in a $\{1\,1\,0\}$ plane. However, if the reacting dislocations are on orthogonal planes, like AOD and COB, their product will lie along an $\langle 0\,0\,1\rangle$ like GH in Fig. 8.16. It is then an edge dislocation with a $\{1\,0\,0\}$ glide plane. The extra plane of this dislocation is parallel to the usual cleavage plane of a b.c.c. crystal, and it has been suggested that the crystal may be cracked open for a short distance beneath the extra plane [8].

The fact that slip is not very rigorously confined to any one plane is strong evidence that the $\frac{1}{2}\langle 1\,1\,1\rangle$ dislocation is not split widely into partials. Nevertheless, it is possible to find plausible-looking stacking faults in a rigid-ball model of the structure. A ball model of the $\{1\,1\,0\}$ planes is shown in Fig. 8.23. The arrows starting at P point out the natural path in which to slide the balls of the upper layer over the lower. This corresponds to the dissociation

$$\tfrac{1}{2}[1\,1\,1] \rightarrow \tfrac{1}{8}[0\,1\,1] + \tfrac{1}{4}[2\,1\,1] + \tfrac{1}{8}[0\,1\,1] \tag{8.56}$$

The fault which is formed on both sides of the $\frac{1}{4}[2\,1\,1]$ dislocation occurs quite naturally in a rigid-ball structure, because it leads to closer packing (for this reason it may be important in the martensitic transformation of a b.c.c. to an h.c.p. structure—see Section 11.8). Thus the $\frac{1}{8}[0\,1\,1]$ which produces the fault drops the ball P from a saddle position into a hollow. However, nearest neighbour relationships are violated and the fault must ordinarily have a rather high energy, since it has never been detected in b.c.c. metals.[†]

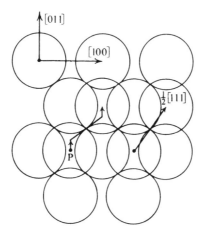

Figure 8.23

[†] Calculations suggest that there are no metastable faulted positions for either $\{1\,1\,0\}$ or $\{2\,1\,1\}$ planes. See Vitek [9].

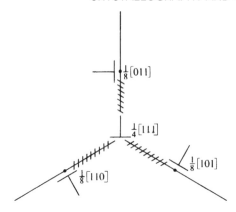

Figure 8.24

If splitting on {1 1 0} can occur, then a screw dislocation can split in an interesting way. Since the screw dislocation is parallel to its Burgers vector, $\frac{1}{2}\langle 1\,1\,1\rangle$, it lies at the intersection of three {1 1 0} planes as shown in Fig. 8.16, and it can reduce its energy by splitting onto all three simultaneously:

$$\tfrac{1}{2}[1\,1\,1] \rightarrow \tfrac{1}{4}[1\,1\,1] + \tfrac{1}{8}[0\,1\,1] + \tfrac{1}{8}[1\,0\,1] + \tfrac{1}{8}[1\,1\,0] \qquad (8.57)$$

The resulting configuration is shown in Fig. 8.24. Before such a screw could glide on any one plane, it would have to be constricted by the applied stress.

A different type of stacking fault is suggested by the fact that twins occur on {2 1 1}. As will be shown in Chapter 10, a b.c.c. twin can be constructed by displacements of $\frac{1}{6}\langle 1\,1\,1\rangle$ on successive {2 1 1} planes. If the energy of the

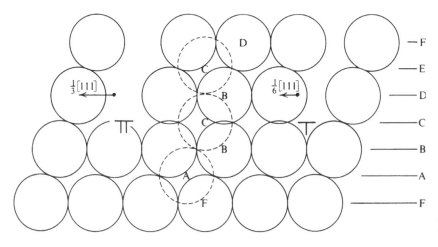

Figure 8.25

{2 1 1} twin boundary is not too high one may suppose that the energy of the fault produced by a single $\frac{1}{6}\langle 1 1 1 \rangle$ displacement will also be moderate, allowing the $\frac{1}{2}\langle 1 1 1 \rangle$ dislocation to split on a {2 1 1} layer as follows:

$$\frac{1}{2}[1\ 1\ 1] \longrightarrow \frac{1}{3}[1\ 1\ 1] + \frac{1}{6}[1\ 1\ 1] \tag{8.58}$$

It is easier to illustrate this fault by means of a {1 1 0} section normal to the fault plane and containing the displacement vector, rather than by a plan of the {2 1 1} planes, because no fewer than six {2 1 1} layers must be stacked on top of one another before the seventh falls directly over the first. This is done in Fig. 8.25, which shows that the stacking sequence ABCDEF A ... is transformed into the faulted sequence ABCBCDE .... . In order to produce this fault, the order of the two partials $\frac{1}{3}[1\ 1\ 1]$ and $\frac{1}{6}[1\ 1\ 1]$ is strictly defined; if they are interchanged, an entirely different fault is formed.

Unfortunately, we do not know the value of the energy of the stacking fault on {2 1 1}. It would be quite wrong to speak of splitting if the stacking fault energy

Table 8.4 Fault energies
(After Teutonico [10])

Element	E_{max} (mJ m^{-2})
Li	81
K	26
Nb	537
Fe	939
Mo	1450

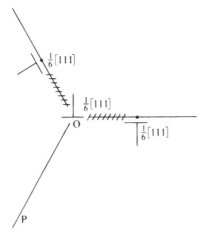

Figure 8.26

is so high that the calculated splitting is only of the order of an interatomic distance. Table 8.4 gives some limiting fault energies, calculated on the basis of a separation of the partials of one lattice spacing, for an edge dislocation splitting on {2 1 1}.

Assuming that splitting on {2 1 1} is possible, it is interesting to see how a screw dislocation can split. It lies on three different {2 1 1} planes and it can lower its energy by splitting on two of them, as in Fig. 8.26. Geometrically, it could split on to all three {2 1 1} planes, but by examining the forces acting on each partial, it is easy to show that after it has split on two, the central partial, at O, will not be forced out along OP, the third {2 1 1}, unless an external stress is applied [11].

8.7 DISLOCATIONS IN SOME COVALENT SOLIDS

In crystals of diamond, Ge and Si, the atoms are held together by strong, directional covalent bonds (Fig. 3.1e). Dislocations in these crystals are immobile at low temperatures. This is because the Peierls stress is high; that is to say the energy of the dislocation varies sharply with its position in the crystal. Extensive slip occurs only at elevated temperatures — above about 60% of the absolute melting temperature in Ge and Si. The slip plane is then {1 1 1} and the slip direction $\langle 1\,\bar{1}\,0 \rangle$.

One can imagine the way in which glide might occur. The dislocation will lie, as much as it can, in a position of low energy, probably along some prominent crystallographic direction. It advances when, aided by thermal activation, a small length of line surmounts its energy barrier and spills over into the next energy trough (Fig. 8.27). The connecting segment which is stranded on the energy hill is called a kink, distinguished from a jog by the fact that it lies wholly in the slip plane (Section 7.6). The kinks can glide apart relatively easily and, as they do so, the dislocation is advanced by the trough separation.

The dislocation lying on a {1 1 1} plane at 60° to its Burgers vector, $\frac{1}{2}\langle 1\,\bar{1}\,0 \rangle$, happens to be easy to draw in projection on to a {1 0 $\bar{1}$} plane normal to the dislocation; it is shown in this way in Fig. 8.28. This figure serves equally well for the sphalerite (zinc-blende) structure. There are two possible {1 1 1} slip planes, type I and type II. The 60° dislocation is shown with its extra plane ending at a type II plane, where it breaks the lesser number of bonds. The one bond per atom

Figure 8.27 Double kink in a dislocation line. The dotted lines represent positions of high energy for the dislocation

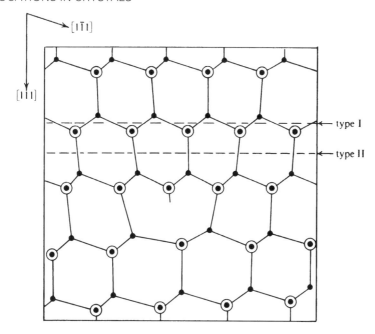

Figure 8.28 A 60° dislocation in sphalerite. The structure of a 60° dislocation in Ge is identical, except that all the atoms are then the same

which is broken has been described as a dangling bond. Intuitively, it seems that slip should occur at the widely spaced type II planes, by the glide of dislocations like that of Fig. 8.28. The situation becomes more puzzling, however, when the possibility of splitting into partials is examined.

There can be stacking faults on the {1 1 1} plane that do not violate tetrahedral bonding and are exactly analogous to intrinsic and extrinsic faults in f.c.c. metals. The stacking sequence of the perfect crystal can be written as $a\,\alpha b\,\beta c\,\gamma a\,\alpha\,\ldots$ (see Sections 3.4 and 3.5). In this sequence, the diatom $a-\alpha$ can be treated as a unit, denoted by A. Type I planes separate sheets of diatoms packed in the f.c.c. sequence ABCA .... An intrinsic fault is formed by passing a Shockley partial, $\frac{1}{6}\langle 2\,1\,1\rangle$, over a type I plane; passing two different Shockley partials over successive type I planes produces the extrinsic fault. The $\frac{1}{2}\langle 1\,1\,0\rangle$ dislocation could split into partials on a type I plane:

$$\tfrac{1}{2}[0\,1\,\bar{1}] = \tfrac{1}{6}[1\,1\,\bar{2}] + \tfrac{1}{6}[\bar{1}\,2\,\bar{1}] \qquad (8.59)$$

It is uncertain whether moving dislocations are split into partials; if they are, then the slip plane is presumably a type I plane.

A rich variety of stacking faults is found in Si films grown epitaxially from the vapour phase [12]. Faults originating at irregularities in the substrate grow

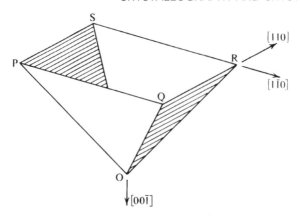

Figure 8.29 Stacking fault in a silicon film. Point O lies at the bottom surface of the film and points P, Q, R and S at the top surface. OP, OQ, OR and OS are stair-rod dislocations

along with the film, sometimes joined together as pieces of polyhedra. Figure 8.29 shows a piece of an octahedral stacking fault in a film, which is parallel to $\{1\,0\,0\}$. The faces of the half-octahedron are alternate intrinsic and extrinsic faults, joined by a $\frac{1}{6}\langle1\,1\,0\rangle$ stair-rod. The configuration at a corner can be derived from the Cottrell–Lomer lock (Fig. 8.13) by gliding one of the Shockleys of the lock to the other side of the stair-rod.

An alternative tetrahedrally bonded structure is that of wurtzite, in which ZnS units are stacked in the hexagonal sequence ABA .... . Stacking faults in sphalerite consist of thin layers of the wurtzite structure. Similarly, stacking faults in wurtzite can occur that are thin layers of the sphalerite structure and are exactly analogous to stacking faults in a hexagonal metal. Dislocations in AlN, which has this structure, are widely split, the fault energy being only about 5 mJ m^{-2} [13].

Neither wurtzite nor sphalerite possess a centre of symmetry, with the interesting consequence that the core structures of dislocations of opposite sign can be distinguished from one another in these structures. Suppose that a 60° dislocation in, say, InSb, which has the sphalerite structure, always lies on a type II plane (Fig. 8.28). Its extra plane then ends either on an Sb atom or on an In atom according as to whether it comes in from the top or the bottom of the crystal. Dislocations of opposite sign can therefore be labelled In and Sb dislocations, and because of their different core structures they may have quite different Peierls stresses.

Tetrahedrally bounded crystals contain a three-dimensional network of strong bonds, but in many crystals the atoms are bonded together strongly only within layers which are themselves connected by relatively weak forces. Graphite is such a crystal. The dislocations and stacking faults on the basal plane of graphite are exactly analogous to those on the basal plane of a hexagonal metal. Figure 8.19b

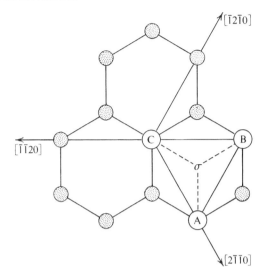

Figure 8.30 Atoms and lattice vectors in the basal plane of graphite

is applicable and Fig. 8.30 shows its relationship to the structure of graphite. Because of the weak interlayer bonding, the energies of the three basic types of stacking fault described in Section 8.5 are very small. One of these faults is produced by the dissociation of a perfect dislocation, as follows:

$$AC \rightarrow \sigma C + A\sigma$$

$$\tfrac{1}{3}[2\,1\,1\,0] \rightarrow \tfrac{1}{3}[\bar{1}\,0\,1\,0] + \tfrac{1}{3}[\bar{1}\,1\,0\,0] \qquad (8.60)$$

By direct measurement of the separation of the $\tfrac{1}{3}\langle 1\,0\bar{1}\,0\rangle$ partials [14], the energy of the fault has been found to be only 0.5 mJ m^{-2}. Figure 8.31 is a side view of a split 60° dislocation, showing the layer of rhombohedral stacking at the fault.

Although slip dislocations are strictly confined to the basal planes, their reactions with one another are a little more complex than might at first be expected. The complexities are a result of the wide separation of the partials and of the fact that, because of this wide separation, the individual partials on nearby planes can, in effect, react with one another. As an example, consider a reaction between the dislocations AB and AC which would simply repel one another if they were not split. Suppose that they lie on adjacent planes, so that if AB splits in the order A$\sigma + \sigma$B, then AC must split in the order of σC + Aσ. If the split dislocations are caused to approach, AB from the left and AC from the right, the partials σB and σC will come over one another and react:

$$\sigma B + \sigma C \rightarrow A\sigma \qquad (8.61)$$

The result is a stable double ribbon, shown in Fig. 8.32.

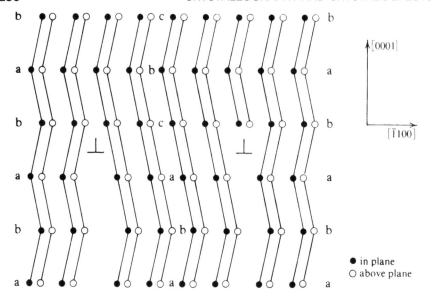

Figure 8.31

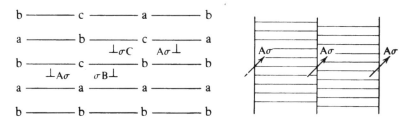

Figure 8.32

The detailed analysis of the geometry of dislocations, which has been carried out up to this point for a number of specific crystal structures, could be repeated for many other crystal structures. It is hoped that the principles employed have now been sufficiently demonstrated for the reader to be able to deal with any case that may be encountered.

PROBLEMS

8.1 In a b.c.c. metal, two dislocations of the Burgers vector $[1\,1\,1]$ and $[1\,\bar{1}\,\bar{1}]$ respectively have reacted over a certain length (as in Fig. 8.11). If the reacting dislocations are pinned at points p, q, r and s, which lie at the corners of a square of side L, make an estimate of the length of the product dislocation (neglect the variation of the energy of a dislocation with the orientation of its line).

8.2 A dislocation is being extruded between small precipitate particles which lie in a row, each particle being a distance l from its neighbours. Determine the force acting on each particle as a function of the shear stress σ applied on the slip plane of the dislocation parallel to the Burgers vector **b** and estimate the maximum force that can be exerted on each particle.

8.3 Using Frank's rule, make a study of reactions between stable perfect dislocations in a body-centred tetragonal crystal. Consider the effect of the value of the c/a ratio on the reactions.

8.4 Compare the number of atoms that are improperly surrounded by nearest neighbours (i.e. as an atom in the c.p.h. structure instead of as in the f.c.c. structure) at intrinsic and in extrinsic faults in f.c.c. metals. What structure is formed if both the Shockley partials in a Cottrell–Lomer lock are forced to glide to the opposite side of the stair-rod dislocation?

8.5 Using a dislocation corresponding to one particular slip plane and slip direction as one of the reacting dislocations forming a Cottrell–Lomer lock, how many different locks can be formed? For the slip system $(1\,1\,1)\,[\bar{1}\,0\,1]$, give the Burgers vector of each partial dislocation in all of the possible Cottrell–Lomer locks.

8.6 If the concentration of vacancies in aluminium at its melting point is 9.4×10^{-4} and if after quenching the aluminium from its melting point the vacancies condense into discs on the close packed plane and form Frank loops, determine the density of dislocations introduced (a) when the loops are 50 Å in radius and (b) when the loops are 500 Å in radius.

8.7 Assuming isotropic elasticity, calculate the separation of the Shockley partials of a screw dislocation in Ag, Al and Au. Use the stacking fault energies given in Table 8.2. For Ag take $\mu = 28$ GPa, $\nu = 0.38$; for Al take $\mu = 26$ GPa, $\nu = 0.34$; for Au take $\mu = 28$ Gpa, $\nu = 0.42$.

8.8 For the first four lattices listed in Table 8.1, find the Burgers vectors of any dislocations whose stability, according to Frank's rule, is doubtful, i.e. dislocations that might dissociate into two perfect dislocations whose Burgers vectors are at 90° to one another.

8.9 Suppose that a screw dislocation in a b.c.c. metal splits according to the equation

$$\tfrac{1}{2}[1\,1\,1] = \tfrac{1}{6}[1\,1\,1] + \tfrac{1}{6}[1\,1\,1] + \tfrac{1}{6}[1\,1\,1]$$

and two of the $\tfrac{1}{6}[1\,1\,1]$ partials move out on two $\{2\,1\,1\}$ planes, leaving the third partial lying along the line of intersection of these planes (Fig. 8.26). Determine the separation of the partials when they are in equilibrium with respect to glide on $\{2\,1\,1\}$, in terms of the energy γ of the stacking fault on {2 1 1}, and show that the third partial will not be forced out along the third {2 1 1}. Assume isotropic elasticity.

8.10 Lead sulphide, PbS, has the same crystal structure as NaCl, but the usual slip system is {1 0 0}⟨0 1 1⟩ instead of {0 1 1}⟨0 $\bar{1}$ 1⟩.
   (a) Compare dislocation reactions that may occur as a result of slip in PbS with those occurring in NaCl.
   (b) Sketch the structure of the edge of the extra plane of an edge dislocation in PbS.

8.11 Describe the dislocation loops that might be produced in zinc by the aggregation of interstitial atoms into discs of atoms parallel to the basal plane. Also describe the dislocation loops that might be produced in graphite by the aggregation of vacant sites into discs parallel to the basal plane.

8.12 Project the structure of silicon on to a (1 0 $\bar{1}$) plane and show, in this projection, the structure of an intrinsic stacking fault and the structure of an extrinsic stacking fault in silicon.

8.13 Show how, in graphite, two dislocation ribbons (i.e. dislocations that are split into widely separated partials) that are on next-nearest neighbour basal planes can interact to form a threefold ribbon.

8.14 An alloy of 50 at % Cu–50 at % Zn contains dislocations of Burgers vector $\frac{1}{2}$⟨1 1 1⟩ and ⟨1 0 0⟩. The alloy is then ordered to produce the $L2_0$ superlattice. Describe the effect of gliding the $\frac{1}{2}$⟨1 1 1⟩ dislocation on {1 $\bar{1}$ 0} and the ⟨1 0 0⟩ dislocation on {0 1 0} after ordering has occurred. From your result, how would you expect two identical $\frac{1}{2}$⟨1 1 1⟩ dislocations to interact in the ordered alloy?

8.15 Use the rule that the observed slip direction is in the direction of the shortest lattice translation vector to deduce the slip directions of (a) mercury and (b) bismuth. Compare your results with Table 6.1. Is the rule obeyed?

SUGGESTIONS FOR FURTHER READING

See the suggestions for Chapter 7 and the following:
Hirth, J. P., and Lothe, J., *Theory of Dislocations*, Wiley, New York and Chichester (1982).
Nabarro, F. R. N. (ed.) *Dislocations in Solids*, Vol. 9, *Dislocations and Disclinations*, North Holland (1992).
Suzuki, T. Takeuchi, S., and Yoshinaga, H., *Dislocation Dynamics and Plasticity*, Springer, Berlin (1991) (Translated from the Japanese edition c. 1985).

REFERENCES

1. A. J. E. Foreman, 'Dislocation energies in anisotropic crystals', *Acta Met.*, **3**, 322 (1955).

2. D. J. H. Cockayne, M. L. Jenkins and I. L. F. Ray, *Phil. Mag.*, **24**, 1383 (1971).

3. P. S. Dobson, P. J. Goodhew and R. E. Smallman, *Phil. Mag.*, **16**, 9 (1967).

4. M. L. Jenkins, *Phil. Mag.*, **26**, 747 (1972).

5. M. Stobbs and C. Sworn, *Phil. Mag.*, **24**, 1365 (1971).

6. W. Thompson, *Proc. Phys. Soc.*, **66B**, 481 (1953).

7. T. H. Alden, 'Extreme latent hardening in compressed lithium fluoride crystals', *Acta Met.*, **11**, 1103 (1963).

8. A. H. Cottrell, 'Theory of brittle fracture in steel and similar metals', *Trans. Am. Inst. Min. Metall. Petrol. Engrs.*, **212**, 192 (1958).

9. V. Vitek, 'Intrinsic stacking faults in body-centred cubic crystals', *Phil. Mag.*, **18**, 773 (1968).

10. L. Teutonico, *acta Met.*, **13**, 105 (1965).

11. A. W. Sleeswyk, '$\frac{1}{2}\langle 1\,1\,1\rangle$ screw dislocations, and the nucleation of $\{1\,1\,2\}\langle 1\,1\,1\rangle$ twins in the b.c.c. lattice', *Phil. Mag.*, **8**, 1467 (1963).

12. G. R. Booker, 'Crystallographic imperfections in silicon', *Disc. Faraday Soc.*, **38**, 298 (1964).

13. P. Delavignette, H. B. Kirkpatrick and S. Amelincx, 'Dislocations and stacking faults in aluminium nitride', *J. Appl. Phys.*, **32**, 1098 (1961).

14. C. Baker, Y. T. Chou and A. Kelly, 'The stacking fault energy of graphite', *Phil. Mag.*, **6**, 1305 (1961).

9

Point Defects

9.1 INTRODUCTION

Disturbances in a crystal that, apart from elastic strains associated with them, extend for no more than a few interatomic distances in any direction are called point defects. Although impurity atoms in solution can be classified as point defects, we shall be mainly concerned with structural point defects, of which there are two elementary types. The first is the vacancy, which consists of an atomic site from which the atom is missing. The second consists of a small region of crystal which contains an extra atom. This is called an interstitial defect, from the idea that the extra atom is squeezed into an interstice between the others.

A point defect differs from a dislocation and from a two-dimensional defect such as a crystal boundary in two important respects. The first is that it is difficult to observe directly, so that in the main it can be detected and studied only through its effect upon some physical property of the crystal. The second distinction is that point defects may be present in appreciable concentrations when the crystal is in *thermodynamic equilibrium*. Whilst dislocations and interfaces always raise the free energy of a crystal, adding a certain number of point defects to an otherwise perfect crystal reduces its free energy to a minimum value. This is because of a gain of entropy which arises from the many possible sets of places in the crystal where the point defects can be put. This configurational entropy is given by

$$S = k \ln W \tag{9.1}$$

where k is Boltzmann's constant (1.38×10^{-23} J K^{-1} or 8.68×10^{-5} eV deg^{-1}) and W is the number of different arrangements of point defects. When n defects are distributed amongst N sites, the number of arrangements is $N(N-1)\ldots(N-n+1)$ if each defect is distinct. In fact, all the defects are identical, and the n labels needed to distinguish them from one another could be distributed in $n!$ ways. Therefore

$$W n! = N(N-1)\ldots(N-n+1). \tag{9.2}$$

Manipulating Eqn (9.2) into a form suitable for the application of Stirling's approximation ($\ln x! \approx x \ln x - x$, for large x) gives

$$W = \frac{N!}{(N-n)!n!}. \tag{9.3}$$

From Eqn (9.1),

$$S = k[N \ln N - (N-n)\ln(N-n) - n \ln n] \tag{9.4}$$

If the presence of one defect increases the internal energy of the crystal by E_f, the change in free energy of a crystal containing n identical defects, at a temperature of T K, is

$$\Delta F = nE_f - T(S + nS') \tag{9.5}$$

Here it is assumed that n/N is so small that the defects do not interact with one another. The extra entropy term nS' in Eqn (9.5) represents the fact that each defect may add a certain entropy S' to the crystal because of its effect on the vibration of the atoms in its neighbourhood. The number of defects producing the minimum free energy n_e is found by setting $d\Delta F/dn = 0$, giving

$$\frac{n_e}{N - n_e} = \exp\left(\frac{S'}{k}\right) \exp\left(-\frac{E_f}{kT}\right) \tag{9.6}$$

If the defect in question is a vacancy, then $N - n_e$ is simply the number of atoms in the crystal. Since in fact $n_e \ll N$, Eqn (9.6) can be written as

$$\frac{n_e}{N} = \exp\left(\frac{S'}{k}\right) \exp\left(-\frac{E_f}{kT}\right) \tag{9.7}$$

In f.c.c. metals, S'/k is of the order of unity, so that the atomic concentration of vacancies, n_e/N, is essentially given by $\exp(-E_f/kT)$. For example, in the f.c.c. metals, Cu, Ag and Au, the energy of formation of a vacancy, E_f, is approximately 1 eV, leading to vacancy concentrations of the order of 10^{-4} at temperatures close to the melting point. The vacancies can be generated at surfaces, grain boundaries or dislocation lines. An atom jumping on to a surface or a grain boundary leaves a vacant site into which deeper atoms can jump, driving the vacancy into the interior (Fig. 9.1). Similarly, an atom can leave a site vacant by jumping on to the edge of the extra plane of an edge dislocation. If an interior atom were to jump into an interstitial position, both a vacancy and an interstitial would be created at the same time. Such a defect, called a *Frenkel defect*, does not occur in appreciable concentration in equilibrated f.c.c. metals because of the high internal energy increase produced by an interstitial. This is not necessarily true for other crystals; e.g. crystals of AgCl and AgBr contain an atomic concentration of Ag in interstitial positions of the order of 10^{-3} and 10^{-2} respectively, at the melting temperatures.

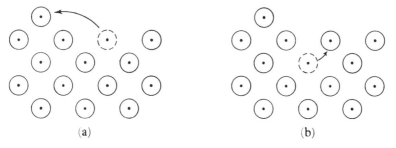

(a) (b)

Figure 9.1 Crystal surface acting as a vacancy source

Since the energy of formation of a point defect determines its concentration in a crystal which is in equilibrium, it would be helpful to be able to calculate this quantity. Unfortunately, this is a very difficult task, although it may be easy to make a crude estimate. For example, the energy needed to form a vacancy can be estimated very roughly, as follows. Assume that the cohesive energy of a crystal can fairly be represented as the sum of the energies of interaction of nearby pairs of atoms. Starting with any one atom, we sum the energies of the bonds made with each of its neighbours. This summation is repeated for a second atom, and so on. The final total sum gives twice the cohesive energy of the crystal, since the energy of each bond is counted twice. If each atom is identically attached to bonds of total energy $2E_c$, the cohesive energy per atom is E_c. To form a vacancy, we remove an atom from the interior of the crystal and place it on the surface. The first step breaks bonds of energy $2E_c$ whilst, the second reforms half the bonds (on average), leaving E_c as the energy to form the vacancy. This estimate is very inaccurate, because it makes no allowance for the possibility of a change in the energies of the bonds between the atoms near the vacant site. In fact, when an atom is removed, the surrounding nuclei will be displaced and the electrons will be redistributed in such a way as to release energy. Consequently, a relaxation energy R must be subtracted from E_c:

$$E_f = E_c - R \qquad (9.8)$$

Since the value of R is of the same order of magnitude as E_c and is very difficult to calculate, the formation energy E_f is correspondingly difficult to obtain accurately. However, a number of careful calculations of the formation energy of a vacancy in copper have been made (e.g. see Ref. [1]), and in agreement with experimental results they cluster about the value of 1 eV. Calculations set a much higher value of about 3 eV on the formation energy of an interstitial defect, in agreement with the lack of evidence for the existence of interstitials in f.c.c. metals which have been brought to equilibrium at high temperatures.

Point defects with a high energy of formation can occur in crystals that are *not* in equilibrium. Both vacancies and interstitials are produced by any radiation that can knock atoms out of their normal sites, since the displaced atom leaves a vacancy behind and itself becomes an interstitial defect. Similarly, plastic deformation can be expected to produce both vacancies and interstitials through the action of dislocations upon one another (Section 7.6).

The ability of a crystal to retain a point defect produced in one of these ways depends upon the mobility of the defect in the crystal, since a highly mobile defect that is not in equilibrium will quickly be lost. Point defects can destroy themselves at a surface or grain boundary, or at a dislocation, or by combining with other point defects. The mobility of a defect, which is present in equilibrium, is also an important property, because atoms can be transported through a crystal by the movement of equilibrium point defects. For example, self-diffusion in f.c.c. metals occurs by the movement of vacancies.

The mobility of a defect is governed by the increase in the free energy of the crystal as the defect passes through the position of maximum energy on the way from one rest position to the next. The frequency with which the defect jumps to a new position is given by

$$\nu = \nu_0 \exp\left(\frac{S_M}{k}\right) \exp\left(-\frac{E_M}{kT}\right) \qquad (9.9)$$

where ν_0 is the 'attempt frequency', or the frequency of vibration of the defect in the appropriate direction. The free energy increase has been divided up into an entropy increase S_M and an increase in internal energy E_M. Table 9.1 shows the number of jumps that can be made in 1 s for various values of the migration energy E_M, assuming that $\nu_0 \sim 10^{13}$ s^{-1} and $\exp(S_M/k) \sim 1$, which are reasonable values for a vacancy in an f.c.c. metal, for example.

The energy of migration of a defect is even more difficult to calculate than its formation energy. When the defect is responsible for self-diffusion, as is the vacancy in f.c.c. metals, the migration energy can be obtained from measurements of rates of self-diffusion. The self-diffusion coefficient in an equilibrated f.c.c. metal is proportional to the equilibrium concentration of vacancies and to their mobility, so that the activation energy for self-diffusion, E_{SD}, is given by

$$E_{SD} = E_f + E_M \qquad (9.10)$$

From Eqn (9.10), using a calculated value of E_f, the migration energy of a vacancy in copper is deduced to be about 1 eV. It can be judged from Table 9.1

Table 9.1 Estimated number of jumps made per second by a point defect having migration energy E_M

E_M (eV)	ν at 77 K	ν at 300 K	ν at 773 K	ν at 1273 K
0.1	3.2×10^6	2.1×10^{11}	2.2×10^{12}	4.0×10^{12}
0.5	$\ll 1$	4.6×10^7	5.8×10^9	1.1×10^{11}
1.0	$\ll 1$	$\ll 1$	3.4×10^6	1.2×10^9
2.0	$\ll 1$	$\ll 1$	1.1	1.4×10^5

that the mobility of vacancies in copper is insignificant at room temperature. An interstitial atom in copper has a lower migration energy; in fact, some calculations suggest that the value may be as low as 0.1 eV. If this estimate is correct, then an isolated interstitial atom in copper is too mobile to be retained at room temperature, since with a migration energy in the range 0.1–0.5 eV it can cover macroscopic distances in 1 s.

9.2 POINT DEFECTS IN IONIC CRYSTALS

So far, examples of point defects have been drawn from f.c.c. metals. Defects in alkali halides have also been studied intensively. They have additional complexities; e.g. both anion and cation defects may occur. Suppose that the formation energy E^+ of a cation vacancy is less than that of an anion vacancy, E^-, as is believed to be the case in sodium chloride. In coming to equilibrium at high temperatures the crystal will initially produce more cation than anion vacancies at surfaces or dislocation lines. This will set up an electric field opposing the issue of further cation vacancies and aiding the formation of anion vacancies. At equilibrium, the crystal will contain very nearly equal numbers n of both types. An equation analogous to Eqn (9.7) can be derived, viz.

$$\frac{n}{N} = \exp\left(\frac{S^+ + S^-}{2}\right) \exp\left(-\frac{(E^+ + E^-)}{2kT}\right) \qquad (9.11)$$

where N is the number of lattice points (equal to the number of ion pairs). The defect consisting of an anion and a cation vacancy (separated from one another) is called a Schottky defect. In sodium chloride a concentration of Schottky defects of 10^{-4} has been found near the melting point, with a temperature dependence indicating that $(E^+ + E^-) = 2.7$ eV (see Table 9.3).

Schottky defects in sodium chloride are formed because of the entropy which they add to the crystal. Ionic crystals frequently contain point defects for a different reason. For example, crystals of ferrous oxide, which has the same crystal structure as sodium chloride, always contain a rather high concentration of cation vacancies. This is a result of the fact that some of the iron is always present in the ferric state. The extra oxygen associated with the ferric iron is accommodated in the normal oxygen ion sublattice, leaving some cation sites unoccupied. In this instance the presence of the point defect is a consequence of a departure from the stoichiometric composition FeO.[†] If the composition is Fe_xO, the concentration of cation vacancies is $(1 - x)$, and since x may be as low as 0.9 the concentration of vacancies is much greater than the equilibrium

---

[†] Point defects due to non-stoichiometry can also occur in metallic systems. For example, when Al atoms are added to the superlattice Ni–Al, which has the CsCl structure, they extend the Al sublattice, leaving Ni sites vacant.

concentration of Schottky defects in a stoichiometric crystal would be. The presence of an impurity can introduce point defects into ionic crystals in a similar way. For example, sodium chloride containing a small number of divalent impurity cations such as Ca^{2+} contains an equal number of vacant cation sites, which compensate for the extra positive charge introduced by the Ca^{2+} ions.

Although pure alkali halides normally have a precisely stoichiometric composition, it is possible to force a crystal to take up an excess of cations by heating it in the vapour of the alkali metal. The extra cations are incorporated in the normal cation sublattice, leaving a number of anion sites vacant. The electrons which maintain the electrical neutrality of the crystal are associated with the anion vacancies. By exciting these electrons, electromagnetic radiation in the frequency range of the visible spectrum is absorbed as it passes through the crystal, and the crystal becomes coloured; e.g. sodium chloride becomes a yellowish-brown. The anion vacancy with its trapped electron is called an F-centre (after the German world for colour, *Farbe*). Exposure of an alkali halide to a damaging radiation also produces F-centres, but various other defects are produced at the same time, so that the colour of an irradiated crystal is not the same as that of a crystal heated in the vapour of the alkali metal.

9.3 POINT DEFECT AGGREGATES

An interstitial atom and a vacant site of the same atom obviously reduce the internal energy of the crystal when they come together, since by doing so they annihilate one another and restore the perfect crystal. Even defects that are not complementary to one another can often lower the internal energy of the crystal by joining together. For example, if an atom next to a vacancy in an f.c.c. metal is removed so as to form a second vacancy, the number of bonds that must be broken is one less than the number that would be broken in forming the second vacancy in an isolated position. Therefore less energy is needed to form a divacancy than is needed to form two isolated vacancies, and the difference in energy is equal to the decrease in internal energy that would result from joining two vacancies together. This energy is defined as the binding energy of the vacancies. The binding energy of a pair of vacancies in gold, for example, is believed to be about 0.3 eV (see Table 9.4).

In general, the energy of formation of a divacancy may be written as ($2E_F - E_b$), where E_F is the formation energy of a single vacancy and E_b the binding energy. The equilibrium concentration of divacancies n_2 is given by

$$n_2 \propto \exp\left(-\frac{2E_f - E_b}{kT}\right) \qquad (9.12)$$

and so, using Eqn (9.7), the equilibrium numbers of vacancies and divacancies, n_1 and n_2 respectively, in a crystal having N atomic sites are related by the equation

$$\frac{n_1^2}{n_2 N} = q \exp\left(-\frac{E_b}{kT}\right) \tag{9.13}$$

The number q depends on the entropies of formation of the defects and on the number of possible divacancy positions in a crystal of N atomic sites. From Eqns (9.13) and (9.7),

$$\frac{n_1}{n_2} = q \exp\left(-\frac{S'}{k}\right) \exp\left(\frac{E_f - E_b}{kT}\right) \tag{9.14}$$

Since $E_f > E_b$, the ratio of single vacancies to divacancies increases as the temperature is lowered. Even at the melting point of an f.c.c. metal, probably fewer than 20% of its vacancies are associated.

Equation (9.14) describes the situation in a crystal that has come to equilibrium. When a crystal is cooled, it may be possible to maintain a 'local' equilibrium between vacancies and divacancies, whilst the total number of vacant sites, n_T, is unable to decrease, because of a scarcity of nearby sinks such as surfaces or dislocations. In this case, Eqn (9.13) still applies, together with

$$n_1 + 2n_2 = n_T \tag{9.15}$$

Dividing Eqn (9.15) by Eqn (9.13) gives

$$\frac{n_2}{n_1} + 2 \left(\frac{n_2}{n_1}\right)^2 = \frac{n_T}{Nq} \exp\left(\frac{E_b}{kT}\right) \tag{9.16}$$

Under these circumstances, the ratio of single vacancies to divacancies decreases as the temperature is lowered. Physically, this is because a fixed number of vacant sites has a higher energy and higher entropy when in the form of single vacancies than when in the form of divacancies.

In general, both interstitials and vacancies which are trapped at low temperatures in excess of their equilibrium number will condense into aggregates of two or more defects. Ultimately, an aggregate may grow into a form that can be observed in the electron microscope, such as a prismatic dislocation loop, or, in the case of vacancies in some f.c.c. metals, a stacking fault tetrahedron (Fig. 8.14). In the intermediate stages of aggregation when the cluster contains only a few defects, it is impossible to observe the cluster directly and it is difficult to decide theoretically which of several alternative configurations should actually be adopted. Whether or not a given configuration occurs may depend on how easily it is nucleated.

9.4 POINT DEFECT CONFIGURATIONS

The positions of the atoms around a point defect can only be observed directly under the special conditions that prevail in the field-ion microscope or in the scanning tunnelling microscope. In both cases the atoms in a surface can be imaged, and it has been possible to identify vacant sites. In general, however, it is necessary to turn to indirect experiments or to theoretical treatments.

In the theoretical treatment of point defect configurations there is an important difference between vacancies and interstitials. The basic configuration of a vacancy is known, whereas that of an interstitial defect is uncertain. In the case of a vacancy, all that needs to be determined is the displacement of the atoms surrounding the unoccupied site. In the case of an interstitial defect, the extra atom might be located at any one of a number of *non-equivalent* positions in the crystal. Several different configurations must therefore be compared in order to discover which has the lowest energy.

These problems have been attacked by treating the energy of a crystal containing the defect as the sum of the energies of interaction of pairs of atoms. A simple function is chosen to describe the dependence of the energy ϕ of two atoms on their separation r. One example is the Morse function

$$\phi = D\{\exp[-2\alpha(r - r_0)] - 2\exp[-\alpha(r - r_0)]\} \tag{9.17}$$

In Eqn (9.17), D is the dissociation energy of the pair of atoms and r_0 their equilibrium separation. The constant α can be derived from the elastic compressibility of the crystal. Other workers have used the Born–Meyer repulsive potential

$$\phi = A \exp\left[-\frac{B(r - r_0)}{r_0}\right] \tag{9.18}$$

together with a pressure applied to the boundaries of the crystal in order to hold the atoms together. Minimum energy configurations have been determined by summing the energies of pairs of atoms for a large number of trial configurations. The various authors of these calculations point out that they are based on an inexact treatment of the energy of the crystal, but that more exact treatments are difficult.

Calculated values of the displacements of the atoms around a vacancy are listed in Table 9.2. The atoms in the innermost shell move inwards, whereas those in the next shell move away from the vacancy. It should be remembered that this reversal may well be merely a consequence of the model used and that it might not occur in a real crystal. The relaxations in b.c.c. metals are considerably larger than those in f.c.c. metals. A relatively small relaxation is to be expected in a 'full' metal like copper, in which the ion cores of the atoms are close to one another, since the inward collapse of the shell of atoms next to the vacancy is very quickly arrested by the repulsive force that the atoms within that shell exert upon one another.

The form of the displacements around a vacancy in a sodium chloride crystal can be deduced by considering the electric charges involved. A positive ion vacancy, being a region deficient in positive charge, draws the surrounding cations inwards and repels the anions, which are its nearest neighbours (Fig. 9.2).

Table 9.2 Relaxation displacements around a vacancy expressed as a percentage of the normal distance from the vacant site. A positive sign indicates a displacement towards the vacancy. (After Girafalco and Weizer [3])

Metal	Shell	
	First	Second
Pb	1.42	−0.43
Ni	2.14	−0.39
Cu	2.24	−0.40
Ca	2.73	−0.41
Fe	6.07	−2.12
Ba	7.85	−2.70
Na	10.80	−3.14

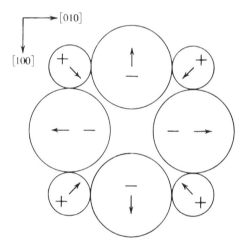

Figure 9.2 {1 0 0} plane of NaCl, showing the sense of the displacements of the ions surrounding a Na$^+$ ion vacancy

According to Mott and Littleton [2], the anions next to a sodium ion vacancy in sodium chloride are displaced outwards by $0.07d$, where d is the interionic distance (equal to half the lattice parameter). In addition to being displaced as a whole, the ions themselves are deformed and polarized by the electric field of the vacancy.

Calculations of the energies of interstitial configurations in metals have yielded some surprising results. Intuitively, one might expect that an extra atom forced into an f.c.c. metal would occupy the centre of a unit cube, where it is surrounded by atoms that lie at the corners of an octahedron. Examination of a ball model of the structure shows that this octahedral interstice is the largest hole available (see Section 3.2). However, a less symmetrical configuration is reported to have a lower computed energy [4]. In this configuration, shown in Fig. 9.3, two atoms are displaced equally from a normal lattice site, each by about $0.3a$ along $a\langle 1\,0\,0\rangle$, where a is the lattice parameter. The energy difference which favours this 'split' interstitial is small, but at least the possibility that this configuration does occur should not be overlooked. Similarly, in a b.c.c. metal an obvious site for an extra atom is the octahedral interstice at the midpoint of an edge or face of the unit cube, but the configuration that is computed to have the least energy is the $\langle 1\,1\,0\rangle$ split configuration, shown in Fig. 9.4 [5].

An interstitial defect that is similar to the split configurations described above has been postulated to occur in b.c.c. alkali metals. An extra atom is supposed to be located between two atoms in a close packed $\langle 1\,1\,1\rangle$ row, where it produces extensive relaxation outwards along the row. Paneth [6] coined the word crowdion for this defect, which consists of six atoms squeezed into a length

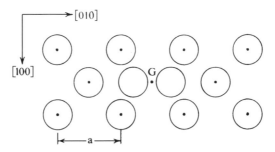

Figure 9.3 Hypothetical split interstitial in an f.c.c. metal. The plane of the diagram is {0 0 1} Reproduced by permission of the American Physical Society

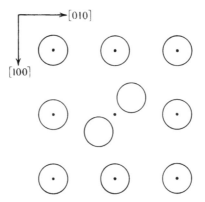

Figure 9.4 Hypothetical split interstitial in a b.c.c. metal. The plane of the diagram is {0 0 1} Reproduced by permission of the American Physical Society

normally occupied by five (Fig. 9.5). The existence of crowdions has never been established.

A defect that resembles the crowdion occurs in alkali halides that have been damaged by ionizing radiation at very low temperatures. In sodium chloride, this defect, called an H-centre, consists of a chlorine *atom* lodged in a close packed $\langle 1\,1\,0 \rangle$ row of Cl⁻ ions (Fig. 9.6). In this instance the electronic properties of the defect provide strong experimental evidence for the configuration.

When a point defect is in the course of moving from one rest position to a neighbouring one, its atomic configuration changes. The accompanying increase in internal energy determines the frequency of movement, through Eqn (9.9). Whilst these changes are difficult to calculate with precision, some interesting qualitative observations can be made. For example, when a b.c.c. metal vacancy moves, two maximum energy configurations must be overcome before it achieves its new rest position. This can be seen from Fig. 9.7, which shows that the atom which changes place with the vacancy must squeeze successively through the

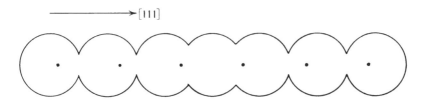

Figure 9.5 Crowdion in an alkali metal, as postulated by Paneth [6]. The line of atoms is $\langle 1\,1\,1 \rangle$ and the dots mark lattice sites

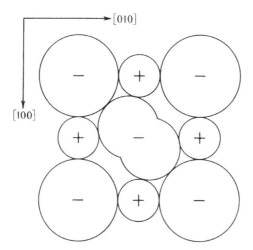

Figure 9.6 H-centre in a NaCl crystal. The plane of the diagram is {0 0 1}. (No attempt is made to depict relaxation displacements)

centres of two triangles of atoms. Between these two triangles, the atom has a little more room, so that a local energy minimum occurs at the midpoint of its jump (displacements of atoms near the vacancy from their normal sites have been neglected in drawing Fig. 9.7, which may be rather unrealistic for alkali metals where such relaxation displacements could be large). When an f.c.c. metal vacancy moves, the replacing atom need only pass through a single rectangle of atoms (Fig. 9.8). By examining a model of close packed spheres, which is appropriate for a 'full' metal like copper, it can be seen that this rectangle is forced to expand to allow the atom to pass through it. The configuration of greatest energy is expected to occur when the rectangle is fully expanded, at the midpoint of the jump.

An interstitial defect that consists simply of an extra atom lying in an interstice between normally situated atoms can move in two basically distinct ways. The extra atom may either jump into a neighbouring, equivalent, interstice or it may replace a normally situated atom which itself moves into an interstice. The replacement mechanism is called an interstitialcy jump. It is believed that interstitial Ag^+ ions move through AgBr crystals by this mechanism.

The mobility of an unsymmetrical defect can be expected to be anisotropic. A crowdion, for example (Fig. 9.5), would move very easily along its own line

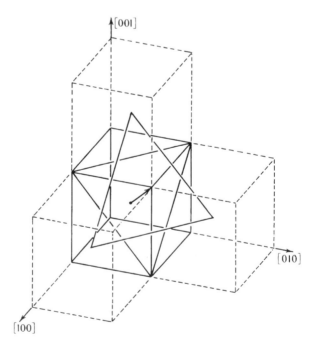

Figure 9.7   Path followed by an atom jumping into a vacant nearest neighbour site in a b.c.c. crystal

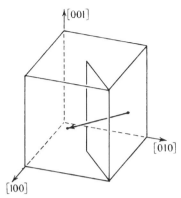

Figure 9.8 Path followed by an atom jumping into a vacant nearest neighbour site in an f.c.c. crystal

by means of a set of quite small atom movements. The mobility of simpler defects may be anisotropic because of crystallographic anisotropy. For example, a vacancy in the graphite structure (Fig. 3.1f) must move more easily within the sheet of strongly bonded carbon atoms which form the basal plane of the crystal structure than from one sheet to another.

9.5 EXPERIMENTS ON POINT DEFECTS IN EQUILIBRIUM

There are various ways of putting point defects into a crystal. The most selective method is to heat the crystal to a temperature at which its point defects can come to equilibrium. Then there may be only one type of defect with a small enough energy of formation to be present in large numbers. Point defects can be produced at low temperatures by irradiating or plastically deforming the crystal. These two methods have the drawback that they introduce several different types of defects whose effects upon the crystal and upon one another are difficult to disentangle. This is especially the case when the defects are being studied through some property of the crystal that does not discriminate between one type of defect and another; experiments that do have some power of discrimination are very valuable.

The least ambiguous experiments are those performed at high temperatures upon a crystal containing an equilibrium number of point defects. Simmons and Balluffi [7] performed experiments of this type by measuring the lengths and lattice parameters of rods of f.c.c. metals as a function of their temperatures. The principle of the experiment is as follows. Suppose that vacancies are formed as the temperature is raised. Each time a vacancy is created by means of an atom jumping out on to a surface, grain boundary or dislocation, one atomic site is added to the metal. The volume of the metal would then increase by the volume occupied by one atom in a perfect crystal were it not for the fact that the atoms around the vacancy move slightly from their normal lattice sites. This relaxation is propagated to the surfaces of the metal as an elastic strain, which causes the surfaces of the metal to move inwards. The total volume change produced by adding n vacancies to N atoms is then

$$\frac{\Delta V}{V} = \frac{n}{N} + \left(\frac{\Delta V}{V}\right)_e \tag{9.19}$$

where $(\Delta V/V)_e$ is the volume change due to elastic strain. Now it happens that the change in lattice parameter measured by X-rays, which results from the elastic strain caused by evenly distributed point defects, is simply that which would be given by the homogeneous dilatation $(\Delta V/V)_e$; i.e. the change in lattice parameter $\Delta a/a$ is given by

$$\frac{\Delta a}{a} = \frac{1}{3}\left(\frac{\Delta V}{V}\right)_e \tag{9.20}$$

The small fractional change in length of the rod, $\Delta L/L$, is given by

$$\frac{\Delta L}{L} = \frac{1}{3}\left(\frac{\Delta V}{V}\right)$$

From Eqns (9.19) and (9.20),

$$\frac{\Delta L}{L} = \frac{n}{3N} + \frac{\Delta a}{a}$$

i.e

$$\frac{n}{N} = 3\left(\frac{\Delta L}{L} - \frac{\Delta a}{a}\right) \tag{9.21}$$

Figure 9.9 shows how the changes in length and lattice parameter diverge as the temperature is raised and vacancies are generated in a rod of Al. Although very precise measurements are needed, the experiments are successful in giving a direct measure of the equilibrium concentration of vacancies in a number of f.c.c. metals; some values are listed in Table 9.3.

Equation (9.21) can evidently be used to describe the effect of any defect whose production adds or removes atomic sites. If ΔN sites are added,

$$\frac{\Delta N}{N} = 3\left(\frac{\Delta L}{L} - \frac{\Delta a}{a}\right) \tag{9.22}$$

The creation of an interstitial atom at a surface, grain boundary or dislocation destroys one atomic site, so that the measured value of $(\Delta L/L - \Delta a/a)$ would be negative if interstitial atoms were the dominant point defects.

Other studies of point defects in equilibrium have been based on measurements of electrical conductivity. This method can be applied more easily to ionic crystals

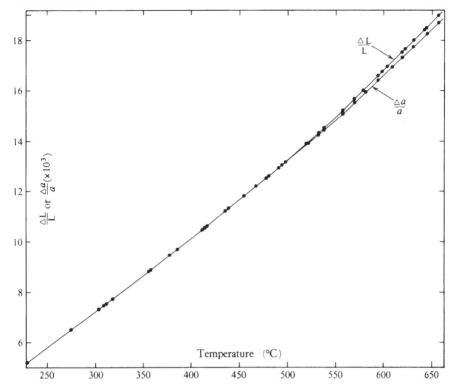

Figure 9.9 Effects of changes in length and lattice parameter with temperature (From Simmons and Balluffi [8])

Table 9.3 Fraction of sites vacant at the melting point, n/N. in equilibrium

Material	n/N at melting point	Method	Ref.
Au	7.2×10^{-4}	Length–lattice parameter	7
Ag	1.7×10^{-4}	Length–lattice parameter	9
Cu	1.9×10^{-4}	Length–lattice parameter	10
Al	9.4×10^{-4}	Length–lattice parameter	8
Pb	2×10^{-4}	Length–lattice parameter	11
Na	7.5×10^{-4}	Length–lattice parameter	12
NaCl	2.8×10^{-4}	Electrical conductivity	13
KCl	1.6×10^{-4}	Electrical conductivity	13

than to metals, since all of the conductivity of an ionic crystal may be due to the mobility of its charged point defects, whereas only a small part of the high-temperature resistivity of a metal is due to the scattering of electrons by point defects. For example, the conductivity of NaCl at high temperatures is due to Schottky defects. If an electric field of 1 V cm^{-1} drives the sodium and chlorine ion vacancies at speeds of μ^+ and μ^- cm s^{-1} respectively, then the conductivity due to a concentration c of Schottky defects, per cm^3, is

$$K = c(\mu^+ + \mu^-)e \qquad (9.23)$$

where e is the electronic charge. It is probable that in all alkali halides except the caesium salts, the cation vacancy is more mobile than the anion vacancy. The immobility of the chloride ion in NaCl has been confirmed by measuring the amounts of material transported to the anode and cathode when a current is passed through a crystal of NaCl. The fraction of the current carried by sodium ions moving to the cathode (the transport number of Na$^+$) is found to be close to unity. Consequently, Eqn (9.23) can be approximated by

$$K = c\mu^+ e \qquad (9.24)$$

Provided that the mobility μ^+ is known, the conductivity K directly measures the concentration of Schottky defects. The mobility μ^+ can be derived from the conductivity of crystals containing a known concentration c_i of a divalent cation impurity, such as Cd^{2+}. Each divalent cation introduces a sodium ion vacancy, and the total of these can be made to exceed by far the equilibrium number of vacancies in a pure crystal. The conductivity is then given by

$$K = c_i\mu^+ e \qquad (9.25)$$

In order to apply Eqn (9.25), the temperature should be sufficiently low that the equilibrium concentration of Schottky defects is small compared to c_i, but sufficiently high that few of the sodium ion vacancies are bound to divalent cations, which attract them because of their opposite charge. Figure 9.10 shows schematically how the conductivity of a slightly impure crystal changes with temperature. In region I, where Schottky defects are abundant, the conductivity is almost identical to that of a pure crystal. Here, the slope of the line gives an activation energy E_I composed of both the Schottky defect formation energy, $(E^+ + E^-)/2$, and the energy barrier to the motion of a cation vacancy, E_M^+, which enters through the mobility μ^+:

$$E_I = E_M^+ + \tfrac{1}{2}(E^+ + E^-) \qquad (9.26)$$

In region II, the conductivity is controlled by the cation vacancies introduced by the divalent impurity, and the slope in this region gives E_M^+ alone. The steepening reduction of conductivity as the temperature is lowered into region III is due to

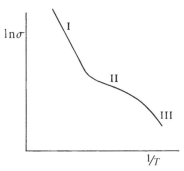

Figure 9.10 Effect of temperature on the electrical conductivity of a NaCl crystal containing a small concentration of a divalent cation (a cation fraction of 10^{-4}, say)

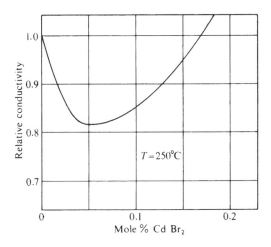

Figure 9.11 Effect of $CdBr_2$ additions on the electrical conductivity of AgBr. (After Tetlow [14])

the association of cation vacancies and impurity ions, and possibly to further clustering of defects.

A striking example of a discriminatory conductivity experiment is provided by measurements on AgBr doped with $CdBr_2$. As Cd^{2+} ions are added, the conductivity at first decreases, but then passes through a minimum as the Cd^{2+} content is increased (Fig. 9.11). This curious behaviour can be explained by supposing that the current in a pure crystal is carried mainly by interstitial Ag^+ ions. The Ag^+ ion vacancies which are introduced by the divalent Cd^{2+} ions destroy a number of Ag^+ interstitials but, at sufficiently high concentration, carry enough current themselves to make up for this loss.

9.6 EXPERIMENTS ON QUENCHED METALS

While measurements made near the melting point provide the most direct evidence for the existence of point defects in equilibrium, many data have been obtained from quenched metals. By cooling rapidly enough, point defects can be 'frozen in' at a low temperature, where they can be studied more conveniently. A practical difficulty is that the population of defects will change during the initial stages of a quench if the quench is too slow, while a quench that is too fast will produce thermal gradients severe enough to cause plastic deformation or fracture. These conflicting conditions have tended to restrict quenching experiments to metals, which are generally less susceptible to thermal shock than non-metals.

The property of a quenched metal that is most frequently studied is its electrical resistivity. This can be measured quickly and accurately, but it has the drawback of being non-discriminatory, since all lattice defects increase the resistivity of metals.

Typically, a fine wire of the metal is quenched into brine, or cold helium, where it cools at a rate of 10^4 or 10^5 °C s^{-1}. An advantage of working at liquid helium temperature is that the resistivity of the perfect crystal is then so low that most of the measured resistivity is caused by defects, whereas at room temperature quenched-in defects add only a few per cent to the resistivity. An activation energy can be derived from the resistivities of a metal quenched from various temperatures, by fitting them to the equation

$$\Delta\rho = A \exp\left(-\frac{E}{kT}\right) \tag{9.27}$$

where $\Delta\rho$ is the extra resistivity measured after quenching from a temperature T. By assuming the presence of one type of defect, which increases the resistivity in direct proportion to its concentration, it is possible to identify the activation energy E in Eqn (9.27) with the formation energy of the defect.

Measurements of this type have been made chiefly in f.c.c. metals, and in these the vacancy is almost certainly the predominant defect. Successful quenching experiments on b.c.c. metals have been rare, partly because of the ease with which b.c.c. metals pick up impurities. It is also commonly believed that the ratio of the formation energy to the motion energy of a vacancy is relatively large for b.c.c. metals, which would make it relatively difficult to retain a substantial concentration in a quench. However, some results have been obtained for b.c.c. metals, which are included in Table 9.4.

Most of the energies of motion listed in Table 9.4 were obtained from studies of the rate at which the resistivity returns to its normal value when the quenched metal is reheated. As soon as the temperature allows the point defects to migrate, they will reduce their concentration towards the equilibrium value by combining with other point defects or destroying themselves at sinks such as surfaces or dislocations. The recovery of the resistivity can be followed in various ways. In

Table 9.4 Energies of formation and migration of vacancy defects in metals and activation energies for self-diffusion: V_1 = single vacancy, V_2 = divacancy, E_b = binding energy of divacancy (values tentative)

Material	Defect	E_f (eV)	Ref.	E_M (eV)	Ref.	E_{SD} (eV)	Ref.
Ag	V_1	1.1	15	0.83	15	1.91	16
	V_2	$E_b = 0.38$	15	0.57	15		
Al	V_1	0.76	17	0.65	18	1.4	19
	V_2	$E_b = 0.17$	20	0.5	20		
Au	V_1	0.98	21	0.82	21	1.81	22
	V_2	$E_b = 0.3$	23	0.7	23		
Cu	V_1	1.14	24	1.08	25	2.04	26
	V_2			0.71	24		
Ni	V_1	1.4	27	1.5	28	2.7	29
Pt	V_1	1.51	30	1.38	30	2.89	31
Na	V_1	0.42	32			0.45	33
Mo	V_1?	2.5	34			4.0	35
W	V_1?	3.3	36	1.93	36	5.23	37
Mg	V_1?	0.89	38	0.52	38	1.39	39
Sn	V_1?	0.51	40	0.68	40	1.04	41

an isochronal anneal the specimen is heated at a constant rate or, for a short, fixed, time, at a succession of increasing temperatures; in an isothermal anneal the recovery is followed at a constant temperature. A third method is to suddenly step up the temperature after a period of isothermal recovery (Fig. 9.12). The ratio of the recovery rates at the intersection of the curves for the two temperatures, T_1 and T_2, can be used to define an activation energy by the equation

$$\frac{(d\rho/dt)_1}{(d\rho/dt)_2} = \exp\left[-\frac{E_M}{k}\left(\frac{1}{T_1} - \frac{1}{T_2}\right)\right] \tag{9.28}$$

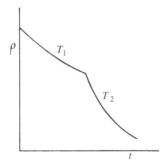

Figure 9.12 Annealing out of the quenched-in resistivity of a metal, with a sudden increase in temperature from T_1 to T_2

Equation (9.28) can easily be derived when recovery occurs by a *single* thermally activated mechanism, such as the diffusion of vacancies to sinks, since in this case

$$\frac{dc}{dt} = f(c)\exp\left(-\frac{E_M}{kT}\right)$$

and

$$\frac{d\rho}{dt} \propto \frac{dc}{dt} \qquad (9.29)$$

The function $f(c)$ depends on the type and number of sinks. The advantage of finding the activation energy E_M by suddenly changing the temperature, rather than by comparing complete curves at different temperatures, is the greater probability that the function $f(c)$ remains the same.

Unhappily, there are grounds for doubting that recovery occurs, in general, by a single mechanism. Unless their concentration is below some value, which is difficult to assign, isolated vacancies will combine to form divacancies or larger clusters during recovery. The process of recovery might then be a complex one, consisting of the combination of vacancies, divacancies and perhaps larger clusters with one another, as well as the removal of some defects at sinks. When this happens, it is difficult to interpret an experimentally determined activation energy. A touchstone in the interpretation of results for f.c.c. metals has been the principle that the motion energy E_M of an isolated vacancy obeys Eqn (9.10):

$$E_f + E_M = E_{SD}$$

This equation is based on the belief that self-diffusion in an f.c.c. metal occurs by the motion of its equilibrium vacancies. Discrepancies with Eqn (9.10) have led to the interpretation of certain results in terms of the motion of divacancies, giving the very tentative values of the energy of binding of the divacancy and of the motion of divacancies given in Table 9.4.

9.7 RADIATION DAMAGE

In the ease with which lattice defects are produced by radiation, metals can be distinguished rather sharply from non-metals. Only particles that can directly knock atoms out of their normal sites will produce lattice defects in metals. Radiation which merely excites electrons does not produce lattice defects unless the excited electrons themselves can displace atoms, as is the case with electrons excited by γ-irradiation. By contrast, many non-metals are severely damaged by an ionizing radiation such as X-rays. This is because of the greater permanence of an electronic disturbance in a non-metal. For example, in X-irradiated NaCl, a chlorine ion may become positively charged through the loss of two of its electrons. This abnormal state may persist for long enough to cause the chlorine ion to escape from its now unfavourable position by jumping into an interstitial site (the Varley mechanism).

Since a displaced atom leaves a vacant site behind it, irradiation normally introduces equal numbers of vacancies and interstitials. A method of adding only interstitials is to pummel the surface of the material with argon ions which are moving to slowly to penetrate it, but quickly enough to jolt atoms into interstitial sites [42]. An interstitial may be produced at some depth by means of *replacement collisions* in which an atom struck at the surface replaces a deeper neighbour, and so on, until finally an atom lacking the energy to replace another becomes lodged in an interstice. This process occurs most effectively along a close packed row of atoms. Even with the help of replacement collisions, interstitials can only be introduced to a depth of about 100 Å by this technique.

Fast neutrons produce damage evenly throughout specimens of ordinary size, because they travel several centimetres between collisions. When a head-on collision does occur, the struck atom is displaced with such violence that it travels for a considerable distance, perhaps as much as a thousand atom spacing, and itself displaces hundreds of other atoms. The majority of displacements occur at the end of the track of this 'primary knock-on', where they produce a very disturbed region called a *spike*. As a result of replacement collisions, interstitials are probably concentrated in the outer regions of the spike, with vacancies clustered in the centre.

A more uniform distribution of vacancy–interstitial pairs can be produced by irradiating a metal foil with electrons that have only enough energy to displace one atom. The maximum energy that an electron of energy E can transfer to an atom of mass M is given by

$$E_{max} = \frac{2E(E + 2mc^2)}{Mc^2} \tag{9.30}$$

where m is the rest mass of the electron and c the velocity of light. The energy that must be imparted to an atom in order to displace it lies between 20 and 40 eV for most metals. Comparison with the energy given by Eqn (9.30) shows that 1 MeV electrons can typically displace one or two atoms.

Despite the relatively simple nature of the damage produced by 1 MeV electrons at low temperatures, its removal by annealing is a complex process. An outstanding feature is that a great deal of recovery occurs at a remarkably low temperature. Fig. 9.13 shows the isochronal recovery of the electrical resistivity of electron irradiated copper in the range 10–80 K. This is called stage I recovery and is evidently composed of five substages. The small extra resistivity left at the end of stage I anneals out in three further stages at successively higher temperatures. Recovery in neutron-irradiated copper follows a broadly similar pattern, with the addition of a fifth stage. It is impossible to deal concisely with the problem of assigning mechanisms to the various stages, but presumably some part of stage I represents the self-annihilation of close vacancy–interstitial pairs.

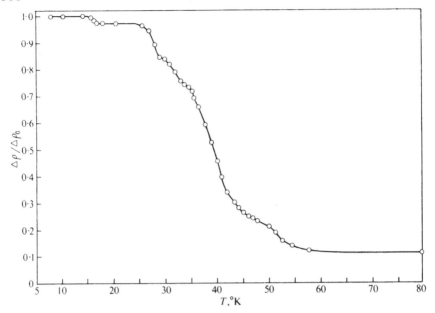

Figure 9.13 Isochronal recovery of electron irradiated copper containing an atomic concentration of Frenkel defects of about 10^{-6}. (After Corbett, Smith and Walker [43])

Stage V probably involves the removal of prismatic dislocation loops produced by the clustering of like defects. Such loops are commonly observed when heavily irradiated materials of all kinds are examined in the electron microscope.

9.8 ANELASTICITY AND POINT DEFECT SYMMETRY

The configuration of atoms about a point defect can be such that its orientation with respect to the crystal axes must be specified. For example, the interstitial defect, centred on the point G with coordinates $(\frac{1}{2}, \frac{1}{2}, 0)$ in Fig. 9.3, can lie along any of the three $\langle 1\,0\,0 \rangle$ directions. In Fig. 9.3 it lies along $[0\,1\,0]$. These three possible orientations are crystallographically equivalent. Ordinarily, the free energies of all crystallographically equivalent configurations are the same, but when a stress is applied to the crystal, their energies may become different. The defects will then tend to reorient themselves so as to increase the proportion of those with less energy. Whether a stress can distinguish between different orientations of a defect or not depends on the symmetry of the configuration of the defect. Consequently, an experiment that detects a reorientation of defects under stress has the power to discriminate between different configurations of the defect.

The best-known example of a defect reorientation induced by stress is the redistribution of interstitial carbon atoms (or nitrogen atoms) present as impurity in iron. Each carbon atom lies at the centre of a squashed octahedron of Fe atoms, as for the site marked F in Fig. 9.14. The symmetry of the atomic arrangement around F is such that the point group symmetry at F is 4/*mmm*; i.e. the site F has tetragonal symmetry. At the site marked F, the fourfold rotation axis is parallel to the z axis. Sites A and B are equivalent position to F in the unit cell and also possess point group symmetry 4/*mmm*, but their fourfold axes are parallel to the x and y axes respectively. This is very easily seen by drawing the orientations of the squashed octahedra of iron atoms forming the neighbours of sites A and B respectively. At each of these sites the two iron atoms that are closest to the carbon atoms lie along the fourfold axis, and so the carbon atom is expected to force them apart. Consequently, an interstitial at the F site will have a smaller energy than one at either the A or B sites when a *tensile* stress is applied along the z axis. Thus some of the carbon atoms in A and B sites will jump into F sites, causing the tensile strain in the z direction to increase. The rate at which this occurs will be determined by the frequency of jumping of the carbon atom. When the stress is removed, the distribution of carbon atoms will tend to become random again, because the F sites are no longer preferred, and so the strain will return to zero. If a tensile stress were applied along $\langle 1\,1\,1 \rangle$ instead of $\langle 0\,0\,1 \rangle$, the three sites A, B and F, being symmetrically disposed about this tensile axis, would remain equal in energy, and there would be no extra strain.

A strain which is time dependent but reversible is called *anelastic*. When the anelastic strain ε_a due to a small constant stress σ is produced by a single simple process like that described above, it is found that ε_a increases with time according

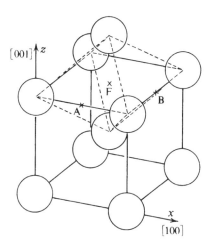

Figure 9.14 Interstitial sites occupied by C or N atoms in Fe. The dotted lines show the squashed octahedron of Fe atoms surrounding the site F

to the equation

$$\frac{d\varepsilon_a}{dt} = \frac{1}{\tau}(\varepsilon_a^\infty - \varepsilon_a) \tag{9.31}$$

This equation states that the strain ultimately approaches the value ε_a^∞. This quantity ε_a^∞ is proportional to the applied stress, so we can write

$$\varepsilon_a^\infty = \sigma\delta s \tag{9.32}$$

The constant δs is called the *relaxation of the compliance* and the quantity τ, with the dimensions of time, is called the relaxation time at constant stress. When the anelastic strain is produced by the motion of single atoms, the relaxation time is proportional to the time in which an atom makes one jump. Integrating Eqn (9.31) and adding the instantaneous elastic strain gives

$$\text{Total strain} = \sigma s + \sigma\delta s \left[1 - \exp\left(-\frac{t}{\tau}\right)\right] \tag{9.33}$$

where s is the elastic compliance and is usually at least an order of magnitude larger than δs. The behaviour is shown in Fig. 9.15.

Anelastic strain in a crystal is often detected experimentally through the viscous drag that it exerts upon any mechanical vibration of the crystal, causing such to die out. This effect is called damping or *internal friction*. The magnitude of the damping experienced by the vibration is a maximum at a certain frequency. When the frequency is very high, the anelastic strain never has time to come into play and there is no damping. On the other hand, if the period of vibration (the reciprocal of the frequency) is very large compared to the relaxation time τ defined in Eqn (9.31) or (9.33), the anelastic strain has ample time to maintain its full value, given by Eqn (9.32), and is therefore indistinguishable from the elastic strain so far as that particular vibration is concerned. Again, there is no

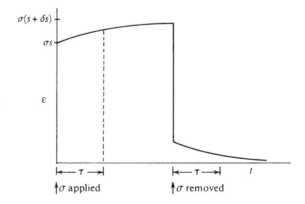

Figure 9.15 Effect of relaxation on the strain caused by a constant stress σ

damping. It can be shown that the maximum damping occurs when the period of the vibration is comparable to the relaxation time or, more precisely, when it is equal to $2\pi\tau$.

The effect of an applied stress on the number of point defects in various different orientations depends upon the strains that the defects produce in the crystal. If the macroscopic strain due to a dispersion of defects of one orientation is such that the forces applied to the crystal do work when this strain occurs, then this orientation will be favoured in comparison to orientations producing strains that do not allow the applied forces to do work. If two different orientations happen to produce identical strains (referred to fixed axes, of course), then they cannot be distinguished from one another by an applied stress of any kind. The strain that a defect can produce is not entirely arbitrary, because it must be consistent with the *symmetry of the defect*. This is defined as follows. Examine an infinite crystal containing a single defect for the presence of point group symmetry operations at the centre of the defect.[†] The group of symmetry elements that is then found defines the symmetry of the defective crystal, which for brevity can be called the *symmetry of the defect*. Evidently it is impossible for the homogeneous, macroscopic strain due to a uniform dispersion of a defect of a given orientation to be less symmetrical than the defect itself, because the symmetry of the defect is limited by the displacements of the atoms around it, which in turn determine the strain. However, the symmetry of the strain may be greater than that of the defect. A homogeneous strain is inherently centrosymmetric (Section 5.2) and so, for example, the strain produced by a dispersion of cation vacancy–anion vacancy pairs in an ionic crystal must be centrosymmetric even though the defect itself has no centre of symmetry. The limitations on the type of strain that can be produced by a defect of any given symmetry can be found by consulting Table 4.2, reading 'defect system' in place of 'crystal system'. For example, from Table 4.2 we see that the only type of strain quadric consistent with cubic symmetry is a sphere (i.e. pure dilatation). Therefore a defect of cubic symmetry can never be a source of anelasticity under homogeneous stress, because its strain quadric is the same for all orientations. To illustrate this, Fig. 9.16 shows a defect of cubic symmetry — a hypothetical tetravacancy in an f.c.c. metal. Its symmetry is $\bar{4}3m$ and it has two distinct orientations (one is shown by solid lines and the other by broken lines in Fig. 9.16). The relative numbers of these two orientations could not be changed by an applied stress of any kind. An example of a defect of lower symmetry is the carbon atom in iron, whose symmetry is that of the site of the type F (Fig. 9.14), provided that the displacements of the surrounding iron atoms conform to the site symmetry. In this case the defect point group is the tetragonal one, $4/mmm$. The strain quadric which is consistent with a tetragonal

---

[†] The centre of the defect, if it is not obvious, may be determined as being that point within the defect volume at which the number of symmetry operations is a maximum. Space group tables such as a those in the *International Tables for X-ray Crystallography* [44] are of great help here, since they list the point group symmetries at different positions in the unit cell.

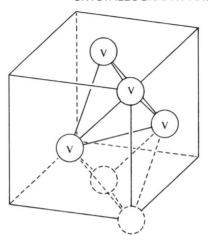

Figure 9.16 Hypothetical tetravacancy in an f.c.c. metal. The dotted lines show the alternative orientation of the tetrahedron of vacancies

defect is a quadric of revolution about the fourfold axis of the defect (Table 4.2). The three orientations of the interstitial carbon (F, A and B in Fig. 9.14) are distinguished by their differently oriented fourfold axis and so a general applied stress will distinguish between them.

Whether a *particular* stress will distinguish between defect orientations, each of which produces a different strain, depends on whether it does a different amount of work on the crystal as the strain is introduced in each case. Sometimes it is possible to pick out by inspection a special stress which cannot produce a change in defect orientation. An example, already mentioned, is the absence of an anelastic response to a tensile stress in the $\langle 1\,1\,1 \rangle$ direction of a crystal of iron containing interstitial carbon atoms. Nowick and Heller [45] have tabulated complete 'section rules' for deciding whether or not there can be an anelastic response to a given tensile stress in a crystal containing a defect of given symmetry.

PROBLEMS

9.1 If the activation energy for self-diffusion in copper is 2.04 eV and the activation energy for the migration of a vacancy is 1.08 eV, determine the ratio of the concentration of vacancies present in equilibrium at 1000 °C to the concentration present in equilibrium at 500 °C.

9.2 Experimental determinations of the concentration of vacancies in copper gave the values of 8×10^{-5} at 980 °C and 16×10^{-5} at 1060 °C. From these values, compute the energy of formation and entropy of formation of a vacancy in copper.

9.3 Compare the concentration of positive ion vacancies in a NaCl crystal due to the presence of a mole fraction of 10^{-4} of $CaCl_2$ impurity with the concentration present in equilibrium in a pure NaCl crystal at $400\,°C$. The formation energy of a Shottky defect $E^+ + E^- = 2.12$ eV and the concentration of Shottky defects at the melting point, $800\,°C$, is 2.8×10^{-4}.

9.4 A crystal of ferrous oxide, Fe_xO, is found to have a lattice parameter $a = 4.30$ Å and a density of 5.72 g cm^{-3}. What is the composition of the crystal (i.e. the value of x in Fe_xO)? State clearly the assumptions you make about the structure of the crystal.

9.5 Simmons and Balluffi observed that when a rod of silver was heated to its melting point, $960\,°C$, the fractional increase in the length of the rod exceeded the fractional increase in its lattice parameter by 5.6×10^{-5}. Assuming that the only defects present are isolated vacancies and taking the entropy of formation of a vacancy to be $1.5\,k$, calculate the formation energy of a vacancy E_f. If in fact there were divacancies present as well as vacancies, and if there were one-tenth as many divacancies as vacancies, what error would this introduce into the value of E_f calculated on the assumption that only isolated vacancies occur?

9.6 Assume that a divacancy in an f.c.c. metal consists of two vacant sites which are nearest neighbours of one another. Determine the number of possible divacancy positions in one unit cell of the f.c.c. lattice. Hence show that, in equilibrium, the number of divacancies per lattice site is given by

$$\frac{n_2}{N} = 6\exp\left(\frac{2E_f - E_b}{kT}\right)$$

Ignore any entropy which a divacancy may add to a crystal by virtue of its effect on the vibrations of the atoms in its neighbourhood.

9.7 A crystal of gold is cooled from its melting point to a temperature of $500\,°C$. Calculate the ratio of the number of vacancies to the number of divacancies in the crystal at $500\,°C$, firstly, on the assumption that complete equilibrium is maintained and, secondly, on the alternative assumption that the total number of vacant sites remains constant but that the vacancies and divacancies achieve an equilibrium amongst themselves. Take the formation energy of a vacancy to be 0.98 eV, the binding energy of a divacancy to be 0.3 eV and the ratio of the number of vacancies to the number of divacancies at the melting point, $1063\,°C$, to be 20.

9.8 Consider an atom in a close packed hexagonal metal jumping into a vacant nearest neighbour site. How many essentially different types of jump are there? In each case, sketch the ring of neighbouring atoms that must be squeezed through on the way to the vacant site and compare this with the rectangle of atoms that must be squeezed through when an atom jumps into a nearest neighbour vacancy in an f.c.c. metal (Fig. 9.8).

9.9 What is the most likely position for an interstitial atom in a crystal of silicon or germanium? Sketch and describe the shape of the ring of atoms which the interstitial atom must pass through in moving from a site of this type to a nearest neighbouring site of the same type.

9.10 A rod of gold is heated to its melting point, at which the concentration of vacancies is 7.2×10^{-4} per atomic site. The rod is then quenched so that all the vacancies generated at the melting point are trapped in the metal. Upon annealing at a slightly higher constant temperature, the rod is observed to shrink as the vacancy concentration is reduced towards its very low equilibrium value. If the total contraction is 1.1×10^{-4}, what change in lattice parameter should be observed during the annealing process?

9.11 What is the point group symmetry of the following defects:
(a) a vacancy pair consisting of an anion vacancy and a nearest neighbour cation vacancy in NaCl,
(b) a vacancy pair consisting of vacant nearest neighbour sites in a b.c.c. metal?
In each case determine by inspection the direction of a tensile stress that will not produce any anelastic strain due to the presence of the defect.

9.12 What is the multiplicity of the general form for the point group of (a) a b.c.c. crystal and (b) an interstitial atom of carbon in iron? What is the relationship between the number of orientations in which a point defect exists and the multiplicities of the general form for the point groups of the crystal and of the defect?

9.13 When a crystal of thoria, ThO_2 (fluorite structure), is doped with calcium, the Ca^{2+} ions occupy Th^{4+} sites, the lack of positive charge being compensated for by O^{2-} ion vacancies. The Ca^{2+} ion and O^{2-} ion vacancy are believed to be associated as a pair on nearest neighbour sites. Determine:
(a) the number of different orientations of the pair,
(b) the orientations that can be distinguished by an electric field applied along [1 0 0],
(c) the orientations that can be distinguished by a tensile stress applied along [1 1 0].

SUGGESTIONS FOR FURTHER READING

Agullo-Lopez, F., Catlow, C. R. A. and Townsend, P. D., *Point Defects in Material*, Academic Press, London (1988).

Cotterill, R. M. K., Doyama, M., Jackson, J. J. and Meshii, M., (eds.), *Lattice Defects in Quenched Metals*, Academic Press (1965).

Damask, A. C. and Dienes, G. J., *Point Defects in Metals*, Gordon and Breach (1963).

Flynn, B. P., *Point Defects and Diffusion*, Clarendon Press, Oxford (1972).

Girafalco, L. A., *Atomic Migration in Crystals*, Blaisdell, New York (1964).

Gittus, J., *Irradiation Effects in Crystalline Solids*, Applied Science, Barking, Essex (1978).
Mott, N. F. and Gurney, R. W., *Electronic Processes in Ionic Crystals*, Oxford (1948).

REFERENCES

1. H. B. Huntington and F. Seitz, 'Mechanism for self-diffusion in metallic copper', *Phys. Rev.*, **61**, 315 (1942).
2. N. F. Mott and M. J. Littleton, 'Conduction in polar crystals', *Trans. Farad. Soc.*, **34**, 485 (1938).
3. L. A. Girafalco and V. G. Weizer, *J. Phys. Chem. Solids*, **12**, 260 (1960).
4. R. A. Johnson and E. Brown, 'Point defects in copper', *Phys. Rev.*, **127**, 446 (1962).
5. R. A. Johnson, 'Interstitials and vacancies in α-iron', *Phys. Rev.*, **134A**, 1329 (1966).
6. H. R. Paneth, 'The mechanism of self-diffusion in alkali metals', *Phys. Rev.*, **80**, 708 (1950).
7. R. O. Simmons and R. W. Balluffi, 'Measurements on the equilibrium concentration of lattice defects in gold', *Phys. Rev.*, **125**, 862 (1962).
8. R. O. Simmons and R. W. Balluffi, *Phys. Rev.*, **117**, 52 (1960).
9. R. O. Simmons and R. W. Balluffi, *Phys. Rev.*, **119**, 600 (1960).
10. R. O. Simmons and R. W. Balluffi, *Phys. Rev.*, **129**, 1533 (1963).
11. F. M. D'Heurle, R. Feder and A. S. Nowick, *J. Phys. Soc. Japan.* **18**, Suppl., 184 (1963).
12. R. Feder and H. P. Charbnau, *Phys. Rev.*, **149**, 464 (1966).
13. R. W. Dreyfus and A. S. Nowick, *J. Appl. Phys.*, **33**, Suppl., 473 (1962).
14. J. Tetlow, *ann. Phys.*, **6**, 68 (1949).
15. M. Doyama and J. S. Koehler, *Phys. Rev.*, **127**, 21 (1962).
16. C. T. Tomizuka and E. Sonder, *Phys. Rev.*, **103**, 1182 (1956).
17. C. Panseri and T. Federighi, *Phil. Mag.*, **3**, 1223 (1958).
18. W. DeSorbo and D. Turnbull, *Phys. Rev.*, **115**, 560 (1959).
19. J. J. Spokas and C. P. Slichter, *Phys. Rev.*, **113**, 1462 (1959).
20. M. Doyama and J. S. Koehler, *Phys. Rev.*, **134**, A522 (1964).
21. J. E. Bauerle and J. S. Koehler, *Phys. Rev.*, **107**, 1493 (1957).
22. S. M. Makin, A. H. Rowe and A. D. Le Claire, *Proc. Phys. Soc. Lond.*, **70**, 545 (1957).
23. J. W. Kauffman and M. Meshii, *Lattice Defects in Quenched Metals.* (eds. R. M. K.Cotterill *et al.*), Academic Press (1965).
24. P. Wright and J. H. Evans, *Phil. Mag.*, **13**, 521 (1966).
25. F. Ramsteiner, G. Lampert, A. Seeger and W. Schüle, *Phys. Stat. Sol.*, **8**, 863 (1965).
26. A. Kuper, H. Letaw, L. Slifkin, E. Sonder and C. T. Tomizuka, *Phys. Rev.*, **98**, 1870 (1955).
27. Y. Nakamura, *J. Phys. Soc. Japan*, **16**, 2167 (1961).
28. D. Schumacher, W. Schüle and A. Seeger, *Z. Naturforsch*, **17a**, 228 (1962).
29. H. Burgess and R. Smoluchowski, *J. Appl. Phys.*, **26**, 491 (1955).
30. J. J. Jackson, *Lattice Defects in Quenched Metals* (eds. R. M. K. Cotterill *et al.*), Academic Press (1965).
31. F. Cattaneo, E. Germagnoli and F. Grasso, *Phil. Mag.*, **7**, 1373 (1962).
32. R. Feder and H. P. Charbnau, *Phys. Rev.*, **149**, 464 (1966).
33. N. H. Nachtrieb, E. Catalano and J. A. Weil, *J. Chem. Phys.*, **20**, 1185 (1952).
34. J. D. Meakin, A. Lawley and R. C. Koo, *Appl. Phys. Lett.*, **5**, 133 (1964).
35. J. Askill and D. H. Tomlin, *Phil. Mag.*, **8**, 997 (1963).

36. H. Schultz, *Acta Met.*, **12**, 761 (1964).
37. W. Danneberg, *Metall.*, **15**, 977 (1961).
38. C. J. Beevers, *Acta Met.*, **11**, 1029 (1963).
39. P. G. Shewmon and F. N. Rhines, *J. Metals*, **6**, 1021 (1954).
40. W. DeSorbo, *J. Phys. Chem Solids*, **15**, 7 (1960).
41. E. Klokholm and J. D. Meakin, *Bull. Am. Phys. Soc.*, **4**, 171 (1959).
42. P. Bowden and D. G. Brandon, 'The generation of dislocations in metals by low energy ion bombardment,' *Phil. Mag.*, **8**, 935 (1963).
43. J. W. Corbett, R. B. Smith and R. M. Walker, *Phys. Rev.*, **114**, 1452 (1959).
44. *International Tables for X-Ray Crystallography* (ed. Theo Hahn), 4th revised edn, published for the International Union of Crystallography by Kluwer academic, London (1995).
45. A. S. Nowick and W. R. Heller, 'Dielectric and anelastic relaxation of crystals containing point defects', *Adv. Phys.*, **14**, 101 (1965).

10

Twinning

10.1 INTRODUCTION

When a crystal is composed of parts that are oriented with respect to one another according to some symmetry rule, the crystal is said to be twinned. The most frequently occurring symmetry rule of twinning, but not the only one, is that the crystal structure of one of the parts is the mirror image of the crystal structure of the other part, in a certain crystallographic plane called the *twinning plane*. Often the plane of contact between the two parts, called the *composition plane*, coincides with the twinning plane.

Twinned crystals are often produced during growth, from the vapour, liquid or solid. Alternatively, a single crystal may be made to become twinned by mechanically deforming it. Twinned crystals of minerals are often found in nature, and then it is not always clear whether the twins were produced by growth or by deformation. In this chapter we shall be dealing mainly with deformation twins, but it is important to realize that this represents only a fraction of the general phenomenon of twinning.

A simple example of twinning which can be produced in a variety of ways is provided by f.c.c. metals. In this case the symmetry rule connecting the differently oriented parts is that one part is the *mirror image* of the other in a (1 1 1) plane. Because the plane of contact between the parts is usually itself a (1 1 1) plane, the structures can be shown as being mirror images of one another in the contact plane, as in Fig. 10.1b. It can be seen that nearest neighbour relations are preserved at the boundary, but that an error in the stacking of the (1 1 1) planes occurs, such that the stacking sequence ABCABC is turned into ABCBAC. It is natural to suppose that such an error might occur during the growth of the crystal. The experimental evidence is that the formation of growth twins is quite sensitive to the conditions of growth. Twins in f.c.c. crystals grown from the melt are not common, but gold films produced by vapour deposition contain a multitude of twins, as do layers of electro-deposited copper, and in either case, the faster the rate of deposition, the higher the density of twins. Twinned grains are often seen in copper or α-brass polycrystals which have been cold-worked

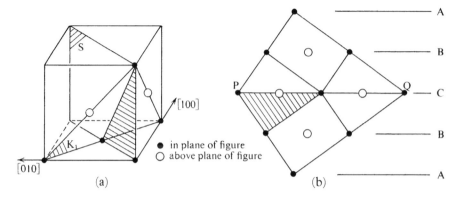

Figure 10.1 Structure of a twin in an f.c.c. metal. The plane S in (a) is the plane of the figure in (b), with the corresponding triangles in (a) and (b) shaded. The plane K_1 is the composition plane, PQ in (b)

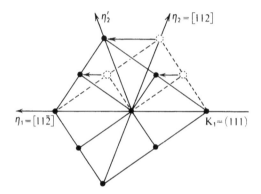

Figure 10.2 Formation of a twin by shear. The dotted lines indicate the structure above the composition plane before the shear occurs

and then heated so as to produce recrystallization.[†] In a recrystallized material, a straight-sided lamella which is in twin orientation to the rest of a grain is called an annealing twin; it can be regarded as a twin produced during crystal growth from the solid state. It is interesting that growth twins in aluminium are rare. This may be because the energy of a twin boundary in aluminium is high.

Examination of Fig. 10.2 shows that the twinned crystal could have been produced by homogeneously shearing part of a single crystal. This immediately suggests that twin formation may be a mechanism of plastic deformation. In fact,

---

[†] When a plastically deformed metal is heated to a temperature at which self-diffusion can occur, new strain-free grains nucleate amongst the deformed grains and grow so as to consume them. This process is called recrystallization.

this mechanism is not very easy to observe in f.c.c. metals, because of the relative ease with which they glide. However, Cu, Ag and Au crystals all twin prolifically during the later stages of tensile tests at low temperatures, and Cu has been found to twin when shock-loaded at ordinary temperatures. The minimum atom movements needed to accomplish twinning are shown in Fig. 10.2. A displacement of $\frac{1}{6}[1\,1\,\overline{2}]$ applied to the upper part of an originally single crystal produces a stacking fault. The same displacement applied at successively higher $(1\,1\,1)$ layers produces a twin. This could be achieved by passing a partial dislocation with the Burgers vector $\frac{1}{6}[1\,1\,\overline{2}]$ over each $(1\,1\,1)$ plane above the composition plane. The geometry of deformation twinning will be described in more general terms in the following section.

10.2 DESCRIPTION OF DEFORMATION TWINNING

Deformation by glide preserves the crystal structure in the same orientation. Deformation by twinning reproduces the crystal structure, but in a specific new orientation. Macroscopically, one can see that the deformed regions that have taken up a twin orientation with respect to the original crystal are lamellae that have undergone a homogeneous simple shear. Before proceeding further, it will be useful to enlarge upon the different levels at which deformation twinning can be described.

Firstly, one can describe the change of shape of the region which becomes a twin lamella. This purely macroscopic description gives the elements of a simple shear. The interface between the twin and the matrix is the plane that is neither distorted nor rotated by the shear. The direction and magnitude of the shear can, ideally, be obtained from the displacements due to the lamella at two surfaces of the crystal, as shown in Fig. 10.3.

Next one can describe the relationship between the orientations of the lattice in the lamella and the lattice in the matrix. Then it can be decided whether the homogeneous simple shear found from the shape change adequately describes the lattice reorientation when the shear is applied to the points of the Bravais lattice. Often, in specific cases, it is found to do so. The effectiveness of the simple shear in describing the lattice reorientation has led to a general description of deformation twins in terms of a simple shear of the Bravais lattice, and this is the treatment we are about to pursue, partly in order to introduce the nomenclature that has grown up around it.

The ultimate level of description is to describe the atom movements that occur during twinning. In contrast to the description of shape change and orientation change, description at this level is ambiguous at present. Further consideration of this level of description will be postponed until actual examples are discussed, but it may be emphasized in advance that a simple shear applied to atom positions, as distinct from lattice points, is *not* always capable of producing all the atom movements that are needed to form a twin.

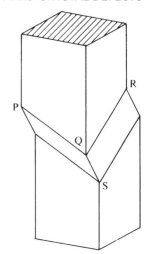

Figure 10.3 Displacements produced by a twin lamella. The traces PQ and QR define the unrotated plane of the shear (K_1), and the magnitude and direction of the shear can be determined from SQ

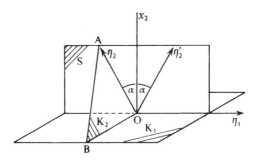

Figure 10.4

Confining ourselves, for the present, to the problem of shearing the Bravais lattice, we will ask the following question: can a homogeneous simple shear be applied to a Bravais lattice so as to change its orientation but not its structure?

Figure 10.4 illustrates the geometry of a simple shear. All points of the lattice on the upper side of a plane K_1 are displaced in the direction η_1 by an amount u_1 proportional to their distance above K_1. Thus

$$u_1 = gx_2 \tag{10.1}$$

where g is the strength of the simple shear. The plane containing η_1 and the normal to K_1 is called the plane of shear S. It can be seen that a vector parallel to η_2 in S will be the same length after the shear has been applied if the angle α which it makes with the normal to K_1 is given by

$$g = 2\tan\alpha \tag{10.2}$$

Evidently all vectors in that plane through η_2, which is normal to S, are unchanged in length, although rotated. This plane, AOB in Fig. 10.4, is conventionally labelled K_2 and called the second undistorted plane. K_1 is neither rotated nor distorted; it is called the twin plane.

So far it has not been specified whether the elements K_1, K_2, η_1, η_2 are rational or not, i.e. whether or not they pass through sets of points of the Bravais lattice. In this respect, the condition that the *lattice is to be reproduced* can be fulfilled in two ways. In the first K_1 is rational. Therefore, we can pick two lattice vectors $\mathbf{l}_1$ and $\mathbf{l}_2$ in K_1 (Fig. 10.5a). These vectors are not affected by the shear. If then η_2 is rational, there is a third lattice vector $\mathbf{l}_3$ parallel to η_2 that is unchanged in length by being sheared to $\mathbf{l}'_3$. Remembering that $-\mathbf{l}_3$ is also a lattice vector (the lattice is centrosymmetric), it can be seen that the new cell formed by $\mathbf{l}_1$, $\mathbf{l}_2$ and $\mathbf{l}'_3$ is a reflection in the twin plane K_1 of the cell $\mathbf{l}_1$, $\mathbf{l}_2$, $-\mathbf{l}_3$ in the unsheared lattice. Provided that this cell is a possible primitive unit cell of the lattice the whole lattice is reconstructed by the shear. If it is not a primitive unit cell, then only a superlattice made up of a fraction of the lattice points is necessarily reconstructed. Twins whose shear elements K_1 and η_2 are rational, while K_2 and η_1 are irrational, are called type 1 twins or reflection twins. A second possibility is shown in Fig. 10.5b. Now η_1 and K_2 are rational but K_1 and η_2 are not. Two lattice vectors $\mathbf{l}_1$ and $\mathbf{l}_2$ can be chosen in K_2 together with $\mathbf{l}_3$ in the rational direction η_1. These vectors form a cell that is reconstructed as $\mathbf{l}'_1$, $\mathbf{l}'_2$, $\mathbf{l}_3$ by the shear. Its orientation relationship to the cell $-\mathbf{l}_1$, $-\mathbf{l}_2$, $\mathbf{l}_3$ in the unsheared lattice is one of $180°$ rotation about η_1. This is called a type 2 twin or rotation twin.

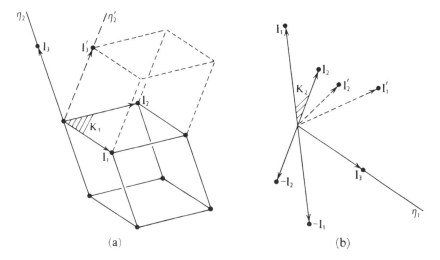

(a) (b)

Figure 10.5 (a) Type 1 twin. (b) Type 2 twin. Lattice vector $\mathbf{l}_1$ is sheared to $\mathbf{l}'_1$ and $\mathbf{l}_2$ to $\mathbf{l}'_2$. The vectors $\mathbf{l}'_1$ and $\mathbf{l}'_2$ are related to $-\mathbf{l}_1$ and $-\mathbf{l}_2$ by a rotation of $180°$ about η_1

Very commonly, all four elements K_1, K_2, η_1 and η_2 are rational and the two types merge. Then the twin may be called compound (or degenerate), which is the usual type in the more symmetrical crystal structures. In a cubic lattice, *only* compound twins are possible. To prove this, consider a type 1 twin for which K_1 and η_2 are rational. In the cubic lattice, the normal to a rational plane is itself rational, so that if K_1 is rational, then the plane of shear S is also rational because it contains two rational directions, viz. η_2 and the normal to K_1. Therefore η_1, being the intersection of two rational planes K_1 and S, is rational also. Finally, K_2 is rational because it contains two rational directions η_2 and the normal to S. A similar argument can be framed starting with K_2 rational, i.e. for a type 2 twin.

Orthorhombic uranium shows, amongst other twins, a type 1 with $K_1 = (1\,1\,2)$, $\eta_2 = [3\,1\,2]$ and a type 2 with $K_2 = (1\,1\,2)$ and $\eta_1 = [3\,1\,2]$. In this example it happens that the elements of the two twins are related by interchange of K_1 and K_2. Two twins related in this way are said to be *reciprocal* to one another. The crystallography of two such twins is distinct, but they have the same shear magnitude since the angle between K_1 and K_2 is the same for each.

A set of lattice points that forms a plane will still form a plane after the twinning shear has been applied to the lattice (Section 5.3), but except in certain special cases the pattern of lattice points in the plane will be changed. The plane therefore becomes a lattice plane of a different kind, so that its Miller indices change. It is useful to be able to write down the new Miller indices in the general case. The transformation equations for doing this were first derived by Mügge [1] and will be stated without proof. They are

$$h' = 2H(Uh + Vk + Wl) - h(UH + VK + WL)$$

$$k' = 2K(Uh + Vk + Wl) - k(UH + VK + WL) \qquad (10.3)$$

$$l' = 2L(Uh + Vk + Wl) - l(UH + VK + WL)$$

where $(h\,k\,l)$ are the Miller indices of the plane before its transformation by twinning and $(h'\,k'\,l')$ are the indices after twinning. The values of h', k' and l' given by Eqns (10.3) may contain a common factor, which must be removed to convert them to Miller indices. If the twin is of type 1, $(H\,K\,L)$ are the indices of K_1 and $[U\,V\,W]$ are the indices of η_2. The equations also apply to type 2 twins, where then $(H\,K\,L)$ are the indices of K_2 and $[U\,V\,W]$ are the indices of η_1. The indices $(h'\,k'\,l')$ are of course referred to the usual crystal axes *in the twin*;[†] these axes are in general *not* the directions into which the original crystal axes are transformed by the twinning shear. The equations for the transformation of the indices of a direction $[U\,V\,W]$ are obtained by simply interchanging $[U\,V\,W]$

[†] In the case of a type 1 twin these axes are the reflection in K_1 of the crystal axes in the matrix; in the case of a type 2 twin they are obtained from the matrix axes by a rotation of 180° about η_1.

and $(H K L)$ and substituting $[U V W]$ for $(h k l)$ in Eqns (10.3). This gives

$$u' = 2U(Hu + Kv + Lw) - u(HU + KV + LW)$$
$$v' = 2V(Hu + Kv + Lw) - v(HU + KV + LW) \qquad (10.4)$$
$$w' = 2W(Hu + Kv + Lw) - w(HU + KV + LW)$$

The transformation equations may be of value in working out how a defect is changed when the lattice is twinned, assuming that the atoms defining the defect move only according to the twinning shear.

A short way of writing down transformation equations for specific cases is to use a transformation matrix (see Section A1.4 in Appendix 1 and Appendix 4). For example, in the case of a twin in an f.c.c. crystal with $K_1 = (1\,1\,1)$, $\eta_2 = [1\,1\,2]$, we can write

$$\begin{pmatrix} h' \\ k' \\ l' \end{pmatrix} = \mathbf{Q} \begin{pmatrix} h \\ k \\ l \end{pmatrix} \qquad (10.5)$$

where the transformation matrix $\mathbf{Q}$ is

$$\mathbf{Q} = \begin{pmatrix} -1 & 1 & 2 \\ 1 & -1 & 2 \\ 1 & 1 & 0 \end{pmatrix}$$

The special planes and directions that do not change the form of their indices, when referred to the usual crystal axes in the twin, can easily be found. For example, in a type 1 twin, the plane K_1 obviously stays the same, and so does any plane in the zone of η_2 because it contains two lattice vectors that do not change, namely η_2 and the vector parallel to its own intersection with K_1. Equations (10.3) confirm this result.

10.3 EXAMPLES OF TWIN STRUCTURES

F.C.C. METALS

The structure of a twin in an f.c.c. metal has already been described (Fig. 10.1). From a study of deformation twin lamellae in Ag and Cu it has been shown that a shear of magnitude 0.707 in the $[1\,1\,\bar{2}]$ direction describes both the macroscopic shear and the lattice reorientation of a twin on a $(1\,1\,1)$ plane. This agreement is not entirely trivial, because the same lattice reorientation could be produced by a shear of double the magnitude, i.e. 1.4, in the reverse sense. For the shear of 0.707 the twinning elements are

$$K_1 = (1\,1\,1), \qquad \eta_1 = [1\,1\,\bar{2}], \qquad K_2 = (1\,1\,\bar{1}), \qquad \eta_2 = [1\,1\,2]$$

These elements are shown in Fig. 10.2. Since the structure of an f.c.c. metal consists of an atom located at each of the lattice points, the atom movements are

most compactly described by the same shear. This amounts to displacing each
$(1\,1\,1)$ layer in the twin by $\frac{1}{6}[1\,1\,\overline{2}]$ over the layer underneath. In this example,
the same simple shear describes the macroscopic shape change, the lattice reori-
entation and the most plausible atom movements needed to accomplish the twin.

B.C.C. METALS

Twinning is a relatively important mode of deformation in many b.c.c. transition metals such as Fe, V, Nb. Twinning is favoured relative to slip by a low
temperature and a high strain rate. Purity appears to be important — for instance
200 ppm of C has been found to suppress twinning in Nb at 77 K. Some solid
solution alloys, such as Mo with 30 wt % Re, twin much more freely than the
pure metal. Only one type of twin is commonly found; the elements are

$$K_1 = (1\,1\,2), \qquad \eta_1 = [\overline{1}\,\overline{1}\,1], \qquad K_2 = (1\,1\,\overline{2}), \qquad \eta_2 = [1\,1\,1]$$

and the magnitude of the twinning shear is 0.707. This twin is depicted in
Fig. 10.6. As in the case of f.c.c. metals, it has been confirmed that the macroscopic shape change agrees with that predicted from the twinning element [2].
The atom movements described by this shear are quite plausible. They corre-
spond to a shift of $\frac{1}{6}[\overline{1}\,\overline{1}\,1]$ on successive $(1\,1\,2)$ planes. In forming the b.c.c.
structure, from horizontal $(1\,1\,2)$ layers, six layers are stacked before the seventh
falls vertically on top of the first; the stacking sequence can be written as
ABCDEF AB.... . The passage of a single partial dislocation with the $\frac{1}{6}[\overline{1}\,\overline{1}\,1]$

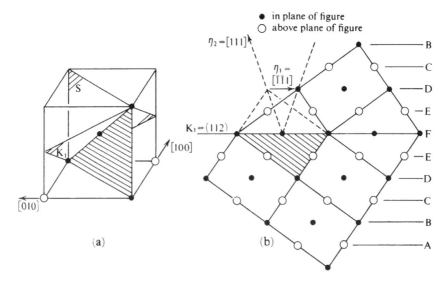

Figure 10.6   Twin in a b.c.c. metal. The scheme of the figure is the same as that
of Fig. 10.1

Burgers vector produces a stacking fault of the form ABCDEFEF AB . . . (Section 8.6). Little is known about the energy of such a fault, except that it is not so low as to permit extended dislocations to be readily observed in the electron microscope. The passage of $\frac{1}{6}[1\,\bar{1}\,1]$ partials on successive planes would produce the twin sequence ABCDEFEDCB . . . .

SPHALERITE (ZINC-BLENDE)

We now turn to structures in which there is more than one atom per lattice point. Twinning in sphalerite, cubic ZnS, is an interesting example. The lattice is f.c.c. and there are two atoms associated with each lattice point, typified by Zn at $(0, 0, 0)$ and S at $(\frac{1}{4}, \frac{1}{4}, \frac{1}{4})$ (Fig. 3.1j). The Zn and S atoms lie on interleaved f.c.c. sublattices and twins in this structure are closely related to twins in f.c.c. metals. The probable structure at a twin–matrix interface is shown in Fig. 10.7, which should be compared with Fig. 10.1. The twinning shear is identical to that shown in Fig. 10.1, namely $K_1 = (1\,1\,1)$, $\eta_1 = [\bar{2}\,1\,1]$, $K_2 = (\bar{1}\,1\,1)$, $\eta_2 = [2\,1\,1]$. The movement of the Zn atoms could be described by this shear, with the plane xx' as origin. The same shear would not place the S atoms correctly, however. The simplest way of producing the necessary atom movements is to apply the twinning shear to *layers* of ZnS. Each layer consists of a sheet of Zn atoms lying on top of a sheet of S atoms, the sheets being parallel to $(1\,1\,1)$. In Fig. 10.7 the Zn atoms are shown connected to the S atoms directly beneath them in the same layer. It can be seen that the structure, as distinct from the lattice, is *not* mirrored in a K_1 plane. A mirrored structure would change the $[1\,1\,1]$ direction to $[\bar{1}\,\bar{1}\,\bar{1}]$ in the twin — two distinct directions in this non-centrosymmetric structure. It has been confirmed that this does not happen and that the structures of twin and matrix are

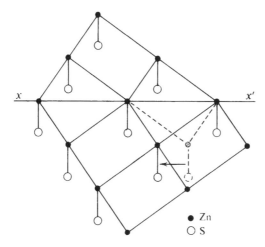

Figure 10.7 Twin in sphalerite

related by a rotation of 180° about the normal to K_1, as shown [3]. The structure at the twin–matrix interface can be described as a fault in the stacking of the ZnS layers described above. The stacking sequence of these layers in sphalerite is the same as the stacking sequence of close packed planes in an f.c.c. metal, viz. ABCA.... At the twin boundary the stacking sequence is ABCBA... , which contains three layers in the sequence of the wurtzite structure, BCB... .

CALCITE

Deformation twinning in calcite crystals has long been recognized and is quite easy to produce. For example, large pieces of a cleaved rhombohedron can be forced into twin orientation by pressing a knife into the proper edge. Referred to the cleavage cell (see Fig. 3.14), the elements of the twinning shears are

$$K_1 = (1\,0\,1), \qquad K_2 = (0\,1\,0), \qquad \eta_1 = [0\,1\,0], \qquad \eta_2 = [1\,0\,1]$$

and the magnitude of the shear is 0.694. Figure 10.8 shows the structure of the twin. The twinning shear moves the Ca^{2+} ions and the centres of the CO_3^{2-} ions to their correct positions in the twin, but leaves the plane of the triangular CO_3^{2-} group at the wrong angle. To complete the reconstruction of the crystal in the twin, each CO_3^{2-} group must be rotated through 52.5° about an axis through its centre and normal to the plane of shear.

HEXAGONAL METALS

Twinning in hexagonal metals is of great technical importance, because the limited nature of the common slip modes in these metals makes twinning a

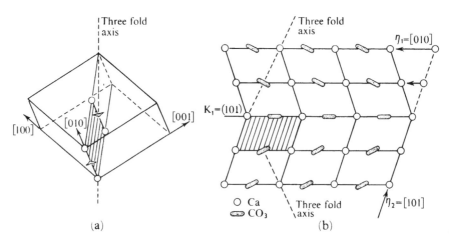

Figure 10.8 Twin in calcite. The scheme of the figure is the same as that of Fig. 10.1

necessary component of their deformation. Many types of twinning are exhibited; to some extent the type can be related to the *c/a* ratio of the metal (see Table 10.1). Broadly speaking, the lower the *c/a* ratio, the greater the variety of modes exhibited. All hexagonal metals twin by the mode

$$K_1 = (1\,0\,\bar{1}\,2), \qquad \eta_1 = [\bar{1}\,0\,1\,1], \qquad K_2 = (1\,0\,\bar{1}\,2), \qquad \eta_2 = [1\,0\,\bar{1}\,1]$$

The magnitude and sense of the shear varies with the *c/a* ratio, but is always small, ranging from 0.175 for Cd (*c/a* = 1.89) to −0.186 for Be (*c/a* = 1.57). Figures 10.9 and 10.10 show (1 0 $\bar{1}$ 2) twins in Zr and in Zn respectively. The sense of the shear in Zn is opposite to the sense of the shear in Zr. It can be seen that the twinning shear is not capable of describing the atom movements. In this case there is no obvious, simple way in which the atom movements can be described. Arrows have been drawn connecting an original position of an atom with that atom in the twin which is closest to it after it has been displaced by the twinning shear. There is no evidence whatever that the atoms actually follow these paths. In Fig. 10.9, only the atoms at P, Q and R are carried directly to their new positions by the twinning shear. A movement that must be added to the twinning shear to describe the displacement of an atom to a position in the twinned crystal is usually termed a *reshuffle*. When assumed atom movements are formally split up into a twinning shear followed by reshuffles, the reshuffles required will depend on the position of that K_1 plane chosen as the origin of the shear. It should be evident that, physically, twinning in hexagonal metals is not well understood.

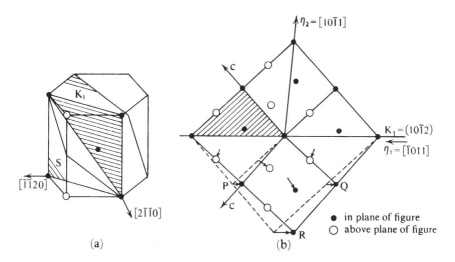

Figure 10.9 The (1 0 $\bar{1}$ 2) twin in zirconium. The scheme of the figure is the same as that of Fig. 10.1

Table 10.1 The twinning elements of various crystals

Material	Lattice	K_1	η_1	K_2	η_2	Magnitude of shear, g
Fe, V, Nb, W, Mo, Cr	B.c.c.	$1\,1\,2$	$\bar{1}\,\bar{1}\,1$	$1\,1\,\bar{2}$	$1\,1\,1$	0.707
Cu, Ag, Au	F.c.c.	$1\,1\,1$	$1\,1\,\bar{2}$	$1\,1\,\bar{1}$	$1\,1\,2$	0.707
Cd	Hex. $c/a = 1.886$	$1\,0\,\bar{1}\,2$	$\bar{1}\,0\,1\,1$	$1\,0\,\bar{1}\,2$	$1\,0\,\bar{1}\,1$	0.171
Zn	Hex. $c/a = 1.856$	$1\,0\,\bar{1}\,2$	$\bar{1}\,0\,1\,1$	$1\,0\,\bar{1}\,2$	$1\,0\,\bar{1}\,1$	0.140
Co	Hex. $c/a = 1.623$	$1\,0\,\bar{1}\,2$	$\bar{1}\,0\,1\,1$	$1\,0\,\bar{1}\,2$	$1\,0\,\bar{1}\,1$	−0.130
		$1\,1\,\bar{2}\,1$	$1\,1\,\bar{2}\,6$	$0\,0\,0\,1$	$1\,1\,\bar{2}\,0$	0.614
Mg	Hex. $c/a = 1.623$	$1\,0\,\bar{1}\,2$	$\bar{1}\,0\,1\,1$	$1\,0\,\bar{1}\,2$	$1\,0\,\bar{1}\,\bar{1}$	−0.130
		$1\,0\,\bar{1}\,1$	$\bar{1}\,0\,1\,2$	$\bar{1}\,0\,1\,3$	$3\,0\,\bar{3}\,2$	0.137
Re	Hex. $c/a = 1.615$	$1\,1\,\bar{2}\,1$	$1\,1\,\bar{2}\,6$	$0\,0\,0\,1$	$1\,1\,\bar{2}\,0$	0.621
Zr	Hex. $c/a = 1.592$	$1\,0\,\bar{1}\,2$	$\bar{1}\,0\,1\,1$	$1\,0\,\bar{1}\,2$	$1\,0\,\bar{1}\,\bar{1}$	−0.169
		$1\,1\,\bar{2}\,1$	$1\,1\,\bar{2}\,6$	$0\,0\,0\,1$	$1\,1\,\bar{2}\,0$	0.63
		$1\,1\,\bar{2}\,2$	$\bar{1}\,\bar{1}\,2\,3$	$1\,1\,\bar{2}\,4$	$\bar{2}\,\bar{2}\,4\,3$	0.225
Ti	Hex. $c/a = 1.587$	$1\,0\,\bar{1}\,2$	$\bar{1}\,0\,1\,1$	$1\,0\,\bar{1}\,2$	$1\,0\,\bar{1}\,1$	−0.175
		$1\,1\,\bar{2}\,2$	$\bar{1}\,\bar{1}\,2\,3$	$1\,1\,\bar{2}\,4$	$\bar{2}\,\bar{2}\,4\,3$	0.218
Be	Hex. $c/a = 1.568$	$1\,0\,\bar{1}\,2$	$\bar{1}\,0\,1\,1$	$1\,0\,\bar{1}\,2$	$1\,0\,\bar{1}\,1$	−0.199
Graphite	Hex.	$1\,1\,\bar{2}\,1$	$\bar{1}\,\bar{1}\,2\,6$	$0\,0\,0\,1$	$1\,1\,\bar{2}\,0$	0.367
Calcite, $CaCO_3$	Trig. R Cleavage cell, p. 116	$1\,0\,1$	$0\,1\,0$	$0\,1\,0$	$1\,0\,1$	0.694
Sapphire, Al_2O_3	Trig. R. Indices as in Table 6.1	$\bar{1}\,0\,1\,2$	$\bar{1}\,0\,\bar{1}\,\bar{1}$	$\bar{1}\,0\,\bar{1}\,4$	$\bar{2}\,0\,2\,1$	0.202
Bi	Trig. R	$1\,0\,1$	$0\,1\,0$	$0\,1\,0$	$1\,0\,1$	0.118
β-Sn	Tetrag. I	$3\,0\,1$	$\bar{1}\,0\,3$	$\bar{1}\,0\,1$	$1\,0\,1$	0.119
In	Tetrag.	$1\,0\,1$	$1\,0\,\bar{1}$	$\bar{1}\,0\,1$	$1\,0\,1$	0.150
U	Ortho C	$1\,3\,0$	$3\,\bar{1}\,0$	$1\,\bar{1}\,0$	$1\,1\,0$	0.299
		I	$3\,1\,2$	$1\,1\,2$	I	0.228
		$1\,1\,2$	I	I	$3\,1\,2$	0.228
		$1\,2\,1$	I	I	$3\,1\,1$	0.329
Plagioclase feldspars $NaAlSi_3O_8$ $CaAl_2Si_2O_8$	Triclinic P	$0\,1\,0$	I	I	$0\,1\,0$	0.151
		I	$0\,1\,0$	$0\,1\,0$	I	0.151

Sources. Most of the data tabulated above may be found in the works listed as further reading at the end of this chapter. For Re see Jeffery and Smith, [5]. For Al_2O_3 see Heuer [6].

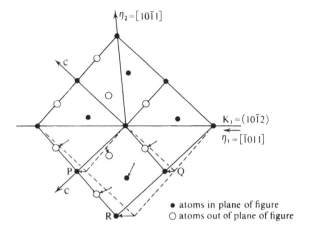

Figure 10.10   The $(10\bar{1}2)$ twin in zinc

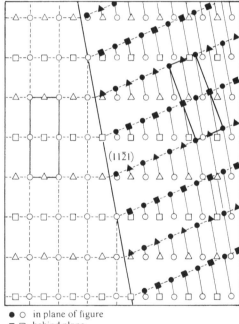

Figure 10.11   Plan view of the $(1\bar{1}00)$ plane in graphite showing the structure produced by applying a twinning shear to the atoms. The outlined cells show that a new structure is produced and that reshuffles must therefore be added to produce the actual structure within the twin. Reproduced by permission of the Royal Society from Freise and Kelly *Proc. Roy. Soc.*, **A264**, 269 (1961)

GRAPHITE

Twinning in graphite offers a beautiful illustration of the distinction that may exist between, on the one hand, the twinning shear, which describes the macroscopic shape change and lattice reorientation, and, on the other, a physically satisfying description of the atom movements. The elements of the twin are

$$K_1 = (1\,1\,\bar{2}\,1), \qquad \eta_1 = [\bar{1}\,\bar{1}\,2\,6], \qquad K_2 = (0\,0\,0\,1), \qquad \eta_2 = [1\,1\,\bar{2}\,0]$$

and the magnitude of the shear is 0.367. The twinning shear reconstructs the Bravais lattice, but not the crystal structure, as shown in Fig. 10.11. Atom reshuffles would be required if a description of the atom movements were based on the twinning shear. Further, any description of a mode of generation based on dislocations moving across $(1\,1\,\bar{2}\,1)$ planes would not be satisfactory, because it would require a disruption of the very strong bonds within the basal plane. These difficulties are elegantly resolved by a description in terms of shear on the basal planes $\{0\,0\,0\,1\}$. A shear displacement of alternate basal planes by $\frac{1}{3}[1\,0\,\bar{1}\,0]$ and

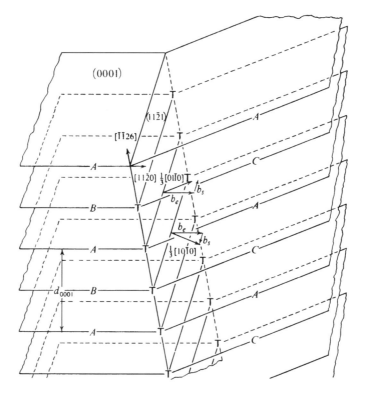

Figure 10.12 The formation of a twin in graphite by a partial dislocation on each basal plane Reproduced by permission of the Royal Society, Freise and Kelly *Proc. Roy. Soc.*, **A264**, 269 (1961)

$\frac{1}{3}[0\,1\,\bar{1}\,0]$ fully describes the atom movements if accompanied by an appropriate rigid body rotation. Partial dislocations with these Burgers vectors are known to exist on the basal planes of graphite. A wall of such alternating partial dislocations at the matrix–twin interface constitutes a tilt boundary, which provides the required rotation (see Section 12.2). The lateral growth of a twin lamella can occur by the glide of the wall of partials into the matrix. This convincingly accounts for experimental observations of twin movements in graphite; the model is illustrated by Fig. 10.12.

10.4 TWINNING ELEMENTS

The twinning elements of various crystals are presented in Table 10.1. Only some of the twinning elements need to be experimentally determined, e.g. K_1, η_1 and the magnitude of the shear, or K_1 and η_2. The remaining elements then follow from the geometry of a simple shear (Fig. 10.3). In the case of the plagioclase feldspars the crystal structure in the twin may not be quite identical with that in the matrix. Cases of this sort should perhaps be classified as pseudo-twins or even martensitic transformations (Section 11.1).

From Table 10.1 it can be seen that crystals of the same structure tend to twin in the same mode. The reason for the choice of mode in a given case is not very well understood. Consideration of Section 10.2 shows that the number of possibilities to choose from is enormous. To define a twinning shear that completely reconstructs the lattice, any three lattice vectors that form a primitive unit cell can be chosen and then from these two must be chosen to define K_1 (type 1) or K_2 (type 2). A commonsense criterion of selection is to make the magnitude of the shear as small as possible, so as to reduce the size of the atom movements needed to form the twin. At first sight, this criterion might not appear to be appropriate, since the shears in Table 10.1 cover a wide range and some are quite large, but when the possibilities for a given structure are carefully examined, it is found that a minimum shear criterion is in fact quite effective [7]. The criterion favours low index twinning planes for type 1 or compound twins. The reason for this can be seen from Fig. 10.13. The higher the index of the plane K_1, the smaller is the separation between K_1 planes. Since the vector l_3, parallel to η_2, cannot be less than the smallest vector of the lattice, the angle that it makes with K_1 must decrease as the spacing of the K_1 planes decreases. This implies that the shear must increase. More precisely, if the vector l_3 defines a lattice point in the nth K_1 plane above O and d is the separation of the K_1 planes,

$$l_3^2 = (nd)^2 + (\tfrac{1}{2}gnd)^2 \qquad (10.6)$$

where g is the magnitude of the shear. If $\mathbf{E}$ is the smallest lattice vector,

$$l_3^2 \geq \mathbf{E}^2 \qquad (10.7)$$

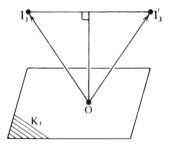

Figure 10.13

and

$$\frac{g^2 + 4}{\mathbf{E}^2} \geq \frac{4}{n^2 d^2}$$

As an illustration of the use of this inequality, consider twinning in the b.c.c. lattice. Any twin in a cubic lattice must be compound; therefore η_2 is rational and Fig. 10.13 applies. Setting $\mathbf{E}^2 = 3a^2/4$, where a is the lattice parameter, and searching for twins with $g \leq 1$ gives

$$\frac{3a^2}{n^2 d^2} \leq 5 \qquad (10.8)$$

In order to secure the reconstruction of the whole lattice or a reasonable fraction of the lattice n should be small. There is no solution for $n = 1$. For $n = 2$ the planes $(1\,1\,0)$, $(1\,0\,0)$ and $(1\,1\,2)$ satisfy the inequality. For example,

$$d_{100} = \frac{a}{2} \qquad (10.9)$$

so that, with $n = 2$,

$$\frac{3a^2}{n^2 d_{100}^2} = 3 \qquad (10.10)$$

However, planes $(1\,1\,0)$ and $(1\,0\,0)$ are rejected as trivial solutions, since they are mirror planes of the lattice (i.e. $g = 0$). The $(1\,1\,2)$ plane is therefore selected as K_1. The direction of the shear on K_1 may be chosen by projecting the crystal structure on to K_1 and selecting the smallest translation needed to set up a mirror image relationship in a K_1 plane. This leads to $\eta_1 = [\bar{1}\,\bar{1}\,1]$ and the twin depicted in Fig. 10.6.

In structures where there is more than one atom per lattice point, such as the hexagonal metals, the twinning shear is not capable of describing all the atom movements that are needed to construct the twin. The choice of mode may still be rationalized in some cases by considering the effects of a shear applied homogeneously to atoms or *small groups* of atoms. A shear is sought that is small, and which demands only a small, plausible reshuffling of the atoms to complete the twin structure, after it has been applied. This shear is then identified with the twinning shear. Some reshuffles are inherently more likely than others; e.g. the rotation of CO_3 radicals in calcite is obviously a more probable physical occurrence than a reshuffle involving the disruption of a CO_3 radical.

Observations of very unusual twinning modes, such as $\{3\,0\,\bar{3}\,4\}$ bands in Mg, may perhaps be explained by the operation of more than one simple twinning mode in the same region of crystal. There is little direct evidence for this interpretation, although in many metals doubly twinned regions are observed where one twin lamella is intersected by another.

The relative importance of twinning as a mode of deformation can sometimes be changed by adding an alloying element. Mo–Re alloys have already been mentioned; another example is that of Be in Fe. Alloys containing 15–30 at % Be, quenched to preserve a solid solution, deform almost entirely by twinning, even at slow strain rates. This alloy exhibits another interesting effect. At the composition Fe_3Be an ordered structure can be formed (Table 3.3). Figure 10.14 shows the effect of the usual b.c.c. twinning shear on this structure. There is a striking feature: the structure in the 'twin' is not the same as that in the matrix. In the Fe_3Be structure the Be atom has Fe atoms as both nearest and next-nearest neighbours, while the twinning shear gives each Be atom two Be atoms as next-nearest neighbours. This shear should perhaps be termed a pseudo-twin because

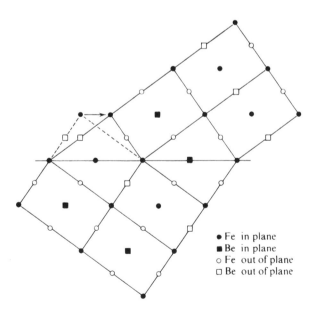

- ● Fe in plane
- ■ Be in plane
- ○ Fe out of plane
- □ Be out of plane

Figure 10.14 A twin, or 'pseudo-twin', in ordered Fe_3Be. The view is of the plane of shear; the K_1 plane is normal to the figure

of the change of structure. The energy to be gained by reverting to the original structure provides a driving force for the reverse shear, and crystals have been found to recover strains as large as 10% by 'untwinning' when the force on them is removed [8].

10.5 THE MORPHOLOGY OF DEFORMATION TWINNING

When a crystal has deformed by twinning, the twinned regions are usually in the form of plates parallel to K_1. Sometimes the plate is very thin and the twin is a lamella whose faces are accurately parallel to K_1. This is common in non-metals such as calcite or graphite where lamellae only a few micrometres thick may run right through the crystal. Iron also twins in this fashion, producing what are called Neumann bands. Measuring the orientations of the intersections of one lamella with two crystal surfaces immediately gives the twin plane K_1, which is accordingly the easiest of the twinning elements to determine. The occurrence of perfect lamellae seems to be associated with the difficulty of deforming the matrix by slip. In metals that deform readily by slip, relatively short slabs of twin with a somewhat irregular interface are common. Twins in calcite assume a shorter, thicker and less regular form at higher temperatures when slip can occur. If a rigid twin is embedded in a perfectly rigid matrix, and if it is everywhere firmly bonded to that matrix, then the only possible interface is the undistorted, unrotated plane K_1. The twin would be a lamella, parallel to K_1, extending right through the matrix. (Possible twin intersections are neglected for the moment.) For a twin of any other form, the matrix has to accommodate itself to the shape change of the twinned region. If the accommodation required is small enough, it may be obtained by elastic strain. Under this condition, a finite lamella must taper to an edge at its sides, and be lens-shaped. The elastic strain field can be represented by an appropriate array of dislocations such as that shown in Fig. 10.15. The Burgers vector b and the vertical spacing of the dislocations d are together defined by the twinning shear g:

$$g = \frac{b}{d} \qquad (10.11)$$

For example, an appropriate model for an f.c.c. twin consists of a twinning partial of the Burgers vector $\frac{1}{6}[1\,1\,\bar{2}]$ on every (1 1 1) plane. The interface contains no dislocations when it is parallel to K_1 and it is then called *coherent*. The boundary becomes incoherent when it deviates from K_1 and its slope is defined by the spacing of dislocations in the K_1 plane. The thickness of the lamella at any point, h, is given by

$$h = nd \qquad (10.12)$$

where n is the number of dislocations between the point in question and the tip of the lamella. The inclination θ of the interface to the twin plane is therefore

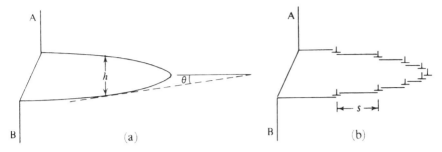

Figure 10.15 (a) Twin lamella intersecting a surface AB. (b) Dislocation model of the same lamella

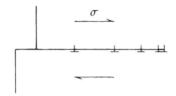

Figure 10.16 Dislocation model of a thin twin lamella

given by

$$\tan \theta = \frac{d}{s} \tag{10.13}$$

where s is the spacing of the dislocations in one interface.

If the lamella is thin and tapered, a pile-up of dislocations on a single plane will represent the stress field adequately at distances large compared to the thickness of the lamella. This model is shown in Fig. 10.16. The shear stress due to a pile-up of n screw dislocations at sufficiently large distances from the head of the pile-up is given by

$$\sigma = \frac{\mu n b}{2\pi r} \tag{10.14}$$

From Eqns (10.11) and (10.12),

$$\sigma = \frac{\mu h g}{2\pi r}$$

The product hg determines the magnitude of the accommodation stresses and strains in this case, and also in the more general case of mixed dislocations where both tensile and shear strains are produced. At the tip of a twin that has been blocked by some obstacle, the tensile strain may be large enough to nucleate a crack. Cracks of this type have been seen in b.c.c. metals, for which g is large ($g = 0.707$).

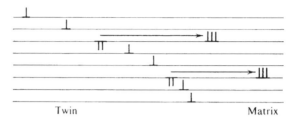

Figure 10.17 Emissary dislocations. The dislocations shown by a single line are $\frac{1}{6}[\bar{1}\,\bar{1}\,1]$ twinning dislocations, the triple lines represent $\frac{1}{2}[\bar{1}\,\bar{1}\,1]$ perfect dislocations and the double lines $\frac{1}{3}[1\,1\,\bar{1}]$ partials

If there are no lattice friction forces opposing the motion of the dislocations in Fig. 10.16, they will run back to the surface when the applied stress is removed. Behaviour of this sort has been observed in calcite, where small twins nucleated by indentation disappear when the load is removed. This phenomenon is called elastic twinning. More generally, twinning is not reversible and twins remain in a crystal after it has been unloaded. The reason is often that accommodation has occurred by slip, relieving the stresses at the edge of the twin or, in terms of the model of Fig. 10.16, neutralizing the net Burgers vector of the pile-up. Under these conditions, blunt twin plates with quite irregular interfaces are possible. An interesting feature of accommodation by slip in b.c.c. metals is that the slip is sometimes concentrated in a band stretching far ahead of the lamella. A model for this process, due to Sleeswyk [9], is shown in Fig. 10.17. Every third twinning dislocation in the incoherent boundary dissociates into a perfect slip dislocation and a complementary partial. The perfect dislocations, called emissary dislocations, can glide ahead of the twin, carrying the twinning shear macroscopically but not twinning the lattice. The long-range stress field of the twin boundary is thereby completely relieved, as can be seen from the fact that the Burgers vectors of its dislocations sum to zero.

Accommodation by slip is usually required at the intersection of two twin lamellae. In rare instances the shears can be matched by twinning alone. In b.c.c. metals this is possible, because for a single shear direction [1 1 1] there is a choice of three {2 1 1} K_1 planes. Zig-zag lamellae are sometimes seen, composed alternately of the twins $K_1 = (2\,\bar{1}\,1)$ and $K_1 = (\bar{1}\,2\,1)$ with common $\eta_1 = [\bar{1}\,\bar{1}\,1]$, for example.

Detailed dislocation mechanisms for the nucleation and growth of a twin lamella have been mainly based on the idea illustrated in Fig. 10.18. A dislocation PQ has a Burgers vector such that it creates a twin by gliding over successive planes parallel to the twin plane K_1. This effect is produced by the intersecting dislocation PP' whose Burgers vector has a component normal to the twin planes which is equal to their spacing. The twin planes are therefore turned into a spiral ramp on which the twinning dislocation PQ glides. The dislocation PP' about

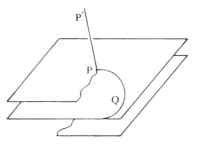

Figure 10.18   Pole mechanism for the growth of a twin

which PQ spirals is called the *pole*. In an f.c.c. crystal PP' may be an AC dislocation on the plane ACD while PQ is a Shockley partial αC, gliding on the plane BCD (see Section 8.3). The Shockley partial may be produced by the dissociation of AC on plane BCD:

$$AC \rightarrow \alpha C + A\alpha$$

The Frank partial Aα cannot glide in the twin plane BCD and so only the Shockley partial αC spirals around the pole.

Some care is needed in specifying the overall deformation of a crystal due to the formation of twins within it. The largest strain that can be obtained from one specific twin is of course the twinning shear itself, which is obtained when the entire crystal has become twinned. However, an unlimited amount of strain can often be obtained by multiple twinning. For example, if part of an f.c.c. twin lamella with $K_1 = (1\,1\,1)$, $\eta_1 = [1\,1\,\overline{2}]$ is retwinned on the same twin plane $(1\,1\,1)$, with $\eta_1 = [\overline{1}\,2\,\overline{1}]$, the orientation of the doubly twinned region becomes the same as that of the original matrix, since the twinning shears add up to the slip displacement $\frac{1}{2}[0\,1\,1]$. In principle, a very large strain could be built up by the repeated application of twinning shears.

In a tensile test, the formation of a lamella of a given twin will contribute to the overall elongation of the specimen provided that the tensile axis lies within the quadrant defined by η_1 and the normal to K_1 (Fig. 10.4) or, in other words, provided that the shear stress on the twin plane K_1, resolved in the direction of the twinning shear η_1, is positive. It should be noted that this condition is not the same as the condition for the tensile axis to be lengthened by the conversion of the entire crystal into its twin [10], because within the twinned region, any vector originally lying in the obtuse sector between K_2 and K_1 is lengthened, while only vectors lying in the acute sector, of angle $(\pi/2 - \alpha)$, are shortened (Fig. 10.4). When the tensile axis lies in the sector of angle α between K_2 and the plane normal to η_1, the formation of a twin *lamella* is inhibited, even though the tensile force would do work on the crystal if it could be *completely* twinned.

PROBLEMS

10.1 One of the ways of describing the orientation relationship of twin and matrix in a type 1 twin is as a tilt about an axis in the twinning plane K_1 and normal to the shear direction η_1. Determine the magnitude of the angle of tilt for the case of a twin in an f.c.c. metal.

10.2 Draw a plan of the K_1 plane in an f.c.c. metal showing two layers of atoms with the positions of the atoms in the upper layer before and after they are sheared to form the first layer of a twin.

10.3 Repeat problem 10.2 for the case of a twin in a b.c.c. metal.

10.4 A twin in a b.c.c. metal has $K_1 = (1\,1\,2)$, $\eta_1 = [\bar{1}\,\bar{1}\,1]$. Determine the Miller indices of the planes into which the following lattice planes are transformed by the twinning shear: $(0\,0\,1)$, $(0\,1\,0)$, $(1\,0\,0)$. Hence write down the matrix for the transformation of planes and determine the plane into which the $(1\,0\,1)$ plane transforms.

10.5 Show that if crystal axes are chosen non-conventionally, with Ox and Oy in the K_1 plane of a type 1 twin and Oz parallel to η_2, then the transformation of any plane $(h\,k\,l)$ is given by $h' = -h$, $k' = -k$, $l' = l$.

10.6 Which lattice rows remain rows of the same type after the lattice has suffered the twinning shear of a type 1 twin? By inspection, or otherwise, determine the lattice directions which the directions $[0\,0\,1]$, $[0\,1\,0]$ and $[1\,0\,0]$ transform into as a result of the shear of a twin in a b.c.c. metal of $K_1 = (1\,1\,2)$, $\eta_2 = [1\,1\,1]$. Hence write down the matrix for the transformation of directions and determine the direction that $[1\,\bar{1}\,1]$ transforms into.

10.7 Twin lamellae parallel to $\{1\,1\,1\}$ planes have been observed in germanium. Sketch the probable structure of this twin.

10.8 For a $(1\,0\,\bar{1}\,2)$ twin in magnesium (see Table 10.1), determine the angle between the c axis in the twin and the c axis in the matrix.

10.9 Supposing that the hexagonal metals could deform only by basal slip or $(1\,0\,\bar{1}\,2)$ twinning (not true of course), divide the metals into those that could be elongated but not compressed parallel to the c axis and those that could be compressed but not elongated in this direction.

10.10 Is the observed twinning mode in f.c.c. metals the mode with the smallest shear (the lattice being completely reconstructed by the shear)?

10.11 The copper–gold alloy containing equal numbers of copper and gold atoms has a superlattice structure such that $(0\,0\,1)$ planes are composed alternately of copper and gold. Show that the passage of a dislocation of Burgers vectors $\frac{1}{2}[1\,0\,1]$ (the indices refer to the f.c.c. lattice of the disordered structure) over successive $(1\,0\,1)$ planes produces a twinned superlattice with $(1\,0\,1)$ as the twin plane.

10.12 The surface of a certain crystal of zinc is a basal plane, $(0\,0\,0\,1)$. If this surface has been interested by $\{0\,1\,\overline{1}\,2\}$ twins (Table 10.1), at what angle do you expect the surface within a twin lamella to be tilted with respect to the surface of the surrounding matrix? If the measured angle of tilt turned out to be less than the expected value, what explanation would you suggest?

10.13 A rod-shaped single crystal of iron has the pole of the $(1\,1\,2)$ plane (which is also the composition plane of the twin) at an angle of $30°$ to the axis of the rod, which lies in the plane defined by the $(1\,1\,2)$ pole and $[\overline{1}\,\overline{1}\,1]$. The crystal is deformed in a tensile test by twinning on the $(1\,1\,2)$ plane over one-third of its length. What is the tensile elongation if the ends of the rod are free to move laterally (i.e. the tensile axis does not rotate in the untwinned part of the crystal)?

10.14 Zinc crystals twin by shear on the plane $(1\,0\,\overline{1}\,2)$ in the direction $[1\,0\,\overline{1}\,1]$. Draw the arrangement of the atoms in the $(1\,\overline{2}\,1\,0)$ plane for several unit cells. This plane is normal to $(1\,0\,\overline{1}\,2)$ and contains the shear direction of the twin. Draw in the trace of one of the $(1\,0\,\overline{1}\,2)$ planes. Suppose the part of your drawing on one side of the trace of the $(1\,0\,\overline{1}\,2)$ plane represents the matrix crystal. Move the atoms situated precisely at the lattice points on the other side of the trace to the lattice points of the twin formed by reflection in $(1\,0\,\overline{1}\,2)$. Find graphically the magnitude of the shear involved and check that it is equal to

$$\left(\frac{c^2}{a^2} - 3\right)\frac{a}{\sqrt{3}c}$$

Note carefully from your drawing that the same shear applied to atoms situated originally at points $(\frac{2}{3}, \frac{1}{3}, \frac{1}{2})$ will not reproduce the crystal structure exactly but that additional displacements of these atoms are required.

SUGGESTIONS FOR FURTHER READING

Bevis, M. and Crocker, A. G., 'Twinning shears in lattices', *Proc. Roy. Soc.*, **A304**, 123–134 (1968).

Bevis, M. and Crocker, A. G., 'Twinning modes in lattices', *Proc. Roy. Soc. Lond.*, **A313**, 509–529 (1969).

Cahn, R. W., 'Twinned crystals', *Adv. Phys.*, **3**, 363 (1954).

Nabarro, F. R. N. (ed.), *Dislocations in Solids*, Vol. 9, *Dislocations and Disclinations*, North Holland (1992).

Pabst, A., 'Transformation of indices in twin gliding', *Bull. Geol. Soc. Am.*, **66**, 897 (1955).

Reed-Hill, R. E., Hirth, J. P. and Rogers, H. C. (eds.), 'Deformation twinning', *Metallurgical Society Conferences*, Vol. 25, Gordon and Breach, New York (1965).

Yoo, M. H. and Wuttig, M. (eds.), *Twinning in Advanced Materials*, TMS, Warrendale, Pennsylvania (1994).

REFERENCES

1. O. Mügge, *Neues Jahrb. Mineral. Geol.*, **11**, 98 (1889).
2. H. W. Paxton, 'Experimental verification of the twin system in alpha-iron', *Acta Met.*, **1**, 141 (1953).
3. M. J. Buerger, 'The plastic deformation of the minerals', *Am. Mineral*, **13**, 35 (1928).
4. E. J. Freise and A. Kelly, *Proc. Roy. Soc.*, **A264**, 269 (1961).
5. R. A. Jeffery and E. Smith, *Phil. Mag.*, **13**, 1163 (1966).
6. A. H. Heuer, *Phil. Mag.*, **13**, 379 (1966).
7. M. A. Jaswon and D. B. Dove, 'The crystallography of deformation twinning', *Acta Cryst.*, **13**, 232 (1960).
8. G. F. Bolling and R. H. Richman, 'Continual mechanical twinning', *Acta Met.*, **13**, 709 (1965).
9. A. W. Sleeswyk, 'Emissary dislocations: theory and experiments on the propagation of deformation twins in α-iron', *Acta Met.*, **10**, 705 (1962).
10. F. C. Frank and N. Thompson, 'On deformation by twinning', *Acta Met.*, **3**, 30 (1955).

11

Martensitic Transformations

11.1 INTRODUCTION

When a steel is quenched from a temperature of, say, 950 °C, at which it is austenite, a face-centred cubic phase, it transforms abruptly to a very hard b.c. tetragonal phase called martensite. Because of its enormous technical importance, this transformation has been studied thoroughly. Its two most striking characteristics are shared by many other transformations, which are accordingly now classified as 'martensitic'. These characteristics are, firstly, that the transformation takes place very rapidly, implying that long-range diffusion plays no part in the transformation, and, secondly, that the shape of a transforming region alters. The transformation of austenite to martensite begins at a certain temperature M_s which decreases as the carbon content of the steel increases. When the steel is quenched to a temperature below M_s, a certain fraction of it transforms almost instantaneously to martensite. There is then little increase in the amount of martensite formed until the temperature is lowered further. The change of shape of a region that transforms can easily be detected as a distortion of the surface of the steel. The shape change is very significant, because it implies that the iron atoms have moved in a regular, systematic way in order to build the new structure.

Some transformations possess one of the above characteristics, but not the other. For example, when a Cu–24 at % Ga alloy is cooled moderately quickly, it transforms rapidly from a b.c.c. phase to a hexagonal phase of the same composition. A sharp interface moves rapidly through the b.c.c. phase, leaving transformed material in its wake. Clearly, no long-range diffusion is involved. However, no change of shape occurs, which suggests that the new structure is not built by systematic atom movements, but rather by atoms moving across a disordered interface in an irregular fashion. This conclusion is reinforced by the fact that a single grain of the new phase can extend across grain boundaries of the old, showing that a single orientation of the new phase can be produced from randomly different orientations of the old. Such transformations are distinguished both from martensitic transformations and from transformations that require diffusion.[†]

Transformations that do require diffusion, which disqualifies them from the title of martensitic, sometimes exhibit a shape change. An example is provided by a Cu–4 wt % Be alloy. The high-temperature f.c.c. phase in this alloy can be preserved at room temperature by quenching. When the alloy is aged, a Be-rich precipitate forms. As a result of the shape change accompanying this precipitation, an alloy crystal may become very distorted.

We shall define transformations that are both distortive and diffusionless as martensitic. We shall discuss only the crystallographic aspects of such transformations.

11.2 GENERAL CRYSTALLOGRAPHIC FEATURES

A crystal that has partly undergone a martensitic transformation often looks somewhat like a crystal containing twins. Plates of the new phase intersect the surface in lines that lie in a few well-defined directions. The surface within the plate is tilted with respect to the rest of the surface, just as it would be in a twin lamella. Even the clicking noise characteristic of twinning may be heard during a martensitic transformation. These observations suggest that the crystallography of martensitic transformations should be approached in the same way as that of twinning. Thus, the question could be asked: can a homogeneous simple shear transform the lattice of the old phase into the lattice of the new phase? In most cases, including the transformation in steels, it is found that a more complex strain than a simple shear is needed to describe the lattice change. The crystallographic theory of a martensitic transformation is concerned with finding a plausible strain that will carry some or all of the lattice points of the old phase into positions where they form part or all of the new lattice. This strain must be consistent with the observable features of the transformation, which will now be defined more carefully.

Assume that a crystal has partly transformed and that the transformed regions are in the form of plates, as is often the case. The most easily measured parameter is the orientation of the plane of the faces of the plate, which is normally expressed as a plane in the matrix (austenite) lattice. This is called the *habit plane*, and it is found in the same way as the composition plane of a twin (Fig. 10.3). The relationship between the orientation of the matrix lattice and the lattice in the plate can be found by getting X-ray reflections from both. The most difficult quantity to determine is the macroscopic strain undergone by the material in the plate. The displacement of fiduciary lines can be used to evaluate

---

[†] They are sometimes called massive transformations, but unfortunately the term 'massive' has also been used by some persons to describe a certain type of martensitic transformation in which the martensite appears in blocks rather than as thin plates.

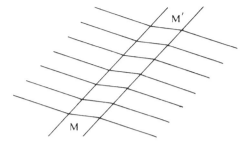

Figure 11.1 Scratched surface intersected by a martensite plate *MM'*

this strain. A corner of the specimen (as in Fig. 10.3), scratches on the surface or slip lines that existed in the matrix before the transformation can all be used as fiduciary lines. Figure 11.1 shows the deflection of scratches in a typical case. The straightness of the scratches within the plate demonstrates that the strain is homogeneous, at least on the macroscopic scale. The fact that the scratches in the matrix remain straight as they approach the plate shows that the matrix around the plate has not been deformed inhomogeneously. This in turn implies that the plate has not been rotated into its present position because this would severely deform the matrix. The continuity of the scratches across the interface implies that the interface, or habit plane, is not distorted. If it were, the scratches would no longer match up at the interface. A plane that is left unrotated and undistorted is called an *invariant plane*. A strain that leaves a certain plane unrotated and undistorted can be described by the following equation for the displacement **u** of a point of position vector **r**:

$$\mathbf{u} = \mathbf{d}(\mathbf{r} \cdot \mathbf{h})$$

where **h** is the unit normal to the invariant plane and **d** is a constant vector in the direction of the displacement. The quantity $\mathbf{r} \cdot \mathbf{h}$ is the perpendicular distance from the point of position vector **r** on to the invariant plane passing through the origin. When **d** is parallel to the invariant plane, the strain is a simple shear. This special case is shown in Fig. 6.7. In general, an invariant plane strain can be resolved into a simple shear given by the component of **d** normal to **h**, coupled with an extension or compression normal to the invariant plane, given by the component of **d** parallel to **h**. To determine **d** in magnitude and direction experimentally, *two* non-parallel fiduciary lines are required.

Although few precise determinations of the macroscopic strain in a martensite plate have been made, and although martensite does not always form well-defined, regular plates, it is generally assumed that the characteristic strain in bulk specimens is at least approximately an invariant plane strain. The belief in an invariant plane strain is undoubtedly reinforced by the consideration that when this type of strain occurs in a very thin plate that is embedded in a matrix, the matrix does not itself have to undergo any strain in order to make way for the strain suffered

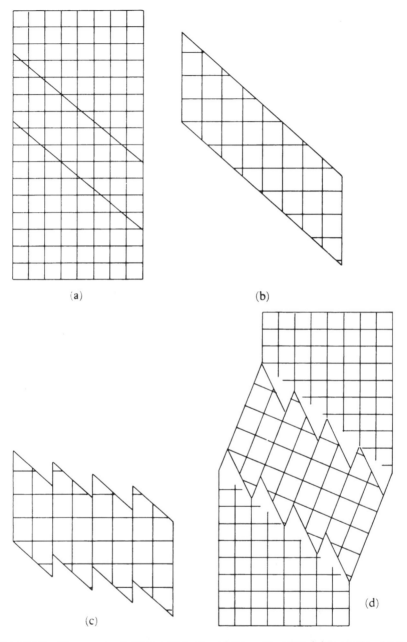

(a)   (b)

(c)   (d)

Figure 11.2 The square lattice within the plate outlined in (a) is strained into a rectangular lattice in (b), lengthening the upper and lower edges of the plate. In (c), slip within the plate has shortened its edges so that macroscopically they match the untransformed matrix, as shown in (d)

by the plate, or, in other words, it has no difficulty in *accommodating* the strain, except at the edges of the plate.

With the understanding that the macroscopic strain is one in which the habit plane is invariant, one may ask whether any such strain, applied *homogeneously* to the points of the matrix (austenite) lattice, would convert this lattice into the martensite lattice. The answer is usually that it would not. The inevitable conclusion is that the lattice strain is *not* the same as the macroscopic strain. Therefore the strain can be homogeneous only on a macroscopic scale; on a fine scale it must be inhomogeneous. In some cases the inhomogeneity can be directly observed, as when the martensite plate can be seen to be *internally twinned*. In other cases the inhomogeneity cannot easily be detected. In cases where the type of inhomogeneity is not known, the theoretical approach has been first to postulate a plausible *homogeneous* strain which *will* transform the matrix lattice into the martensite lattice and then to postulate an *added* deformation in the form of slip or twinning, which preserves the martensite lattice but alters the macroscopic strain and brings it to the invariant plane strain which is observed. This treatment is illustrated, in two dimensions, by Fig. 11.2. The test of the theory is that the final macroscopic strain must be identical with the observed strain and that the orientation relationship between the martensite and matrix lattices must be predicted correctly.

The limitations of such a theory should be obvious. The actual paths of the atoms are not predicted. The theory predicts only the final atom positions, given the initial ones. In principle, this prediction can be tested directly: either the atoms do end up in the specified positions or they do not. In practice, it is only possible to test the theory indirectly, typically through the habit plane and orientation relationship. This leaves open the possibility that some other strain, based on different postulates, would also be compatible with the observed features of the transformation. There is no absolute guarantee of uniqueness.

The general considerations of this section will now be illustrated by a more detailed discussion of a number of specific transformations. Although the transformation in steels is by far the most important case, the complexity of this particular transformation makes it easier to develop the details of a crystallographic theory in other cases. This consideration has governed the choice of the transformations to be discussed in the next three sections; however, the section on transformations in steels can be read on its own without difficulty.

11.3 TRANSFORMATION IN COBALT

Cobalt is f.c.c. at high temperature; on cooling, it transforms at $420\,°C$ to a c.p.h. structure. The crystallography of this transformation is particularly simple, since it merely produces a change in the stacking sequence of close packed planes from ABCAB ... to ABABA .... This change can be achieved by passing a partial dislocation of Burgers vector $\frac{1}{6}\langle 2\,1\,1\rangle$ across *every other* {1 1 1} plane. If each $\frac{1}{6}\langle 2\,1\,1\rangle$ displacement is in the same $\langle 2\,1\,1\rangle$ direction, there accumulates a macroscopic simple shear of magnitude g given by

$$|\tfrac{1}{6}\langle 2\,1\,1\rangle| = g \times 2d_{(1\,1\,1)} \tag{11.1}$$

where $d_{(1\,1\,1)}$ is the spacing between $(1\,1\,1)$ planes. Therefore,

$$g = \sqrt{2}/4 \tag{11.2}$$

The simple shear on the $(1\,1\,1)$ plane leaves this plane unrotated and undistorted, and so the $(1\,1\,1)$ is expected to be the habit plane.

It is interesting to see how this result might be described more formally. It is obvious that no homogeneous strain, applied to the atoms, can turn an f.c.c. structure into an h.c.p., because the atoms lie at the points of a space lattice in the f.c.c. structure whereas only half of them do so in the h.c.p. structure. Any homogeneous strain of the f.c.c. structure would generate a new lattice with one atom at each lattice point. However, a homogeneous simple shear of magnitude $\sqrt{2}/4$ in the $\langle 2\,1\,1\rangle$ direction on the $\{1\,\bar{1}\,\bar{1}\}$ plane does generate the correct hexagonal lattice from *half* the points of the f.c.c. lattice. Additional atom movements, producing no change of shape, are needed to complete the structure change.

Experimental study of the transformation has been hindered by the difficulty of observing partially transformed cobalt crystals. Striations parallel to the traces of $\{1\,1\,1\}$ are observed after transformation, suggesting that $\{1\,1\,1\}$ is the habit

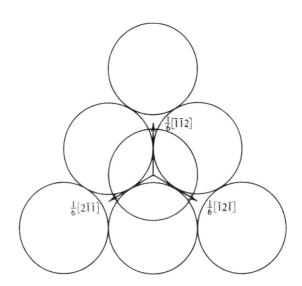

Figure 11.3 Reprinted from Gaunt and Christian, *Acta Met.* **7**, 529 (1959), with permission from Elsevier Science

plane, as predicted. By alloying with Ni, the transformation temperature may be reduced so that some cubic phase is retained at room temperature. On such specimens, the close packed planes and directions of the two phases have been found to be parallel to one another, with high precision (better than $\pm\frac{1}{2}^{\circ}$) [1]. However, the shape change is quite irregular, and well-defined uniform tilts of macroscopic areas of surface are seldom observed. This can be accounted for by assuming that the direction of the $\langle 2\,1\,1\rangle$ displacement varies amongst the three possibilities, which exist for any given $\{1\,1\,1\}$ plane, as shown in Fig. 11.3. If the three possibilities were employed equally overall, then the macroscopic shear would disappear entirely.

It is observed that a very small decrease in volume accompanies the transformation of f.c.c. cobalt to c.p.h. cobalt. Expressed as a fraction, the volume change is -3.6×10^{-3}. It follows that the glide of $\frac{1}{6}\langle 2\,1\,1\rangle$ partial dislocations does not by itself quite complete the transformation, because it does not provide this change of volume.

11.4 TRANSFORMATION IN ZIRCONIUM

Zirconium and titanium both undergo a transformation from a high-temperature b.c.c. phase, β, to a c.p.h. structure, α. Lithium has a similar transformation at about 70 K.

The first step in considering these transformations is to compare the b.c.c. and c.p.h. lattices, in order to pick out a small strain that will convert the one into the other. Figure 11.4 shows the b.c.c. lattice with a distorted hexagonal cell picked out in it. The $(0\,1\,1)_\beta$ forms the basal plane of the distorted hexagonal cell, while the close packed $\langle 1\,\bar{1}\,1\rangle$ directions in this plane correspond to the close packed $\langle 1\,1\,\bar{2}\,0\rangle_\alpha$ directions. This accounts for only four of the six $\langle 1\,1\,\bar{2}\,0\rangle$ directions in the hexagonal lattice. The other two $\langle 1\,1\,\bar{2}\,0\rangle$ directions are derived from an edge of the b.c.c. cell, i.e. from $[1\,0\,0]$ and $[\bar{1}\,0\,0]$ in Fig. 11.4. In the particular case of zirconium, the dimensions of the hexagonal cell can be brought to their correct values by contracting the b.c.c. lattice by 10% along $[1\,0\,0]_\beta$, which becomes $[2\,\bar{1}\,\bar{1}\,0]_\alpha$, expanding by 10% along $[0\,1\,\bar{1}]_\beta$, which becomes $[0\,1\,\bar{1}\,0]_\alpha$, and expanding by 2% along $[0\,1\,1]_\beta$, which becomes $[0\,0\,0\,1]_\alpha$. This strain we will call the pure lattice strain, S_{ij}. Referred to the directions $[1\,0\,0]_\beta$, $[0\,1\,\bar{1}]_\beta$ and $[0\,1\,1]_\beta$ as x, y, z axes respectively gives

$$S_{ij} = \begin{pmatrix} -0.10 & 0 & 0 \\ 0 & 0.10 & 0 \\ 0 & 0 & 0.02 \end{pmatrix} \qquad (11.3)$$

The only justification for choosing the particular correspondence between the b.c.c. and c.p.h. lattices shown in Fig. 11.4 as a basis for a theory of the transformation is that the strain **S** given by Eqn (11.3) is quite small. Any number of

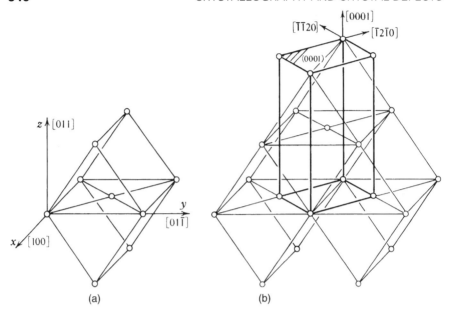

Figure 11.4 (a) Unit cell of the b.c.c. lattice, drawn with (0 1 1) horizontal. (b) Distorted hexagonal cell picked out of the b.c.c. lattice, oriented as in (a)

other correspondences could be postulated because any cell in the b.c.c. lattice containing two lattice points can be chosen and strained until the points at its corners form one of the possible unit cells of the hexagonal lattice. (Because the b.c.c. structure has one atom per lattice point while the h.c.p. structure has two atoms per lattice point, half of its atoms have to be reshuffled after the homogeneous lattice strain has been applied, in order to complete the structure change.) In practice, all but a few correspondences are entirely implausible, because of the large strain that they demand, and it is possible to pick out the most plausible correspondence by inspection. A lattice correspondence can be conveniently represented by a matrix $\mathbf{C}$. If $[p\,q\,r]$ is a vector in the original lattice which becomes the vector $[u\,v\,w]$ in the transformed lattice,[†] (see Section A-1.4 in Appendix 1 and Appendix 4), then

$$\begin{bmatrix} u \\ v \\ w \end{bmatrix} = \mathbf{C} \begin{bmatrix} p \\ q \\ r \end{bmatrix} \tag{11.4}$$

where $\mathbf{C}$ is a 3×3 matrix. By substituting for $[p\,q\,r]$ in turn the vectors $[1\,0\,0]$, $[0\,1\,0]$ and $[0\,0\,1]$, it can be seen that the columns of $\mathbf{C}$ are respectively the

[†] The vector $[p\,q\,r]$ is referred to the usual crystal axes in the original lattice and the vector $[u\,v\,w]$ to the usual crystal axes in the martensite lattice.

vectors that these particular vectors become in the transformed lattice. For the correspondence illustrated in Fig. 11.4,

$$C = \begin{pmatrix} 1 & \frac{1}{2} & -\frac{1}{2} \\ 0 & 1 & -1 \\ 0 & \frac{1}{2} & \frac{1}{2} \end{pmatrix} \tag{11.5}$$

where vectors in the hexagonal lattice are written in the three-index system.

Having chosen a lattice correspondence, the next step is to compare the strain that it predicts with the observed strain of a transformed region of crystal. The effect of the pure lattice strain can be illustrated by showing how it would deform a spherical crystal. Any homogeneous strain deforms a sphere into an ellipsoid (Section 5.3). A unit sphere of b.c.c. zirconium oriented as shown in Fig. 11.5 and strained in accordance with the lattice correspondence given above becomes the ellipsoid

$$\frac{x^2}{0.90^2} + \frac{y^2}{1.10^2} + \frac{z^2}{1.02^2} = 1 \tag{11.6}$$

All the vectors whose lengths are not changed by the strain are given by the intersection of this ellipsoid with the original sphere, i.e. with

$$x^2 + y^2 + z^2 = 1 \tag{11.7}$$

The curve of intersection is shown in projection on the xy plane in Fig. 11.5. Vectors from the origin to this line define a *cone* of elliptical cross-section, and not a plane. However, in the actual transformation it is observed that one plane *does* remain undistorted on the macroscopic scale, namely the habit plane. In the complete theory, the next step is to formally add a slip or twinning deformation

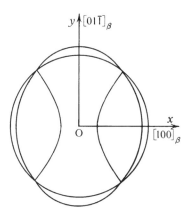

Figure 11.5　A section through a sphere of zirconium and the ellipsoid developed from the sphere by the pure lattice strain. The section is normal to $[0\,1\,1]_\beta$

which of course leaves the lattice unchanged but alters the macroscopic strain so as to produce a plane that is undistorted on the macroscopic scale.

In the case of zirconium, the 2% strain along $[0\,1\,1]_\beta$ is small enough to make an approximate treatment worth while. If this principal strain were zero, the sphere and the ellipsoid would touch at the z axis, and the unlengthened vectors would lie in a pair of planes, as shown in Fig. 11.6. In fact, the necessary and sufficient condition for there to be an undistorted plane is that one of the principal strains should be zero and the other two of opposite signs. Because this condition is nearly satisfied for zirconium, the amount of slip or twinning that needs to be added to the pure lattice strain to produce an invariant plane strain is quite small and rough agreement with the observed features of the transformation can be obtained by taking the strain S'_{ij} to be the lattice transformation strain, where

$$ S'_{ij} = \begin{pmatrix} -0.1 & 0 & 0 \\ 0 & 0.1 & 0 \\ 0 & 0 & 0 \end{pmatrix} \tag{11.8} $$

This strain would be correct if the $[0\,1\,1]_\beta$ lattice vector were exactly equal in length to the c dimension of the hexagonal unit cell. The strains $\varepsilon_x = -0.1$ and $\varepsilon_y = 0.1$ correctly deform the $(0\,1\,1)_\beta$ lattice plane into the basal plane of the hexagonal cell. Because ε_x and ε_y are equal and opposite and small, the strain S' is close to being a pure shear (Section 5.3). Now when the proper rotation is added to a pure shear, the net effect is equivalent to a simple shear. The plane on which this simple shear occurs is neither rotated nor distorted, so this plane must be the habit plane. In the present case, because the magnitude of the (approximate) pure shear is small, the amount of rotation is small and, depending on the sense of the rotation, the plane on which the simple shear occurs is close

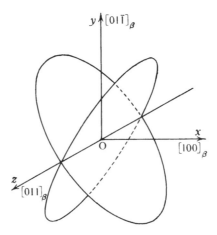

Figure 11.6 Undistorted planes of the strain S'

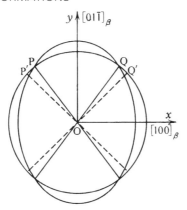

Figure 11.7 Rotation suffered by the undistorted planes of the strain S'

to one or other of the planes lying at an angle of $45°$ to $(1\,0\,0)_\beta$ and containing $[0\,1\,1]_\beta$.

To be more precise, the rotation must be taken into account. Figure 11.7 shows the effect of the strain S' on a sphere, viewed along the z axis (it corresponds to Fig. 11.5). Although the planes OQ, OP have not been distorted by the strain, they have been rotated from their initial positions OQ', OP'. Therefore to produce an unrotated, as well as undistorted, habit plane, a rotation about the z axis must be added to the pure lattice strain S' in order to return one of these planes to its initial position (OQ to OQ' for example). The *total* lattice strain is then an invariant plane strain which can be identified directly with the observed strain.

We will choose the plane passing through OQ' as the habit plane (Fig. 11.7) and compute the angle between this plane and $[1\,0\,0]_\beta$, which is parallel to the x axis. Let Q' be the point (x, y). The coordinates of Q are then found by adding the displacement $\mathbf{u}$ due to the strain S'. This is given by the matrix multiplication

$$\begin{pmatrix} u_x \\ u_y \\ u_z \end{pmatrix} = S' \begin{pmatrix} x \\ y \\ 0 \end{pmatrix} \tag{11.9}$$

whence

$$u_x = -0.1x$$

$$u_y = 0.1y \tag{11.10}$$

Therefore Q is the point $(0.9x, 1.1y)$. The ratio x/y is now given by the condition that OQ' = OQ, or

$$x^2 + y^2 = (0.9x)^2 + (1.1y)^2$$

Therefore

$$x/y = 1.05$$

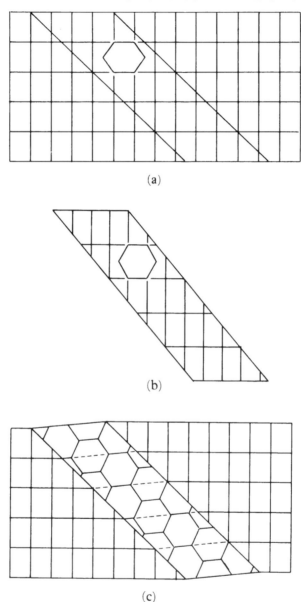

(a)

(b)

(c)

Figure 11.8 Approximate crystallography of a plate of martensite in titanium or zirconium. In (b) the lattice strain is applied to produce the basal plane of the hexagonal structure from the {0 1 1} plane of the b.c.c. structure. After a small rotation, the transformed plate fits into the matrix as in (c). In an actual transformation the pure strain and rotation would occur together rather than in the two distinct stages shown here

The habit plane therefore makes an angle $\tan^{-1} 1.05 (\approx 46.5°)$ with the cube plane of the β phase $(1\,0\,0)_\beta$ and contains the $[0\,1\,1]_\beta$ direction, which becomes the c axis of α, the hexagonal martensite. This is not a rational plane of the β lattice. It is worth noting that even when *no* slip or twinning needs to be added to the lattice strain to produce an invariant plane, the habit plane *still* will not in general be a rational plane. The alternative habit plane, through OP' (Fig. 11.7), is crystallographically equivalent to the habit plane through OQ', because the plane normal to the x axis is a mirror plane.

The orientation relationship between the martensite and the matrix can be determined from the lattice correspondence and the *total* lattice strain, i.e. the pure strain **S'** followed by the rotation which takes OQ' back to OQ. Because the axis of rotation is normal to the $(0\,1\,1)_\beta$, which becomes the basal plane of the martensite, these two planes $(0\,1\,1)_\beta$ and $(0\,0\,0\,1)_\alpha$ remain exactly parallel to one another. The rotation turns the $[1\,0\,0]_\beta$ about $3°$ from the $[2\,\bar{1}\,\bar{1}\,0]_x$ to which it corresponds. The corresponding close packed directions, i.e. $\langle 1\,1\,1 \rangle_\beta$ and $\langle 2\,\bar{1}\,\bar{1}\,0 \rangle_x$, are about $2.5°$ from being parallel.

The production of a plate of martensite by lattice strain alone is illustrated in Fig. 11.8.

Experimentally, the orientation relationship has been found by using faceted crystals, grown by decomposition of the iodide [2,3]. The external symmetry of these crystals depicts the orientation of the high-temperature phase, β, which cannot be retained at room temperature. Titanium as well as zirconium crystals have been studied in this way, and they give essentially the same results. The closest packed planes $(0\,0\,0\,1)_x$ and $\{0\,1\,1\}_\beta$ and close packed directions $\langle 2\,\bar{1}\,\bar{1}\,0 \rangle_x$ and $\langle 1\,1\,1 \rangle_\beta$ are always within a degree or two of being parallel to one another. Groups of parallel striations lie on the surface at room temperature. If presumed to be habit plane traces, these striations show that the habit planes are always nearly normal to the basal plane of the α, as predicted above. Specifically, habit planes close to $\{\bar{4}\,3\,\bar{3}\}_\beta$ have been reported. All these results agree satisfactorily with the predictions of the approximate theory described above.

11.5 TRANSFORMATION OF INDIUM–THALLIUM ALLOYS

The structure of indium can be described as f.c. tetragonal (see Section 3.3) with the ratio $c/a = 1.08$. Alloying with thallium decreases the c/a ratio until at 23 at % Tl the solution becomes f.c.c. The c/a ratio also decreases with temperature and in the region of 20 at % Tl a martensitic transformation from f.c.c. to f.c.t. occurs on cooling at $60°C$.

This transformation is of interest for two reasons. Firstly, the f.c.t. phase consists of parallel, twin-related lamellae, and it will be easy to see how this inhomogeneity produces the macroscopically undistorted habit plane which could not be produced by the homogeneous lattice strain alone. Secondly, a single interface

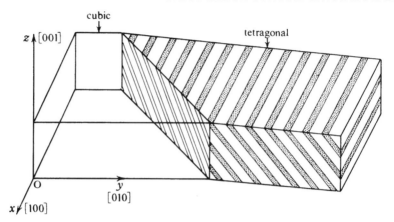

Figure 11.9 A partly transformed In–Tl alloy. The dark and light bands in the tetragonal phase (the martensite) are twin-related lamellae

between the f.c.c. and the f.c.t. regions can be made to sweep reversibly through a crystal [4]. The theoretical approach of searching for a single macroscopically undistorted, unrotated interface plane can be applied very confidently to this case. A partly transformed crystal is sketched in Fig. 11.9.

The f.c.c. and f.c.t. lattices of In–Tl are so nearly alike that the choice of lattice correspondence is obvious. The f.c.t. lattice of a 20.7 at % Tl alloy at the transition temperature is produced from the f.c.c. lattice by expanding one $\langle 0\,0\,1 \rangle$ direction $\frac{2}{3}\%$ and contracting the other two by $\frac{1}{3}\%$. The pure lattice strain referred to the crystal axes is

$$S_{ij} = \begin{pmatrix} -\varepsilon & 0 & 0 \\ 0 & 2\varepsilon & 0 \\ 0 & 0 & -\varepsilon \end{pmatrix}, \quad \varepsilon = 0.003 \quad (11.11)$$

with the y axis chosen for the c axis of the tetragonal lattice. Because the principal strains are so small, a simplified treatment makes it possible to understand the main features of the transformation.

The homogeneous strain **S** does not produce an undistorted plane. The intersection of the unit sphere with the ellipsoid which it becomes after applying **S** gives a circular cone for the locus of unlengthened vectors. However, a macroscopically undistorted plane can exist because the martensite is not in fact homogeneous, but consists of lamellae that are twins of one another. The twin relationship is shown in Fig. 11.10. It can be produced by transforming two different cube axes into the c axis in the two twin-related regions. If $S_1 = (-\varepsilon, -\varepsilon, 2\varepsilon)$ and $S_2 = (-\varepsilon, 2\varepsilon, -\varepsilon)$, then S_1 and S_2 produce the two twin-related regions shown in Fig. 11.10, apart from a small rotation of one region with respect to the other through the angle $\phi \approx 3\varepsilon$. Because ε is so small, this rotation will be neglected. The macroscopic strain $\overline{S}$ due to volume fractions x and $1 - x$ of regions 1 and

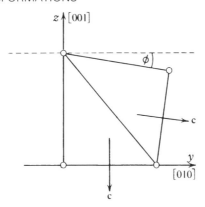

Figure 11.10   The twin relationship of the lamellae shown in Fig. 11.9

2 respectively is then

$$\overline{S} = x S_1 + (1-x) S_2 \tag{11.12}$$

or

$$\overline{S} = (-\varepsilon,\, 2\varepsilon - 3\varepsilon x,\, -\varepsilon + 3\varepsilon x) \tag{11.13}$$

By choosing $x = \frac{1}{3}$, $\overline{S}$ becomes the strain $(-\varepsilon,\, \varepsilon,\, 0)$. One of the macroscopic principal strains is now zero while the other two are opposite in sign, satisfying the condition for there to be an undistorted plane. (**A** similar solution is obtained for $x = \frac{2}{3}$.) In fact, because ε is so small, the macroscopic strain $\overline{S} = (-\varepsilon,\, \varepsilon,\, 0)$ is a very close approximation to a pure shear strain, which after adding a small rotation ε is equivalent to a simple shear on a plane at very nearly 45° to the axes Ox and Oy (Section 5.3). The habit plane will be parallel to the plane on which this simple shear occurs, in other words very close to either $(1\,1\,0)$ or $(1\,\overline{1}\,0)$. The $(1\,\overline{1}\,0)$ habit plane is shown in Fig. 11.9.

A feature that has important consequences is the fact that a given habit plane may be the invariant plane of either of two opposite macroscopic strains. Thus the plane $(1\,\overline{1}\,0)$, as well as serving as the habit plane for a martensite plate whose macroscopic strain is $(-\varepsilon,\, \varepsilon,\, 0)$ as described above, may also serve as the habit plane for a plate that is internally twinned on the $(1\,0\,1)$ plane and therefore has the macroscopic strain $(\varepsilon,\, -\varepsilon,\, 0)$. Obviously, by stacking these parallel plates having opposite strains alternately, it is possible to build up a volume of martensite within which the average strain is zero (Fig. 11.11). Experimentally, it is observed that the transformation often occurs by the growth of such a stack of martensite plates into the matrix, the plates having alternating types of internal twinning as shown in Fig. 11.11.

Although the theory presented above is inexact, it does show clearly how a shear which preserves the lattice (in this case a twinning shear) can produce a macroscopically undistorted plane to match with the untransformed matrix.

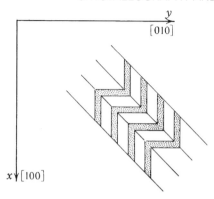

Figure 11.11 Three parallel plates of martensite having alternating shear strains

Because ε is so small, the habit plane as determined by a precise calculation is only $0.5°$ from $\{1\,1\,0\}$, and the tetragonal axes are within $2°$ of being parallel to the cube axes.

The absolute thicknesses of the twin lamellae are not specified by the theory. The thinner the twins, the finer the scale in which matching is achieved at the interface. For twins of average thickness h, elastic strains of order ε will spread from the interface to a distance of order h, giving an interfacial energy proportional to $\varepsilon^2 h$. The tendency to reduce interfacial energy by making h small is presumably limited by the extra volume energy of the martensite due to the twin boundary energy γ, which is of order γ/h per unit volume. The total energy will be a minimum at some finite spacing.

An applied stress has an interesting effect on the transformation. For example, it can be seen that a tensile stress along the y axis will favour a transformation of the particular form shown in Fig. 11.9 because this transformation increases the length of the crystal in the Oy direction. The M_s temperature is increased by such an applied tensile stress, so that in a tensile test carried out just above the normal M_s, deformation by transformation occurs at a critical stress. The transformation reverses when the load is removed.

11.6 TRANSFORMATIONS IN STEELS

Research on martensitic phenomena has always been stimulated by the great practical importance of martensite in steels. Martensitic transformations in steel have themselves continued to provide many problems, due to their complexity and their great variety in steels of different compositions.

The one feature that is basic to the theory of all forms of tetragonal martensite in steels is the lattice correspondence, which was proposed by Bain in 1924 [5]. This is depicted in Fig. 11.12, which shows how a b.c.t. lattice can be picked out

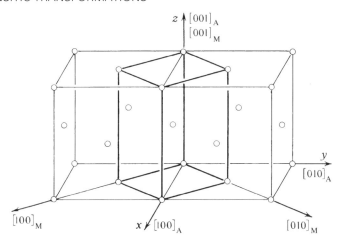

Figure 11.12   The f.c.c. lattice with a b.c. tetragonal cell picked out of it. (After Bain [5])

of an f.c.c. lattice. The correspondence matrix $\mathbf{C}$ for lattice vectors is given by

$$\mathbf{C} = \begin{pmatrix} 1 & -1 & 0 \\ 1 & 1 & 0 \\ 0 & 0 & 1 \end{pmatrix} \qquad (11.14)$$

for the choice of the c axis shown in Fig. 11.12. The correspondence matrix for planes, $\mathbf{K}$, is defined by

$$\begin{pmatrix} H \\ K \\ L \end{pmatrix}_M = \mathbf{K} \begin{pmatrix} h \\ k \\ l \end{pmatrix}_A \qquad (11.15)$$

where subscripts M and A denote martensite (b.c.t.) and austenite (f.c.c.) respectively. By substituting in turn $(1\,0\,0)_A$, $(0\,1\,0)_A$ and $(0\,0\,1)_A$ for $(h\,k\,l)_A$ and inspecting Fig. 11.12 to see which martensite planes $(H\,K\,L)_M$ these become, we find that

$$\mathbf{K} = \tfrac{1}{2} \begin{pmatrix} 1 & -1 & 0 \\ 1 & 1 & 0 \\ 0 & 0 & 2 \end{pmatrix} \qquad (11.16)$$

The tetragonality of the b.c.t. lattice as it stands in Fig. 11.12 is much greater than that of martensite. To achieve the correct lattice parameters, the $[0\,0\,1]$ direction must be contracted by about 20% and the $(0\,0\,1)$ plane expanded uniformly by about 12%. This pure lattice strain, referred to either A or M axes, can be written as

$$B_{ij} = \begin{pmatrix} \eta_1 & 0 & 0 \\ 0 & \eta_1 & 0 \\ 0 & 0 & \eta_3 \end{pmatrix} \qquad (11.17)$$

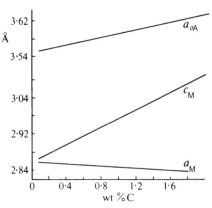

Figure 11.13 Lattice parameters of austenite and martensite as a function of the carbon content of the steel. (After Roberts [6]. Reproduced by permission of the American Institute of Mining, Metallurgical and Petroleum Engineers)

If the lattice parameter of austenite is a_0 and the lattice parameters of martensite are c and a, then

$$\eta_1 = \frac{\sqrt{2}a}{a_0} - 1 \sim 0.12$$

and $\eta_3 = c/a_0 - 1 \sim -0.20$. The exact values of these principal strains depend upon the carbon content of the steel, which affects the tetragonality of the martensite as shown in Fig. 11.13. An increase of c/a with carbon content is just what would be expected from the Bain correspondence and the fact that C atoms in austenite occupy octahedral interstices such as $(0, 0, \frac{1}{2})_A$ (Section 3.2). If C atoms are trapped in these sites during the transformation, they will oppose the contraction along $[0\,0\,1]$. Considering for a moment the Bain correspondence between the f.c.c. lattice and the b.c.c. lattice ($c/a = 1$), it is important to realize that *all* the octahedral interstices in the f.c.c. lattice correspond to *just one of the three* differently oriented squashed octahedral interstices in the b.c.c. lattice, i.e. one of the three sites F, A and B in Fig. 9.14. Referring to Fig. 9.14, if the $[0\,0\,1]_M$ corresponds to $[0\,0\,1]_A$, then the f.c.c. octahedral interstices correspond to F type sites; if the $[0\,1\,0]_M$ corresponds to $[0\,1\,0]_A$, then they correspond to B-type sites; and if $[1\,0\,0]_M$ correspond to $[1\,0\,0]_A$ they correspond to A-tpe sites. As a result, the C atoms which are distributed at random in the austenite expand the transformed lattice in one particular direction and produce its tetragonality.

The strain **B** will change a unit sphere into the ellipsoid

$$\frac{x^2}{(1+\eta_1)^2} + \frac{y^2}{(1+\eta_1)^2} + \frac{z^2}{(1+\eta_3)^2} = 1 \qquad (11.18)$$

All the vectors whose length has not been changed by **B** lie on the circular cone defined by the intersection of this ellipsoid with the unit sphere. Since the change

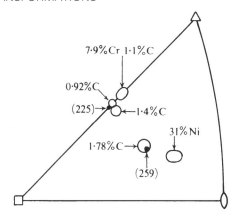

Figure 11.14 Habit plane normals of martensite in various steels

in shape of a transforming region is believed to leave one plane macroscopically undistorted and unrotated, to match with the matrix, the problem is to find an additional deformation that will produce such a plane, without changing the lattice. All the principal strains of **B** are quite large, so that approximate arguments cannot be applied. The first successful detailed theories were produced by Wechsler, Lieberman and Read and by Bowles and Mackenzie. [7,8].[†] Although the principles of their theories are quite straightforward the mathematical details are rather complex, and we shall not reproduce them here.

Experimentally determined habit plane normals for various steels are given in Fig. 11.14 plotted as regions rather than points, because the habit plane in a particular steel really does vary a little. The various orientation relationships have in common that closest packed $\{1\,1\,1\}_A$ and $\{1\,1\,0\}_M$ are roughly parallel. Various directions in these planes may be parallel; in general, the relation is somewhere between $\langle 1\,1\,0\rangle_A \| \langle 1\,1\,1\rangle_M$ (Kurdjumow–Sachs relationship) and $\langle 2\,1\,1\rangle_A \| \langle 1\,1\,0\rangle_M$ (Nishiyama relationship).

Martensites having a habit near $\{2\,5\,9\}$ are well understood. They occur in Fe–Ni–C and Fe–Ni alloys and in Fe–C alloys of relatively high C content. A tolerable prediction of their habit and orientation relationship can be derived from the postulate that the lattice invariant shear strain which is added to the Bain strain is in the $\langle 1\,1\,1\rangle_A$ direction on a $\{1\,2\,1\}_M$ plane that is derived from a $\{1\,1\,0\}_A$ plane. This shear corresponds to twinning in a b.c.c. crystal, and it was proposed by Wechsler, Lieberman and Read and by Bowles and Mackenzie [7,8] that the martensite plate consists of a stack of twin related laths. Under the optical microscope there was little sign that this was the case, but when such steels could

---

[†] M. S. Wechsler, D. S. Lieberman and T. A. Read, 'On the theory of the formation of martensite', *Trans, Am. Inst. Mining Metall. and Petrol. Engrs*, **197**, 1503 (1953). J. S. Bowles and J. K. Mackenzie, 'The crystallography of martensite transformations', *Acta Met.*, **2**, 224 (1954).

be examined with the electron microscope, the proper arrays of very thin twins were found [9]. This completed a triumph for the formal crystallographic theory.

To explain habits which are not near $\{2\,5\,9\}_A$, and even to explain the wide scatter in the region of $\{2\,5\,9\}_A$, some adjustment of this theory is needed. One approach is to postulate a different lattice invariant shear from the $\{1\,2\,1\}\langle1\,1\,1\rangle_M$, which is appropriate for internally twinned plates. This is certainly reasonable for those alloys in which untwinned or only partly twinned plates are seen. Alloys with a relatively high M_s (lower Ni or C contents) seem to fall into this category.

Particular difficulty has been experienced in understanding martensites whose habit plane is close to $\{2\,2\,5\}_A$. This habit plane might possibly be the result of a lattice invariant shear which changes in kind as the plate develops, e.g. from twinning to slip. There is experimental evidence that such a change can occur, and an attempt has been made to incorporate this feature in a rigorous crystallographic theory of the transformation [10].

11.7 TRANSFORMATIONS IN COPPER ALLOYS

Several copper alloys, Cu–Al, Cu–Sn, Cu–Ga and Cu–Zn, behave very similarly within a certain composition range. In each case there exists a high-temperature phase, β, which is b.c.c. If slowly cooled, the b.c.c. phase decomposes into two phases; if cooled rather quickly, it changes structure without changing composition, but also without changing its shape; if cooled very quickly, it undergoes a martensitic transformation. The martensite is a close packed structure, basically either f.c.c. or at the low-Cu end of the composition range, h.c.p. The reasons for calling the martensite structures only basically f.c.c. and h.c.p. are rather interesting. The first point is that the alloys usually order before transforming, so that the corresponding pattern of order is built into the martensite. (β-brass, Cu–Zn, differs from the remaining alloys in having the CsCl ordered structure instead of the Fe_3Al structure; See page 121) A second complication to the basically f.c.c. structure is that the plates of martensite are heavily faulted. The stacking faults can be seen under the electron microscope [11] and their effect on diffraction patterns shows that there is one fault on about every third close packed plane, each produced by the same $\frac{1}{6}\langle2\,1\,1\rangle$ displacement. The assertion that the displacements producing the stacking faults supply an undistorted, unrotated plane which forms the face of the martensite plate can be checked by the formal crystallographic theory.

The lattice correspondence is assumed to be the inverse of the Bain correspondence, and a shear on $\{1\,1\,1\}_M$ in the direction $\langle2\,1\,1\rangle_M$ is chosen for the complementary strain. These postulates lead to the correct habit plane, near $\{1\,3\,3\}_\beta$, and a fault density close to that which is observed.

Plates of the basically h.c.p. martensite of Cu–Al, Cu–Ga and Cu–Sn are internally twinned and not faulted. Here, too, the formal theory agrees well with the observations.

11.8 TRANSFORMATIONS IN NON-METALS

Little can be said about martensitic transformations in non-metals. There are many instances of a rapid change in structure occurring at a certain temperature, but very few instances in which the morphology of the change has been observed carefully. Without this vital observation, it is difficult to decide whether a transformation is martensitic or not.

A transformation whose morphology has been studied occurs in ZrO_2. The lattice of ZrO_2 is monoclinic at low temperatures and tetragonal above $1100\,°C$. When the monoclinic phase is heated under a microscope, at about $1100\,°C$ striations are seen to appear suddenly on the surface [12]. The microstructure of the resulting tetragonal phase is typical of a martensite. This transformation and the reverse of it which occurs on cooling are technically important, because they disrupt a zirconia body in which they occur and thereby limit the usefulness of an otherwise excellent refractory material (melting point of $2680\,°C$) in its pure state.

Also of technical importance is the transformation that occurs at the Curie temperature of a well-known ferroelectric material. Barium titanate is approximately cubic above $120\,°C$, with the perovskite structure (Fig. 3.1o). Below $120\,°C$, down to room temperature, it is tetragonal with a spontaneous electric polarization in the direction of the c axis. In this ferroelectric condition a crystal of $BaTiO_3$ has a domain structure which is often similar to the structure of the tetragonal In–Tl phase (Fig. 11.9). Stacks of twin lamellae occur in which an individual lamella constitutes a ferroelectric domain. The direction of polarization follows the c axis and therefore rotates through approximately $90°$ on passing through a twin boundary.

Mineralogists use the apt word 'displacive' to describe changes in structure that are probably martensitic. For example, if a structure change in a silicate can be achieved by a small distortion of the framework of Si–O bonds, the transformation is called *displacive*. If, on the other hand the new structure could be obtained only by breaking and remaking Si–O bonds, the transformation is called *reconstructive*.

11.9 CRYSTALLOGRAPHIC ASPECTS OF NUCLEATION
AND GROWTH

The nucleation and growth of a martensite plate may well be a complex process which takes place in several distinct stages. The atom movements that produce a nucleus of martensite are likely to be determined by the nature of some defect at which the nucleus forms. As the nucleus grows, its change of shape must be accommodated by the matrix, and it may be imagined that at some stage the constraint of the matrix forces the nucleus to undergo the heterogeneous deformation by slip, faulting or twinning, which allows the new phase to grow as a plate whose faces match the matrix on a macroscopic scale. The special sites at which nucleation occurs may have an arrangement of atoms that is similar to that in the new phase. A stacking fault in f.c.c. cobalt is an obvious example, for at the fault the stacking sequence of close packed planes becomes that of the hexagonal phase. It is natural to suppose that similar nucleation sites may exist in other structures. For example, the transformation of a b.c.c. metal to an f.c.c. or h.c.p. structure could conceivably be nucleated by a fault on the {0 1 1} plane. Figure 11.15 shows two (0 1 1) layers of a b.c.c. crystal. If the atoms in the upper layer are shifted by $\frac{1}{8}[0\,1\,1]$ to positions above the triangular gaps in the layer beneath, the stacking of the two layers becomes more like that of successive close packed planes of the f.c.c. or h.c.p. structure. It is conceivable that a dislocation with the usual $\frac{1}{2}[1\,\bar{1}\,1]$ Burgers vector could provide such a displacement by dissociating on a (0 1 1) plane, according to Eqn (8.57), which is

$$\tfrac{1}{2}[1\,\bar{1}\,1] \rightarrow \tfrac{1}{8}[0\,\bar{1}\,\bar{1}] + \tfrac{1}{4}[2\,\bar{1}\,1] + \tfrac{1}{8}[0\,\bar{1}\,1]$$

Another line of argument leads to the idea that $\{0\,1\,1\}\langle 0\,\bar{1}\,1\rangle$ shear has a particular significance for the nucleation of a close packed structure from a b.c.c. structure. Referring to Fig. 11.15, it can be seen that a b.c.c. structure made from rigid balls would have no resistance to $\{0\,1\,1\}\langle 0\,\bar{1}\,1\rangle$ shear. The balls in the upper (0 1 1) layer are perched up on saddle points and are unstable because the slightest push in a $\langle 0\,\bar{1}\,1\rangle$ direction will topple them into the positions shown in Fig. 11.15b. The b.c.c. structure is a mechanically unstable one for hard spheres. The stiffness of a real b.c.c. crystal with respect to $\{0\,1\,1\}\langle 0\,\bar{1}\,1\rangle$ shear is given by the elastic modulus $(c_{11} - c_{12})/2$, and the ratio of $(c_{11} - c_{12})/2$ to c_{44} is a measure of its stability with respect to a close packed structure. An example of a crystal with the CsCl structure which transforms to a close packed structure is β-brass, for which $(c_{11} - c_{12})/2c_{44} = \frac{1}{18}$ at room temperature. The absolute value of $(c_{11} - c_{12})/2$ is low (9.3 GPa), and $(c_{11} - c_{12})/2$ decreases as the temperature falls towards M_s,

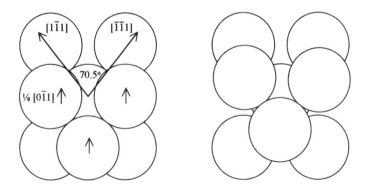

Figure 11.15 The (0 1 1) plant of a b.c.c. metal (a) before and (b) after a displacement of the atoms in the upper layer by $\frac{1}{8}[0\,\bar{1}\,1]$

indicating that the resistance of the lattice to $\{0\,1\,1\}\langle 0\,\bar{1}\,1\rangle$ shear is small at the temperature of its transformation to a close packed structure [13].

Once nucleated, martensite usually develops as a plate. The edgewise growth of the plate is often very rapid; the velocity may be an appreciable fraction of the velocity of sound. In general, the faces of the plate match the untransformed matrix only on a macroscopic scale. The local distortions can be represented by a set of dislocations in the interface whose long-range stress field vanishes. As the plate thickens, the glide of these dislocations produces the slip, faulting or twinning, which when added to the lattice distortion produces an average matching at the interface. Of course, this set of dislocations does not completely solve the problem of accommodating a finite plate in untransformed material. There is a shear on the interface plane which must be accommodated at the edges of the plate. The constraint of the matrix causes the edges of the plate to be tapered, just as in the case of a twin lamella (Fig. 10.15). In addition, there is in general an expansion or contraction normal to the faces of the plate which must be accommodated. If these shear and tensile strains are small, they may be accommodated by the elastic deformation of the plate and matrix. It is then possible for a plate to be in equilibrium, such that a small drop in temperature would allow the plate to grow by supplying enough chemical free energy to offset the energy of its elastic strain field, whereas a small rise in temperature would allow the elastic strain field to shrink the plate. This situation is analogous to elastic twinning and martensite of this type is called *thermoelastic*. More commonly, as in steels, the strains are too large to be accommodated elastically and plastic deformation occurs. Since plastic deformation dissipates energy instead of storing it, the growth of a plate is then no longer reversible. In addition, the damage that plastic deformation does to the lattice of the matrix may inhibit the further transformation of a partially transformed crystal.

Another way of partly accommodating the strain in one plate is to nucleate another plate of different strain nearby. This effect leads to the characteristic zigzag arrangement of martensite plates in high C steels. If one plate nucleates more than one other, a chain reaction develops which may transform a good part of the whole crystal in one fantastic burst. In the case of In–Tl alloys, plates with the same habit may choose between two strains which are, at least approximately, equal and opposite shears. A stack of plates of alternating shear forms a volume with no overall strain. Such a volume can have quite arbitrary boundaries, of course.

As a final brief summary it may be said that much of the crystallography of a martensitic transformation can be understood in terms of the interaction of two factors. The first factor is a tendency to transform by means of the lattice strain which most simply converts a small unit of the parent lattice into a unit of the martensite lattice. The second is the constraint that is imposed upon a finite transforming region by its surroundings.

PROBLEMS

11.1 A single crystal of f.c.c. cobalt is transformed martensitically to the hexagonal phase. How many orientations of c.p.h. cobalt may be present after the transformation? If the reverse transformation, c.p.h. to f.c.c., is also martensitic, how many orientations of the f.c.c. phase could be produced from a single grain of the hexagonal phase? If the f.c.c. crystal is to transform by means of a pole mechanism (Section 10.5, Fig. 10.18), what type of pole dislocation must be available?

11.2 Show that when two different invariant plane strains are applied, one after the other, there always remains a line which is neither rotated nor changed in length and a plane which is not rotated.

11.3 When a crystal of b.c.c. titanium transforms to the c.p.h. phase, how many lattice vectors may transform into the c axis of the hexagonal unit cell? Is this number equal to the number of different orientations in which the c.p.h. phase originating from a single b.c.c. crystal may be found? Derive the matrix equation which gives the transformation of the indices of planes for the lattice correspondence shown in Fig. 11.4 (see Appendix 4).

11.4 The lattice strain S given by Eqn (11.3) produces the correct hexagonal lattice from half the points of the b.c.c. lattice. Describe how the atoms at the remaining points of the b.c.c. lattice must be reshuffled in order to place them at the correct positions in the c.p.h. structure.

11.5 Prove that the atoms on the $(0\,1\,1)$ plane of a b.c.c. metal crystal can be brought into a close packed array by means of a compressive strain of 10% along $[1\,0\,0]$ together with a tensile strain of 10% along $[0\,\bar{1}\,1]$. What is the c/a ratio of the hexagonal unit cell that is derived from the b.c.c. lattice by means of these two strains alone? Compare this value with the c/a ratio of zirconium.

11.6 Calculate approximately the angle of the tilt that should be observed on the $(1\,0\,0)$ surface of a cubic In–Tl crystal which is transforming by the mode shown in Fig. 11.9. Assume that the pure lattice strain is a tensile strain of 0.006 along that cube axis which transforms into the c axis of the tetragonal cell and a compressive strain of 0.003 along the other two cube axes. What is the tensile strain in the Oy direction that can be produced by this mode of transformation? If the twin boundaries within the tetragonal phase are mobile, how may this strain be increased?

11.7 In the transformation of cubic In–Tl to tetragonal In–Tl, interfaces of $\{1\,1\,0\}$ orientation in the cubic phase divide this phase from the twinned tetragonal phase. How many different interfaces may bound:
    (a) a region with a given twin structure,
    (b) any region in which the thicker of the two twin lamellae contains a given c axis — say parallel to $[0\,0\,1]$ of the cubic phase,

(c) any region in which either of the twin lamellae contains a given c axis — say parallel to $[0\,0\,1]$ of the cubic phase.

11.8 Sketch the atomic structure on the planes normal to Ox and Oz in Fig. 11.9 at the places where they are intersected by the martensite habit plane. Indicate the form of the local distortions and show how the macroscopic distortion vanishes in both planes.

11.9 The Bain lattice correspondence for the formation of martensite in steels requires a lattice strain of principal components η_1, η_2 and η_3, as defined by Eqn (11.17) and Fig. 11.12. Show that the cone of vectors which are not changed in length by this strain has a semi-angle ϕ_i when the cone is defined by the initial position of these vectors, where

$$\tan \phi_i = \left(\frac{-2\eta_3 - \eta_3^2}{2\eta_1 + \eta_1^2} \right)^{1/2}$$

Show that, when defined by the position of these vectors after the strain has been applied, the cone has a semi-angle ϕ_f where

$$\tan \phi_f = \left(\frac{1 + \eta_1}{1 + \eta_3} \right) \tan \phi_i$$

Calculate approximate values of the angles ϕ_i and ϕ_f.

11.10 Show that the lattice correspondence for a b.c.c. to f.c.c. transformation (the Bain correspondence) and the lattice correspondence for the b.c.c. to c.p.h. transformation (e.g. in zirconium) can each be regarded as a special case of a correspondence for a b.c.c. to orthorhombic transformation. Derive a matrix for the transformation of the indices of directions, according to this correspondence (see Appendix 4).

SUGGESTIONS FOR FURTHER READING

Bilby, B. A. and Christian, J. W., 'The crystallography of martensitic transformations', *J. Iron Steel Inst.*, **197**, 122 (1961).

Christian, J. W., *The Theory of Transformations in Metals and Alloys*, Pergamon, Oxford (1975).

Greninger, A. B. and Troiano, A. R. 'Crystallography of austenite decomposition', *Trans. Am. Inst. Mining Metall. and Petrol. Engrs*, **140**, 307 (1940).

Lieberman, D. S., 'Martensitic transformations and determination of the inhomogeneous deformation', *Acta. Met.*, **6**, 680 (1958).

Nabarro, F. R. N., (ed.), *Dislocations in Solids*, Vol. 3, *Moving Dislocations*, North Holland (1975).

Porter, D. A. and Easterling, K. E., *Phase Transformations in Metals and Alloys*, Chapman and Hall, London (1992).

Wayman, C. M., *Introduction to the Crystallography of Martensitic Transformations*, Macmillan (1964).

REFERENCES

1. P. Gaunt and J. W. Christian, 'The cubic-hexagonal transformation in signle crystals of cobalt and cobalt–nickel alloys', *Acta Met.*, **7**, 529 (1959).
2. W. G. Burgers, 'On the process of transition of the cubic body centred modification into the hexagonal close packed modification of zirconium', *Physica*, **1**, 561 (1934).
3. P. Gaunt and J. W. Christian, 'The crystallography of the $\beta-\alpha$ transformation in zirconium and in two titanium–molybdenum alloys', *Acta Met.*, **7**, 534 (1959).
4. M. W. Burkart and T. A. Read, 'Diffusionless transforamtion in the indium–thallium system', *Trans. Am. Inst. Mining Metall. and Petrol. Engrs*, **197**, 1516 (1953).
5. E. C. Bain, *Trans. Am. Inst. Mining Metall. and Petrol Engrs.* **70**, 25 (1924).
6. C. S. Roberts, *Trans. Am. Inst. Mining Metall. and Petrol. Engrs*, **197**, 203 (1953).
7. M. S. Wechsler, D. S. Lieberman and T. A. Read, 'On the theory of the formation of martensite', *Trans, Am. Inst. Mining Metall. and Petrol. Engrs*, **197**, 1503 (1953).
8. J. S. Bowles and J. K. Mackenzie, 'The crystallography of martensite transformations', *Acta Met.*, **2**, 224 (1954).
9. P. M. Kelly and J. Nutting. 'The morphology of martensite,' *J. Iron Steel Inst.*, **197**, 199 (1961).
10. D. S. Lieberman, 'The phenomenological theory of composite martensite,' *Acta Met.*, **14**, 1723 (1966).
11. H. Warlimont, 'Microstructure, crystal structure and mechanical properties of martensite phases in copper alloys', in *Physical Properties of Martensite and Bainite*, The Iron and Steel Institute (1965).
12. L. L. Fehrenbacher and L. A. Jacobson, 'Metallographic observation of the monoclinic–tetragonal phase transformation in ZrO_2', *J. Am. Ceram. Soc.*, **48**, 157 (1965).
13. W. D. Robertson, *Physical Properties of Martensite and Bainite*, The Iron and Steel Institute (1965), p. 26.

12

Crystal Interfaces

12.1 THE STRUCTURE OF SURFACES AND SURFACE FREE ENERGY

Every real crystal must have at least one imperfection — its surface. A surface that is parallel to a prominent crystallographic plane is easy to visualize as being a smooth sheet of atoms whose pattern is the same as that on a parallel plane inside the crystal, with perhaps a very slightly different interplanar spacing. This simple picture is likely to be correct for metals; on the other hand, it is quite probable that the atoms in the outer layer of directionally bonded crystals like silicon are substantially rearranged. Since little is known about the real atomic structure of surfaces we shall restrict ourselves mainly to a discussion of some simple geometrical models.

We can formally define a flat surface parallel to any given rational plane by removing all the atoms whose centres lie to one side of a plane of this orientation located within the crystal. In crystal structures with one atom per lattice point, the location of the defining plane cannot affect the structure of the surface, because each atom has identical surroundings. In crystal structures with more than one atom per lattice point, the structure of the surface may depend upon the location of the plane. For example, a $\{1\,1\,1\}$ surface in a NaCl crystal will consist of either a sheet of Na^+ ions or a sheet of Cl^- ions, depending upon the location of the defining plane or the side from which the atoms are removed. In this structure the same is true for all surfaces $\{h\,k\,l\}$ for which h, k and l are all odd. Similarly, there are two alternative structures for some of the surfaces of a c.p.h. metal. An example is the $(1\,0\,\bar{1}\,0)$ surface shown in Fig. 12.1. It is easy to see from Fig. 12.2 that in wurtzite there are *four* alternative surfaces parallel to a closest packed plane. It is interesting to note that any two of these surfaces bounding a crystal must have different structures (a consequence of the absence of a centre of symmetry).

The atoms at a surface are deprived of some of their neighbours. Since the binding of an atom to its neighbours contributes a negative term to the energy of a crystal, we can attribute some excess energy to the presence of the surface. Imagine two identical surfaces to have been created within a single crystal by

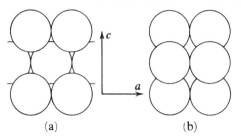

Figure 12.1 The two alternative $\{10\overline{1}0\}$ surfaces of a hexagonal metal. The surface is parallel to the plane of the figure

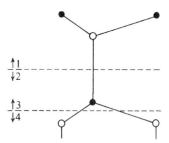

Figure 12.2 The four alternatives for a surface parallel to $(000\,1)$ in wurtzite. The surface is normal to the plane of the figure, intersecting it at one or other of the dotted lines

breaking the atomic bonds through which a plane passes. The surface energy is then equal to half the energy of the broken bonds. (This equation assumes that energies of those bonds which are left do not change.) The idea of broken bonds is useful in discussing how the energy of a surface will change as it is rotated out of a low-index orientation.

Suppose a $\{1\,1\,1\}$ surface of an f.c.c. metal is rotated through a small angle θ about a $\langle 1\,\overline{1}\,0\rangle$. Figure 12.3 shows that it then contains a density of steps ρ per unit length:

$$\rho = \frac{\sin\theta}{h} \tag{12.1}$$

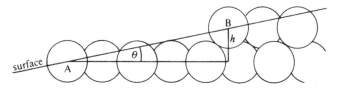

Figure 12.3 Surface at a small angle θ to a $\{1\,1\,1\}$ plane of an f.c.c. metal. The surface is normal to the plane of the figure

where h is the spacing of the $\{1\,1\,1\}$ lattice planes. If the angle θ happens to be such that the points A and B in Fig. 12.3 lie on $\langle 1\,\overline{1}\,0\rangle$ rows of atoms, then the steps can be evenly spaced. For other surfaces of rational orientation the steps can occur in evenly spaced groups, but if the surface is irrational there must be irregularities in the arrangement of steps. An atom on a step lacks more neighbours and so has more broken bonds than an atom in the flat $\{1\,1\,1\}$ surface; consequently the steps introduce an extra energy which is proportional to the number of steps as long as they are so far apart that they do not interact with one another. If each step contributes an energy β per unit length, then the total energy of unit area of surface is

$$E = E_0 \cos\theta + \frac{\beta \sin |\theta|}{h} \tag{12.2}$$

where E_0 is the energy of unit area of $\{1\,1\,1\}$ surface. The surface energy therefore increases as the surface is rotated from its low-index orientation, in either sense. The plot of energy as a function of θ shows a cusp at which $dE/d\theta$ changes discontinuously from $-\beta/h$ to $+\beta/h$ (Fig. 12.4).

Similar arguments can be applied to a small rotation of the $\{1\,1\,1\}$ surface about any axis; therefore a cusp exists in a three-dimensional plot of surface energy against orientation. Such a plot may take the form of a polar diagram in which the energy of a surface is represented by a vector which is normal to the surface and long in proportion to the energy of the surface. The energy plot can be expected to exhibit a number of cusps at the orientations of various low-index planes.

The energy of a surface is closely related to its surface free energy, defined as the work that can be obtained from the destruction of unit area of surface. The surface free energy is

$$\gamma = E - TS \tag{12.3}$$

and although the energy E is usually its more important component, the entropy term TS can be significant. For example, the steps upon a surface that is slightly off a low-index plane will introduce a configurational entropy, because their straightness and spacing may vary. Therefore, at a finite temperature the cusp in the surface free energy plot will not be as sharp as that in the energy plot and it may even disappear for higher-index planes.

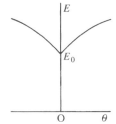

Figure 12.4

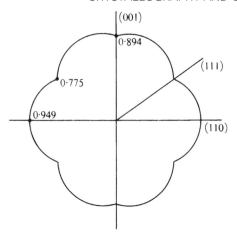

Figure 12.5 Possible {1 1 0} section through the γ-plot of an f.c.c. metal. Surface tension values are Values given are in units of the tension of a {2 1 0} surface. Reprinted with permission from Mackenzie *et al., J. Phys. Chem.*, **23**, 185 (1962). Copyright 1962 American Chemical Society

A polar diagram of surface free energy is called a γ-plot. Figure 12.5 shows a {1 1 0} section through the γ-plot of an f.c.c. metal, as computed from a simple nearest neighbour bond model.

The meaning of the surface free energy of a solid can be illustrated by describing an experiment that has been used to measure it. A very fine wire is loaded at a temperature which is close to its melting point. By finding the rate of extension as a function of load one finds the value of the load W which will just counteract the tendency of the wire to reduce its surface area by shrinking. Considering an infinitesimal increase in length dx at this equilibrium, 'zero-creep' condition, we have

$$\gamma \, dA - W \, dx + \gamma_b \, dA_b = 0 \tag{12.4}$$

In this equation $\gamma \, dA$ is the increase in free energy due to the creation of an area dA of new surface (Fig. 12.6). The term $\gamma_b \, dA_b$ is the change in free energy due to a change dA_b in the area of grain boundary within the wire, which in practice is polycrystalline. The whole equation expresses a condition for the free energy of the system to be a minimum. It is important to realize that the wire is conceived to change length by a frictionless flow at constant volume, not by elastic deformation, so that the increase in surface area, dA, is obtained by creating a new surface, not by stretching the old one. For a wire of radius r containing n transverse grain boundaries per unit length (Fig. 12.6), Eqn (12.4) gives

$$\gamma = W/\pi r + \gamma_b nr \tag{12.5}$$

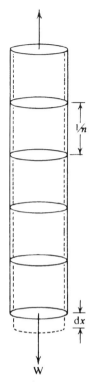

Figure 12.6   W

If the grain boundary energy γ_b is not known, the use of an estimated value in Eqn (12.5) does not seriously affect the accuracy of the value of γ. Since the orientation of the surface varies around the circumference of the wire, Eqn (12.5) gives an average surface free energy.

Some surface free energies obtained by this and other methods are listed in Table 12.1. The surface free energy of the cleavage plane of many brittle crystals has been found from the force needed to split a pre-cracked crystal of convenient shape, such as that shown in Fig. 12.7. The basic equation defining the critical force P_c is the balance of energy,

$$2P_c\,\mathrm{d}y = \mathrm{d}E + 2\gamma w\,\mathrm{d}L \qquad (12.6)$$

where $\mathrm{d}y$ is the deflection of each of the points of application of P and $\mathrm{d}E$ the increase in the elastic energy of the crystal; $\mathrm{d}y$ and $\mathrm{d}E$ accompany an infinitesimal growth $\mathrm{d}L$ of the crack, during which the force P is held constant. To apply this equation one must be sure that the applied force P does not have to do work in plastically deforming the crystal as the crack grows.

Table 12.1   Surface free energies of solids

Crystal	Surface	Environment	Temperature (K)	Surface tension (mJ m^{-2})	Method	Ref.
Ag	Average	He	1200	1140 ± 90	Zero creep	2
Au	Average	He	1300	1400 ± 65	Zero creep	2
Cu	Average	Vacuum	1250	1650	Zero creep	3
CaF$_2$	(1 1 1)	Liquid N$_2$	77	450	Cleavage	4
Calcite	(1 0 $\bar{1}$ 0)	Liquid N$_2$	77	230	Cleavage	4
Fe (δ)	Average	A	1700	1950 ± 200	Zero creep	5
Fe 3% Si	(1 0 0)	Liquid H$_2$	14	1360	Cleavage	4
KCl	(1 0 0)	Air	298	110 ± 5	Cleavage	6
LiF	(1 0 0)	Liquid N$_2$	77	340	Cleavage	4
MgO	(1 0 0)	Air	298	1150 ± 80	Cleavage	7
	(1 0 0)	Vacuum	77	1280		8
Mo	Average	A	2623	1960	Zero creep	9
NaCl	(1 0 0)	Vacuum	77	283 ± 30	Cleavage	8
	(1 0 0)	Liquid N$_2$	77	317 ± 30		
Nb (Cb)	Average	Vacuum	2520	2100	Zero creep	10
Ni	Average	A		1725	Zero creep	11
Si	(1 1 1)	Liquid N$_2$	77	1240	Cleavage	4
Sn	Average	Vacuum	490	685	Zero creep	12
Zn	(0 0 0 1)	Liquid N$_2$	77	105	Cleavage	4

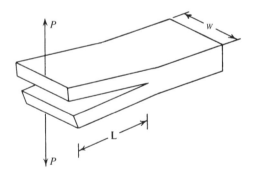

Figure 12.7

12.2 STRUCTURE AND ENERGY OF GRAIN BOUNDARIES

A crystalline solid is usually found in the form of a polycrystal, i.e. an aggregate of randomly oriented single crystals, called grains. Even so-called single crystals usually contain regions called subgrains, which have slightly different orientations.

Perhaps the simplest type of grain boundary to visualize is the symmetrical *tilt boundary*, where the two grains on either side are related by symmetrical

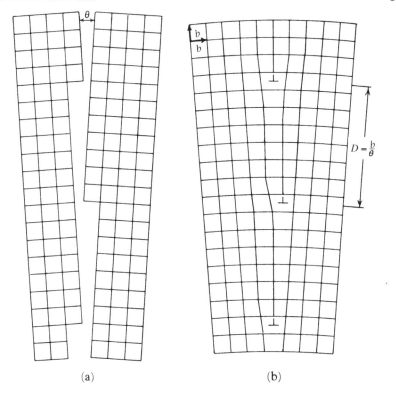

(a)   (b)

Figure 12.8 Low angle symmetrical tilt boundary in a simple cubic lattice. The boundary is normal to the plane of the figure. (From Read [13]. Copyright 1953. Reproduced by permission of McGraw-Hill Book Company)

rotations about an axis lying in the boundary. Figure 12.8 shows a tilt boundary in a simple cubic structure. This boundary could be formed by joining two crystals having surfaces which are rotated from a cube plane by small angles $+\theta/2$ and $-\theta/2$ about a $\langle 1\,0\,0\rangle$. When the two surfaces are joined, the steps in them become edge dislocations of the Burgers vector equal to the step height. Setting $h = b$ in Eqn (12.1), the number of dislocations per unit length of boundary becomes

$$\frac{1}{d} = \frac{2\sin\theta/2}{b} \qquad\qquad (12.7)$$

or when θ is small,

$$\frac{1}{d} \approx \frac{\theta}{b} \qquad\qquad (12.8)$$

This boundary is, in fact, a 'polygon wall' of dislocations of the type mentioned in Section 7.6. Read and Shockley [14] have calculated the energy of such an array, situated in an infinite medium of shear modulus μ and Poisson's ratio ν. They found the energy per unit area to be

$$E = E_0\theta(A_0 - \ln\theta) \qquad (12.9)$$

where

$$E_0 = \frac{\mu b}{4\pi(1-\nu)}$$

and

$$A_0 = 1 + \ln\left(\frac{b}{2\pi r_0}\right)$$

The length r_0 in Eqn (12.9) is related to the energy of the core of one of the boundary dislocations. Its definition is that an integration of an expression for the elastic strain energy such as Eqn (8.1) down to a radius r_0 gives the *total* energy of the dislocation, including the core energy. A high core energy decreases r_0 and increases A_0. Since from Eqn (12.8) the density of dislocations in the boundary is θ/b, for small θ, the first term in Eqn (12.9) depends on the total core energy in unit area of the boundary. The elastic energy of the boundary enters into the second term. The stresses due to the dislocations largely cancel one another at distances from the boundary that exceed the spacing of the dislocations, b/θ. The elastic energy within a cylinder of radius b/θ about an edge dislocation of core radius r_c is

$$E = \frac{\mu b^2}{4\pi(1-\nu)}\ln\frac{b}{\theta r_c}$$

which accounts for the form of the second term in Eqn (12.9).

Equation (12.9) can be applied only to boundaries having a small angle of tilt, θ, such that the cores of the dislocations do not overlap. Measurements of the free energy of tilt boundaries in copper show that at angles greater than about $8°$ the energy is greater than that predicted by fitting Eqn (12.9) to the lower angle results, to which it properly applies. (The entropy term in the free energy of a low-angle boundary is expected to be small compared to its energy, so that energy and free energy can be identified with one another.)

A plot of Eqn (12.9) is shown in Fig. 12.9. The cusp at $\theta = 0$, where the boundary disappears, is very sharp: $dE/d\theta$ becomes infinite at $\theta = 0$. Cusps in surface energy plots are relatively blunt, because the energy of a surface step is always localized at the step and is limited in amount, whereas a far-reaching stress field is set up when an isolated step is pressed into another crystal to make a grain boundary dislocation. Physically, the reason why the energy of a tilt boundary rises so steeply as its angle increases from zero is that the strain field of each dislocation spreads out to a very large distance when the dislocations are widely separated. The increase in energy becomes less steep as the angle of tilt

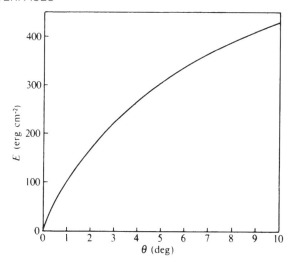

Figure 12.9 Energy of a tilt boundary as a function of the tilt angle θ. Values are given by Eqn (12.9) with $E_0 = 1450 \, \mathrm{erg \, cm^{-2}}$ (appropriate for Cu) and $A_0 = 0$

increases further because the stress fields of the dislocations cancel as they come closer together.

Shallower cusps in the grain boundary energy must exist at angles of tilt at which the dislocations are evenly spaced.[†] In deriving Eqn (12.9) rigorously, it is assumed that the dislocations are uniformly spaced. Evidently this is possible in the simple cubic lattice only when θ is such that

$$\cot \theta/2 = 2n \tag{12.10}$$

where n is an integer. For example, when $\cot \theta/2 = 14$, there is one dislocation on every seventh cube plane. To slightly increase the angle of tilt by $\delta\theta$ the separation must occasionally be decreased to six planes. Suppose that the decrease in spacing, if it could be achieved uniformly, would be δd, the original separation being d. Then the spacing of the *actual* disturbances in separation is

$$D = \left(\frac{b}{\delta d} \right) d \tag{12.11}$$

or, since $d \approx b/\theta$, for small θ,

$$D \approx \frac{b\theta}{\delta\theta} \tag{12.12}$$

If the angle of tilt is decreased, instead of increased, the same equation gives the spacing of the occasional eight-plane separations. This spacing is less than

---

[†] Most of these cusps will be very small and of theoretical rather than practical interest.

the spacing of dislocations in a tilt boundary of angle $\delta\theta$ would be, by the factor $\theta(\theta < 1)$. The extra elastic energy due to an increase or decrease in tilt angle of $\delta\theta$ can be guessed, by comparing Eqn (12.12) with Eqn (12.8), to be roughly that of a $\delta\theta$ boundary composed of dislocations of the small Burgers vector $b\theta$ or

$$\Delta E \sim -E_0\theta\delta\theta \ln \delta\theta \tag{12.13}$$

The form of this expression indicates that a small sharp energy cusp exists at the special angle θ, since $d\Delta E/d(\delta\theta)$ becomes infinite as $\delta\theta \to 0$. Physically, this is because each disturbance in the regular structure of the boundary produces a very far-reaching stress field when the disturbances are widely separated.

Cusps can be expected to occur also at special high angles of tilt. For example, when $\cot\theta/2 = 2$ (or $\theta \approx 53°$), the structure of the boundary is neat and regular, as shown in Fig. 12.10. The plane of the boundary is a $\{2\,1\,0\}$ plane in either grain, and one grain can be described as a twin of the other, with $K_1 = (2\,1\,0)$. (According to the definition of a twin given in Section 10.2, any symmetrical tilt boundary can be described as a twin boundary when it is formed by joining surfaces of rational orientation, i.e. in the simple cubic lattice, surfaces for which $\cot\theta/2 = m/n$, where m and n are integers.) Since the atoms in the twin boundary lie on the lattices of both grains, we may expect the energy to be particularly small when the density of atoms on this boundary is high, i.e. when K_1 is a low-index plane. Since $(2\,1\,0)$ is the closest packed twin plane that can be formed in the simple cubic lattice when the tilt axis is $[1\,0\,0]$, the deepest of the energy cusps should occur here, at the angle $\theta \approx 53°$.

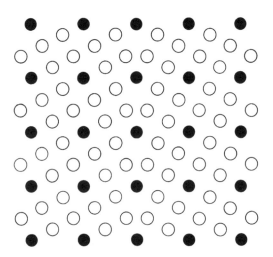

Figure 12.10 Tilt boundary of good fit. The boundary is normal to the plane of the figure. The dark circles represent atoms that lie on points of the lattices of both grains. (From Shewmon [15])

It must be admitted that measurements of the free energies of [1 0 0] tilt boundaries in copper have not revealed cusps at twin angles, perhaps simply because they are too shallow — less than, say, 10% below the maximum energy.

The symmetrical tilt boundary is a special type of boundary which can be specified by the axis of tilt and the single angle θ. In general, a grain boundary has five degrees of freedom, that is to say, five numbers are required to define it. One way of allocating these numbers is to specify the rotation that brings the lattice of one grain parallel to the lattice of the other, which requires two numbers to define the axis of rotation and one to define the angle of rotation, and then to specify the orientation of the boundary with respect to one of the grains, which uses another two numbers.

We will examine some of the consequences of varying the orientation of the grain boundary in a simple cubic crystal whilst keeping $\langle 1\,0\,0\rangle$ as the axis of rotation of one grain with respect to the other.

The tilt boundary of Fig. 12.8 can be turned out of its symmetrical orientation by rotating it about the tilt axis. Figure 12.11 shows the effect of a rotation of χ. Edge dislocations with extra planes that are normal to those of the original set are introduced. It can be shown [14] that their density is given by

$$p_x = \frac{2}{b}\sin|\chi|\sin\frac{\theta}{2} \qquad (12.14)$$

while the density of the original set is reduced to

$$p_y = \frac{2}{b}\cos\chi\sin\frac{\theta}{2} \qquad (12.15)$$

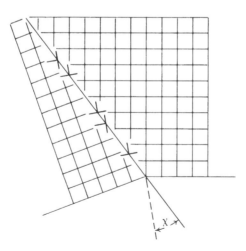

Figure 12.11 Unsymmetrical tilt boundary

The new dislocations must increase the energy of the boundary sharply as χ increases from zero, because at first they are far apart and therefore have far-reaching strain fields. Read and Shockley [14] showed that the energy of the boundary has the same form as Eqn (12.9), viz.

$$E = E'_0\theta(A - \ln\theta) \tag{12.16}$$

but that now

$$E'_0 = \frac{\mu b}{4\pi(1-\nu)}(\cos\chi + \sin|\chi|)$$

while

$$A = A_0 - \frac{1}{2}\sin 2|\chi| - \frac{\sin|\chi|\ln(\sin|\chi|) + \cos\chi\ln(\cos\chi)}{\sin|\chi| + \cos\chi}$$

According to these equations a sharp energy cusp does indeed exist at the symmetrical orientation $\chi = 0$.

A qualitatively similar cusp should exist for a high-angle tilt boundary which is a twin plane, since rotating such a boundary out of its symmetrical position destroys the good fit of the atoms upon it.

A more drastic change in boundary structure is produced by rotating the tilt boundary of Fig. 12.8 through 90° about an axis in the plane of the boundary and normal to the tilt axis. The boundary is then normal to the $\langle 100\rangle$ axis about which the two grains are rotated, relative to one another, and is called a *twist boundary*. It consists of a grid of screw dislocations, as shown in Fig. 12.12. The deformation due to one of the two orthogonal sets of screws is such that its pure shear component cancels that of the other set at large distances from the boundary, whereas its rotational component adds to that of the other set and produces the necessary relative rotation of the grains.

The density of dislocations needed to produce a twist of angle θ can be specified by the following procedure. Choose a mean lattice from which the two grains can be generated by rotations of $\theta/2$ and $-\theta/2$ respectively. In Fig. 12.13 the axis of rotation passes through the point O. Choose a vector **OA** which lies in the grain boundary, which consists of an array of dislocations in the mean lattice. This array produces the required rotations. Since **OA** cuts through a set of dislocations, a circuit A to O in grain 1 and O to A in grain 2 will surround these dislocations, so that a corresponding circuit taken in a reference lattice will fail to close (Section 7.1). Choosing the reference lattice to be parallel to the mean lattice, the reference path corresponding to A to O in grain 1 can be found by rotating grain 1 back through $\theta/2$ so as to align it with the reference lattice. This places the starting point at $a^{(2)}$ (Fig. 12.13). The back-rotation of grain 2 places the finishing point of the circuit at $a^{(1)}$. The closure failure is then $a^{(1)}a^{(2)}$, of magnitude

$$|\mathbf{B}| = 2|\mathbf{OA}|\sin\theta/2 \tag{12.17}$$

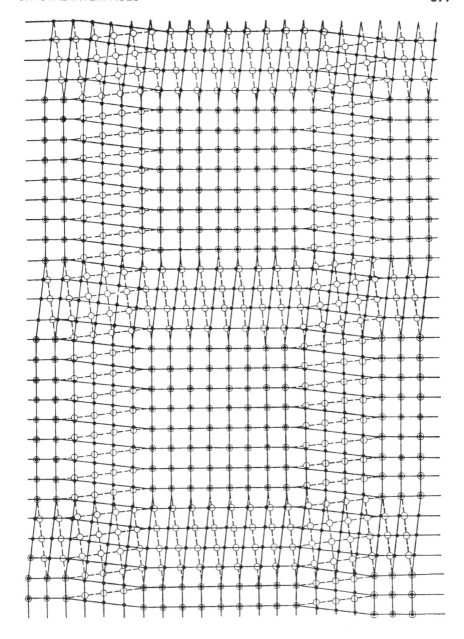

Figure 12.12 Twist boundary in a simple cubic lattice. The boundary is parallel to the plane of the figure. (From Read [13]. Copyright 1953. Reproduced by permission of McGraw-Hill Book Company)

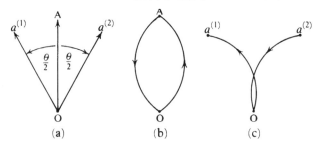

Figure 12.13 (a) Generation of grains 1 and 2 by opposite rotations of $\theta/2$. (b) Burgers circuit through the two grains. (c) Corresponding circuit in the reference lattice

The vector **B** is equal to the sum of the Burgers vectors of all the dislocations intersected by **OA**. Taking **OA** parallel to a $\langle 1\,0\,0\rangle$ in the mean lattice, the number of $\langle 0\,1\,0\rangle$ screw dislocations that it intersects, per unit length, is then

$$p = \frac{2\sin\theta/2}{b} \tag{12.18}$$

This is the same as the density of dislocations in a symmetrical tilt boundary of the same angle, and again for small θ the dislocation spacing d is

$$d \approx \frac{b}{\theta} \tag{12.19}$$

The energy of a twist boundary will increase with the angle of twist in the same general way as the energy of a tilt boundary increases with the angle of tilt. The energy should be cusped at those angles of twist at which the atoms fit together well at the boundary. For example, the rotation of 53° about [1 0 0] at which a symmetrical tilt boundary in a cubic lattice has a good fit also produces a twist boundary, normal to [1 0 0], on which the atoms fit together neatly. This is shown by Fig. 12.14, in which the atoms lying on the lattice points of both grains are picked out. The net of lattice points common to both grains in Fig. 12.14 and the similar net of coincident points lying on the corresponding tilt boundary (Fig. 12.10) are each one of the planes of a single *coincidence lattice*.

A coincidence lattice is a lattice formed by those lattice points that lie on top of one another after two initially superimposed lattices have been rotated with respect to another. A rotation about [1 0 0] of approximately 53°, or more economically of 37°, causes one-fifth of the lattice points of a cubic crystal to coincide. This is true for b.c.c. and f.c.c. lattices, as well as for simple cubic. Figure 12.15 shows the b.c. tetragonal lattice of coincident points produced from an f.c.c. lattice. The plane that has the highest density of coincident lattice points is the {1 1 0} in the coincidence lattice, which corresponds to a {2 1 0} plane in one of the f.c.c. lattices. Table 12.2 lists some other coincidence lattices formed from

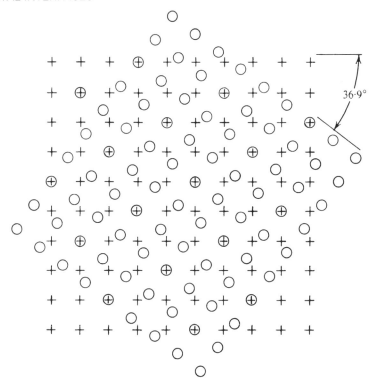

Figure 12.14 Twist boundary of good fit in a simple cubic lattice. The boundary is parallel to the plane of the figure. (From P. G. Shewmon [15])

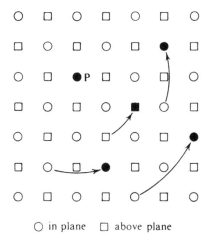

○ in plane   □ above plane

Figure 12.15 Part of the coincidence lattice produced from an f.c.c. lattice by a rotation of 36.9° about a [0 0 1] at P, the plane of the figure being (0 0 1)

Table 12.2   Some coincidence lattices for f.c.c. and b.c.c. crystals.
(After Brandon, Ralph, Rananathan and Wald [16])

Fraction of lattice points on coincident lattice	Rotations producing coincidence lattice		Closest packed plane of coincidence lattice written as plane of parent lattice	
	Axis	Angle	F.c.c.	B.c.c.
1:3	110	70.5	111	112
	111	60		
	210	131.8		
	211	180		
	311	146.4		
1:5	100	36.9		
	210	180	210	310
	211	101.6		
	221	143.1		
	310	180		
	311	154.2		
	331	95.7		
1:7	111	38.2		
	210	73.4	123	123
	211	135.6		
	310	115.4		
	320	149		
	321	180		
	331	110.9		

cubic lattices by rotations about prominent crystallographic directions. Because of symmetry, each coincident lattice can be produced by several different rotations. A grain boundary that contains some points common to the lattices of both grains can be formed by choosing any plane of a coincident lattice and removing one of the component lattices on one side of this plane and the other on the other side. The coincident lattice therefore provides a useful way of *grouping together* various boundaries of good fit between grains having the various equivalent orientation relationships that produce the same coincidence lattice. The degree of fit is striking only when the coincidence lattice has a high density of points and the boundary orientation is that of a prominent plane in this lattice.

The possibility of grouping together boundaries of good fit on the basis of coincidence lattices is really a consequence of crystal symmetry. In less symmetrical crystals such a treatment may not be possible. For example, Fig. 12.16 shows a twin boundary of excellent fit in a monoclinic crystal. The orientation relationship can be described as a rotation of π about the normal to the twin plane OA. In general, OA is irrational and no lattice of coincident sites exists.

The method employed to determine the dislocation content of a twist boundary can be generalized to cover any type of boundary. We can produce any grain

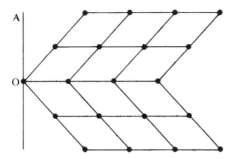

Figure 12.16 Twin boundary in a monoclinic lattice. The boundary is normal to the plane of the figure

boundary by means of an array of dislocations in a lattice which we may call the source lattice. The total Burgers vector of the dislocations intersected by a vector **R** which lies in the boundary is given by

$$\mathbf{B} = \mathbf{r}_2 - \mathbf{r}_1 \qquad (12.20)$$

The vectors $\mathbf{r}_1$ and $\mathbf{r}_2$ are vectors generated from **R** by rotations equal in magnitude but opposite in sense to those required to generate grains 1 and 2 respectively from the source lattice. (The reversal of sense arises from the use of a closure failure in a reference lattice which is parallel to the source lattice.) In a real crystal, the vector **B** must then be allotted to possible crystal dislocations. For a small rotation, expressed in magnitude and direction by the vector θ,

$$\mathbf{B} = \mathbf{R} \times \theta \qquad (12.21)$$

as is easily seen from Fig. 12.17.[†]

Although it is possible to determine a dislocation array that produces the correct orientation relationship for any grain boundary, the array will have little physical significance unless the dislocations within it are so far apart that their core structures do not overlap. Only then will it be possible to relate the properties of the grain boundary to the properties of individual dislocations.

[†] When θ is large the rotation cannot be simply described by a vector and (12.21) becomes

$$\mathbf{B} = (\mathbf{R} \times \mathbf{l})2 \sin \theta/2$$

where **l** is a unit vector in the mean lattice (source lattice) such that grain 1 is generated from the source lattice by a rotation of $-\theta/2$ and grain 2 by a rotation of $+\theta/2$ about **l**. **B** and **R** have the same meanings as for very small θ and so **B** and **R** are *defined in the source lattice* (see, for example, Bollmann [17]. The analysis was first developed by F. C. Frank).

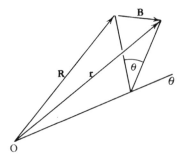

Figure 12.17

12.3 INTERFACE JUNCTIONS

Interface junctions occur when a grain boundary comes to a surface for example, or when three grains are in contact with one another inside a polycrystal. From the equilibrium positions of the interfaces at such junctions it is possible to deduce something about the relative magnitudes of their interfacial free energies or about the variation of the energies with interface orientation.

When the energy of each intersecting interface is independent of orientation, the equilibrium configuration is easily deduced. In effect, each boundary, by trying to reduce its area, exerts a force upon the junction and the total force must vanish. When a grain boundary meets a surface, it pulls the surface down into a groove until its own tension is balanced by the surface tensions (provided that the temperature is high enough to permit atoms to move and thereby accomplish the equilibrium configuration — to form the groove, material must move from the intersection out to the sides, where it forms hills, as shown in Fig. 12.18).

The equilibrium configuration for the symmetrical case shown in Fig. 12.18 can be deduced as follows. Consider an infinitesimal displacement of the junction in

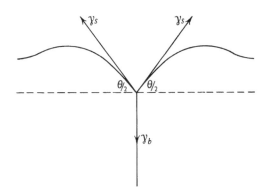

Figure 12.18   Grain boundary groove, seen in cross-section

a direction parallel to the grain boundary. The total free energy change accompanying this displacement from equilibrium must be zero. If unit length of junction is displaced by dx, the area of grain boundary decreases by dx while the area of surface increases by $2dx \sin \theta/2$. Therefore

$$-\gamma_b\, dx + 2\gamma_s\, dx \sin \theta/2 = 0$$

or

$$\gamma_b = 2\gamma_s \sin \theta/2 \tag{12.22}$$

where γ_b and γ_s are the grain boundary and surface free energies. Equation (12.22) can also be obtained by balancing the tensions which are, in effect, exerted by the interfaces and this shows that the magnitudes of these tensions are γ_b and γ_s. By measuring the angles of grain boundary grooves, the ratio γ_b/γ_s has been found for several materials. Some values are listed in Table 12.3, along with values of γ_b deduced from the ratio γ_b/γ_s together with independent measurements of γ_s.

Unfortunately, the assumption that the free energies of the interfaces are independent of their orientations is not usually correct. If a piece of interface can lower its free energy by turning into a different orientation, then in order to prevent it from doing so, a couple must be applied to it. This results in the addition of so-called *torque terms* to the balance of forces at a junction.

The forces that would have to be applied to the edges of a segment of mobile interface in order to hold it in equilibrium are given by the condition for the free energy to be a minimum, or

$$dG = 0$$

where dG is the change in free energy of the system which accompanies an infinitesimal displacement from equilibrium. By considering the creation of an extra length dx of the interface of length l shown in Fig. 12.19,

$$-F_x\, dx + \gamma\, dx = 0$$

Table 12.3 Energies of high-angle grain boundaries

Material	Temperature (°C)	γ_{gb}/γ_s	γ_{gb} mJ m^{-2} (using γ_s from Table 12.1)	Method	Ref.
Fe (δ)	1500	0.24	470	Grooving	18
Fe (γ)	1100	0.40	780	Grooving	19
Cu	850	0.37	610	Grooving	19
Au	850	0.25	350	Grooving	19
Ni	1350	0.40	690	Grooving	19
Nb	2250	0.36	760	Grooving	19
Sn	213	0.235	160	Grooving	19
NaCl	~600	—	266 ± 20%	See Ref.	20
LiF	~600	—	398 ± 20%	See Ref.	20

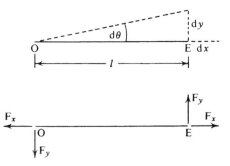

Figure 12.19 A segment of an interface, OE, held in equilibrium by forces F_x, F_y

Therefore

$$F_x = \gamma \tag{12.23}$$

By considering the displacement dy,

$$-F_y \, dy + l \frac{d\gamma}{d\theta} \, d\theta = 0$$

Since $dy \approx l d\theta$,

$$F_y = \frac{d\gamma}{d\theta} \tag{12.24}$$

The force F_x must be applied to ends of the boundary to stop it from shrinking and the force F_y to stop it from turning into an orientation of lower energy. If the point E, for example, is an interface junction, then the boundary OE exerts forces upon the junction which are equal and *opposite* to F_x and F_y.

If the interface happens to be at the orientation of a cusp in free energy, where $d\gamma/d\theta$ changes sharply from, say, $-\alpha$ to $+\alpha$, then the situation becomes indeterminate. No force F_y is then needed to stop the boundary from turning, since any rotation of the boundary out of its cusp orientation increases its energy. In fact, a force F_y of any value between the limits $-\alpha$ and $+\alpha$ could be applied without causing the boundary to rotate.

The existence of energy cusps at special grain boundary orientations can lead to odd-looking configurations. For example, Fig. 12.20 shows two boundaries of the same twin joining at right angles. It is supposed that both are special boundaries; e.g. boundary 1 may be the K_1 plane of the twin so that the atoms upon it are common to the lattices of both twin and matrix, and boundary 2 may be some other plane of good fit. The tilt and twist boundaries of Figs. 12.10 and 12.14 are possible instances. All that can be deduced from this configuration is that the tension of boundary 2 is not great enough to pull boundary 1 out of its special orientation:

$$\gamma_2 < \frac{d\gamma_1}{d\theta} \tag{12.25}$$

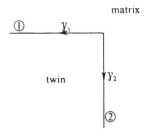

Figure 12.20

and similarly

$$\gamma_1 < \frac{d\gamma_2}{d\theta} \tag{12.26}$$

where, in each case, $d\gamma/d\theta$ is the magnitude of the slope at the cusp in the plot of boundary free energy against the angle of rotation θ about an axis parallel to the line of intersection of boundaries 1 and 2.

The case of a triple interface junction is shown in Fig. 12.21. If it is assumed that none of the intersecting interfaces lies in a special orientation, two equations of equilibrium can be derived by allowing the junction to be infinitesimally displaced in two orthogonal directions and by developing the condition that the resultant change in free energy must be zero. The same equations can be easily written down from the force diagram of Fig. 12.21b. The forces in Fig. 12.21b are forces that the boundaries exert upon the junction (i.e. they are opposite in sense to the forces F_x and F_y shown in Fig. 12.19, which are forces exerted upon the boundary). It is often convenient to use the three symmetrical equations, only two

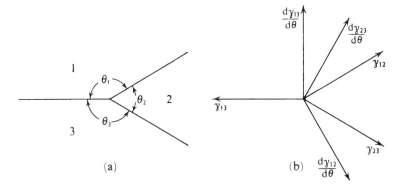

Figure 12.21 The junction of the interfaces between three grains. Each interface is normal to the plane of the figure: (a) geometry of junction and (b) force diagram. The positive sense of $d\theta$ is taken to be a *clockwise* rotation of the interface concerned

of which are independent, which are obtained by setting the sum of the components of force in the directions of the three interfaces equal to zero. These are

$$\gamma_{13} + \gamma_{12}\cos\theta_1 + \gamma_{23}\cos\theta_3 - \frac{d\gamma_{12}}{d\theta}\sin\theta_1 + \frac{d\gamma_{23}}{d\theta}\sin\theta_3 = 0$$

$$\gamma_{12} + \gamma_{23}\cos\theta_2 + \gamma_{13}\cos\theta_1 - \frac{d\gamma_{23}}{d\theta}\sin\theta_2 + \frac{d\gamma_{13}}{d\theta}\sin\theta_1 = 0 \qquad (12.27)$$

$$\gamma_{23} + \gamma_{13}\cos\theta_3 + \gamma_{12}\cos\theta_2 - \frac{d\gamma_{13}}{d\theta}\sin\theta_3 + \frac{d\gamma_{12}}{d\theta}\sin\theta_2 = 0$$

These equations were first derived by Herring [21].

Returning to the case of a grain boundary intersecting a surface, we see that if there are important, unknown, torque terms $d\gamma/d\theta$, it is generally impossible to find the ratio of grain boundary energy to surface energy from the angles at a single junction. However, in the special case of the intersection of a twin lamella with the surface of an f.c.c. crystal, it has been possible to simplify Eqns (12.27) and apply them in an elegant way.

The simplification follows from the relative magnitudes of the various terms. The energy of the {1 1 1} twin boundary is expected to be small; indeed, it may well be even smaller than the magnitude of the torque term $d\gamma_s/d\theta$ arising from the anisotropy of the surface free energy. As a result, the torque terms acting at a junction between a twin boundary and the surface can overcome the tension of the twin boundary and pull the surface up into a ridge, as shown in Fig. 12.22.

The occurrence of a ridge, instead of a groove, demonstrates most strikingly the existence of the torque arising from the tendency of the surface to rotate into a more favourable orientation. Because the twin boundary is itself, at a special orientation where there is a cusp in free energy, the only one of Eqns (12.27)

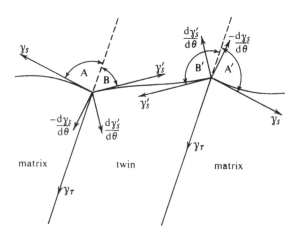

Figure 12.22 Twin boundary grooving, seen in cross-section

that can be applied is the one that does not contain a torque term for the twin boundary. In other words, we can only balance the forces acting parallel to the twin boundary. At the two junctions shown in Fig. 12.22, we have, respectively,

$$\gamma_T = \gamma_s \cos A + \gamma'_s \cos B + \frac{d\gamma_s}{d\theta} \sin A - \frac{d\gamma'_s}{d\theta} \sin B$$

and   (12.28)

$$\gamma_T = \gamma_s \cos A' + \gamma'_s \cos B' - \frac{d\gamma_s}{d\theta} \sin A' + \frac{d\gamma'_s}{d\theta} \sin B'$$

Adding Eqns (12.28) we obtain, with the assumption that $\gamma_s \approx \gamma'_s$,

$$2\gamma_T = \gamma_s (\cos A + \cos A' + \cos B + \cos B')$$
$$+ \frac{d\gamma_s}{d\theta}(\sin A - \sin A') + \frac{d\gamma_s}{d\theta}(\sin B' - \sin B)$$

Now by observation, $A + A' \approx 180°$, $B + B' \approx 180°$ so that $\sin A - \sin A'$ and $\sin B - \sin B'$ are small quantities. Since $d\gamma_s/d\theta$ and $d\gamma_s/d\theta$, although larger than γ_T, are themselves expected to be much smaller than γ_s, we can write

$$2\gamma_T \approx \gamma_s (\cos A + \cos A' + \cos B + \cos B')$$ (12.29)

Hence the ratio of the free energies of the twin boundary and surface, γ_T/γ_s, can be found. This technique was first used by Mykura [22] to measure twin boundary free energies (knowing the surface energy) and to build up a picture of the anisotropy of the surface free energy in f.c.c. metals. The main conclusions are that the free energy of a {1 1 1} boundary of a twin in an f.c.c. metal is very small compared with a surface free energy and that the surface free energy does not vary by more than 10% with change in orientation. Provided that the surfaces are clean, it is found that the close packed {1 1 1} surface has the smallest free energy, as might be expected. Some values of twin boundary free energies are listed in Table 12.4.

Table 12.4   Twin boundary free energies

Metal	Temp. (°C)	Atmosphere	γ_T/γ_s	$\gamma_T/\gamma_{g.b.}$	γ_T (mJm^{-2})	Ref.
Ag	900	air	0.0051	—	1.8	23
Co	1290	vac.	0.0069	—	—	24
Cu (OFHC)	1000	H$_2$	0.027 ±0.01	—	47	25
Cu (high purity)	1000	H$_2$	0.007		12.2	25
Cu	950	—	—	0.045	28	19
Ni	1000	vac.	±0.015 0.005	—	26	22
Pt	1080	vac.	0.053	—	150	24
Pb	220	—	—	0.05	10	19

12.4 THE SHAPES OF CRYSTALS AND GRAINS

Although the shape of a crystal is ordinarily a consequence of the way in which it grew, or perhaps of its cleavage, there must nevertheless be some equilibrium shape that might be reached in practice only by an unconstrained small crystal or by a small void within a crystal. This shape is determined by the surface free energy γ and is such as to minimize the total free energy, i.e.

$$\int_A \gamma \, dA = \text{a minimum} \qquad (12.30)$$

If the surface free energy is isotropic, then the equilibrium shape is a sphere; i.e. the surface area of a given volume is minimized. If the surface free energy varies with orientation, then the equilibrium shape can be derived from the γ-plot by means of a theorem due to Wulff, which will be stated without proof.[†] The equilibrium shape is that of the inner envelope of planes normal to, and passing through the ends of, the vectors representing surface free energies on the γ-plot ('Wulff planes').

If the γ-plot is deeply cusped then the equilibrium shape will be a polyhedron. Figure 12.23 illustrates this situation for the hypothetical γ-plot of an f.c.c. metal first shown in Fig. 12.5. If the cusps are less pronounced, then the equilibrium shape may contain rounded regions, although it must still possess flats at the cusp orientations. The question of whether a surface of given orientation will appear on the equilibrium shape, or 'Wulff body', can be studied by means of a construction due to Herring [27].

Consider some orientation, OA in Fig. 12.24, O being the origin of the γ-plot, which is smoothly curved at A. If the surface which is normal to OA is to appear

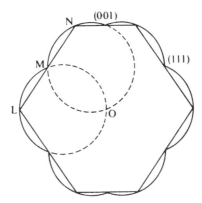

Figure 12.23 The γ-plot of Fig. 12.5, showing the equilibrium shape (or Wulff body) of the crystal

---

[†] For a discussion of the proof of Wulff's theorem, see Mullins, [26].

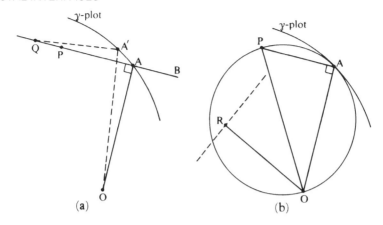

Figure 12.24 Construction due to Herring [27]

on the Wulff body then it must do so in a region of the body that touches the Wulff plane AB. Where on the plane AB does this happen? This question is answered by considering points in the neighbourhood of A, such as A'. These points represent the free energies of surfaces of slightly different orientations to the surface whose free energy is represented by A, and so these surfaces must appear on the Wulff body close to the point at which the surface is normal to OA. As the points are brought closer and closer to A their Wulff planes must touch the Wulff body at points closer and closer to the point at which the Wulff plane at A touches it. The plane which is normal to OA' intersects AB in a line at Q. As various points A', A'', ... around A are brought in towards A, the lines of intersection of their Wulff planes tend towards an intersection at a single point, P. If the Wulff body exhibits the surface which is normal to OA, then it does so at the point P. Now in two dimensions, the locus of points defined by the property that the normals to the radius vector at each point intersect in a common point P' is a circle through the origin O and having OP' as diameter (the diameter of a circle subtends an angle of $90°$ at all points on the circumference). There is a corresponding result in three dimensions, from which it follows that the γ-plot at A touches a sphere having OP as diameter, as shown in Fig. 12.24b. An important result follows immediately. Consider any other direction through O. If it meets the γ-plot inside the sphere of diameter OP, as at R, then its Wulff plane will cut off the point P, which cannot then lie on the *inner envelope* of Wulff planes. Evidently the surface that is normal to OA cannot appear if the γ-plot passes anywhere inside the sphere through O which is tangent to the γ-plot at A. If, on the contrary, the γ-plot is everywhere outside this tangent sphere, then access to P cannot be cut off and the surface that is normal to OA must appear on the Wulff body. By similar reasoning, it follows that an orientation at which there is an energy cusp will appear on the Wulff body if and only if a sphere having the line from O to the cusp point as diameter lies entirely inside the γ-plot.

An interesting special case arises when the γ-plot coincides with a portion of the tangent sphere. Figure 12.23 illustrates this situation, in two dimensions. There, the section of the γ-plot consists of circles through the origin, two of which are shown. OL is a diameter of the circle LMO so that all Wulff planes to points on the γ-plot in the section L $\rightarrow$ M touch the Wulff body at the corner L. Similarly, ON is a diameter of the circle OMN and all Wulff planes to points on the γ-plot from M to N touch the Wulff body at the corner N. In fact, every Wulff plane from the γ-plot touches a corner of the two-dimensional Wulff body, so that, strictly speaking, surfaces of all orientations appear on it, although only surfaces of the cusp orientations such as the {1 1 1} surface LN have any extension, other surfaces being crowded into a corner such as N. If the cusp points of Fig. 12.23 are kept fixed but the rest of the γ-plot is expanded, as in Fig. 12.25, the Wulff body remains the same, but now Wulff planes from general points do not touch it, even at a corner. This distinction has some significance in connection with thermal faceting, which is defined below.

An ordinary-sized crystal of arbitrary shape never reaches its equilibrium shape in practice, because of the large redistribution of material required and the small amount of energy to be gained thereby. It is more likely that the surface of a crystal will be able to reduce its energy by means of the relatively small atom movements needed to break up the surface into small sections, one or more of which have an orientation of lower energy. This process is called thermal faceting (Fig. 12.26). Herring has shown that a surface of an orientation not represented on the Wulff body can always reduce its total surface free energy by faceting. A surface parallel to a Wulff plane which touches the Wulff body cannot reduce

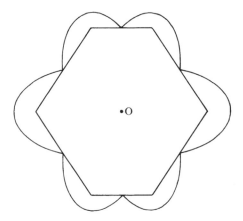

Figure 12.25 The same equilibrium shape as shown in Fig. 12.23, arising from a different γ-plot

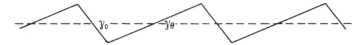

Figure 12.26 A surface that has reduced its energy by breaking up into facets, one of which is a low-index plane having a small surface tension γ_0. The surface is normal to the plane of the figure

its energy by faceting, even when the plane touches only a corner of the Wulff body, as in Fig. 12.23.

Crystals seldom appear in their equilibrium shape; it is much more common for their shape to be the result of the way in which they grew. When the growth rate of a crystal is determined only by the rate at which its surfaces can accept atoms, then the rate of advance of a surface will depend only on its crystallographic orientation. In this way, the shape of a salt crystal grown from aqueous solution may be determined simply by the relative growth rates at its various surfaces. The faster-growing orientations will tend to grow themselves out, leaving the crystal with its more slowly growing surfaces. The steady state shape of the growing crystal can be derived from a polar diagram of the growth rate by means of the same construction used to derive the equilibrium shape from the γ-plot. The analogue of the Wulff theorem is that the crystal becomes trapped in that shape which grows at the slowest possible rate.

Often the rate of advance of the surface of a growing crystal does not depend solely on its orientation. Instead, the growth rate may be controlled by the rate at which material can be transported to the surface or by the rate at which heat can be removed. Under these conditions an unstable form of growth sometimes occurs and produces a spiky or 'dendritic' morphology. Dendrites are observed frequently when a crystal has grown from a liquid, and occasionally in solid state transformations.

To illustrate how growth instabilities can occur, consider, for example, a plane surface of a crystal growing into a supersaturated solution. Suppose that the growth rate is limited only by the rate of diffusion of solute atoms to the surface. Then there must be a concentration gradient from the supersatured value in the solution to the concentration in equilibrium with the crystal at the surface. If now by chance a bump starts to develop, the concentration gradient at its nose must become steeper, because it is thrust in towards the region of high solute concentration. The steepening of the gradient increases the rate of diffusion to the bump and tends to amplify it. Opposing this is the action of surface tension, which tends to flatten the bump and sets up a flux of solute from the bump out to the sides. Provided that the bump is not too sharp, its amplitude will show a net increase. The growth of the bump into a spike, followed by the development of bumps upon the spike, and so on, produces a dendrite. Often crystallographic factors apparently combine with conditions of unstable growth to produce a dendritic crystal whose dendrite arms have crystallographic orientations.

The arrangement of the grains within a polycrystal is observed to be quite similar to that of bubbles within a froth. This suggests that the pattern of grains is governed by grain boundary tensions just as the pattern of bubbles in a froth is governed by the surface tension of the liquid between the bubbles. In drawing this analogy the anisotropy of the grain boundary free energies is neglected. The two principles that govern the shapes of the grains are then, firstly, that they must fill space and, secondly, that grain boundary tensions must balance at grain edges and corners.

The only edges that should occur are those along which three boundaries meet, and the only corners those at which four edges meet, because other junctions can reduce the grain boundary area by decomposing after the manner shown in Fig. 12.27. (In fact, polycrystals, unlike froths, have been found to contain a small proportion of abnormal junctions, perhaps because of some anisotropy in the boundary free energy.) The polyhedron which can be stacked to fill space under this constraint is the cubo-octahedron, shown in Fig. 12.28a. However, this polyhedron, when stacked, does not fulfil the requirement that the boundary tensions should balance at its edges and corners. Kelvin showed that to make the three faces meet at 120° and the four edges at 109.47° each hexagonal face must acquire complex curvatures, as shown in Fig. 12.28 b.

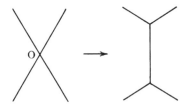

Figure 12.27   The instability of four interfaces meeting along a line through O, normal to the figure

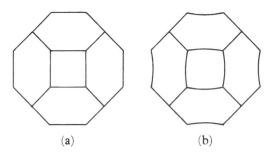

(a)   (b)

Figure 12.28   (a) Cubo-octahedron and (b) distorted cubo-octahedron which, when stacked, meets the requirement that the surface tensions balance at all junctions. (After Lord Kelvin [28])

Of course, neither polycrystals nor froths ever attain the ideal regularity of an array of equal distorted cubo-octahedra. Grains vary in size, in the number of faces that they possess and in the shapes of their faces. Nevertheless, their average properties are found to correspond closely to those of the cubo-octahedron. For example, the average number of faces per grain is close to 14, while the average number of edges per face is close to $5\frac{1}{7}$. A face with five edges occurs more frequently than any other [29].

In a froth, the faces of bubbles with fewer than average faces have to bow out so as to increase the angles at which they meet to 120°. To maintain this convexity, the gas in the bubble must be at a higher pressure than the gas in neighbouring bubbles. If the gas can diffuse through the bubble wall it will leak out and the bubble will shrink. The ultimate loss of such bubbles increases the average bubble size. Grain growth also occurs in polycrystals at high temperatures. The tendency of a curved grain boundary to reduce its area whilst at the same time maintaining a balance of tensions at its edges causes it to migrate towards its centre of curvature.

12.5 BOUNDARIES BETWEEN DIFFERENT PHASES

Like a grain boundary, the boundary between two different crystals can be specified by describing the orientation relationship between the lattices of the two crystals and the orientation of the boundary itself. Corresponding to grain boundaries of special orientation, such as twin boundaries, there are special boundaries between two different crystals in a specific orientation relationship; corresponding to high angle grain boundaries, there are interfaces between randomly oriented crystals of different kinds.

The free energy of a random interphase boundary relative to that of a grain boundary can be obtained from the angles between interfaces in a two-phase mixture. Figure 12.29 shows a particle of a phase B situated at a grain boundary of the phase A. Assuming that the interfacial energies do not vary with orientation (i.e. neglecting torque terms), the 'dihedral angle' θ is given by an equation similar to Eqn (12.22), viz.

$$\cos \frac{\theta}{2} = \frac{\gamma_{AA}}{2\gamma_{AB}} \qquad (12.31)$$

Table 12.5 lists some results derived from the measurement of dihedral angles. It can be seen that the tensions of random interphase boundaries do not differ greatly from the tensions of high-angle grain boundaries in the same system, but are often slightly less.

At the opposite extreme to the random interphase boundary is a boundary between two crystals that have different atoms but identical structures, so that boundaries of any orientation between parallel crystals are fully coherent. This situation occurs when a precipitate is formed by ageing a supersaturated solid

Table 12.5 Relative energies of random interphase boundaries

System	Temperature (°C)	Phase A	Phase B	γ_{AB}/γ_{AA}	γ_{AB}/γ_{BB}	Ref.
Fe–C	825	α b.c.c. 0.01% C	γ f.c.c. 0.22% C	0.93 ± 0.02	0.90 ± 0.02	30
Fe–Cu	825	α b.c.c.	Cu	0.74	0.86	31
	1000	γ f.c.c.	Cu	0.61	0.87	
Cu–Zn	700	α f.c.c.	β b.c.c.	0.83	1.00	31
Cu–Sn	750	α f.c.c.	β b.c.c.	0.76	0.93	31
NaCl–NaF	600	NaCl	NaF	0.90	0.78	20
NaCl–LiF	600	NaCl	LiF	1.13	0.76	20
NaCl–NaI	550	NaCl	NaI	1.08	0.78	20
LiF–CsCl	550	LiF	CsCl	1.38	0.65	20

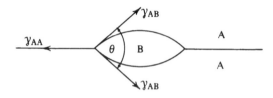

Figure 12.29

solution of Ag in Al. The Ag atoms collect into small spherical clusters on a continuous f.c.c. lattice. Another possibility is that the crystal structures of a precipitate and matrix differ, but possess a common sublattice. For example, magnesioferrite, $MgFe_2O_4$, contains an f.c.c. sublattice of oxygen ions which is almost identical to that of MgO (Section 3.6). Small $MgFe_2O_4$ particles precipitating within MgO preserve the oxygen ion sublattice, and an interface passing through oxygen ions, such as a suitably located {1 1 1}, is common to both structures. A third possibility is that the two crystal structures possess only a plane in common. For example, the close packed planes of an f.c.c. and an h.c.p. crystal match exactly if the interatomic distances are equal; the martensitic transformation in cobalt illustrates this case. In general, of course, it is impossible to find identical planes in two different crystal structures, and arbitrarily chosen crystals cannot be joined by a coherent boundary.

Usually, even the corresponding planes of two specially related crystals match only approximately. An exact fit could then be achieved by means of a uniform elastic strain or, alternatively, the misfit could be localized at dislocation lines. A boundary in which a small misfit is taken up by an array of dislocations is called a partially coherent boundary; it is analogous to a low-angle grain boundary. A difference is that an array of dislocations which constitutes a low-angle boundary is always such that the strain vanishes at large distances, whereas a partially coherent boundary array would produce a long-range strain if it were located within a single homogeneous medium. The *actual* strain within the pair of different crystals can be thought of as being the resultant of a homogeneous elastic strain which gives an exact fit and a strain due to the array of dislocations at the interface, the two strains cancelling at large distances from the interface. For example, Fig. 12.30 shows the case of a misfit in one direction in an interface between two orthorhombic crystals. The junction can be thought of as having been made by first stretching the upper crystal in the c direction to give a perfect fit, and then introducing parallel edge dislocations that cancel this strain at large distances, but at the expense of producing local regions of mismatch at the boundary.

The number of extra planes to be accommodated in the upper crystal of Fig. 12.30, in unit distance, is

$$\rho = \frac{1}{a_1} - \frac{1}{a_2} \tag{12.32}$$

The dislocation spacing $p = \rho^{-1}$ is therefore

$$p = \frac{a_1 a_2}{a_2 - a_1} \tag{12.33}$$

The dimension of a small particle precipitating from solid solution might be much less than the dislocation spacing given by Eqn (12.33). In this case, that particle would be fully coherent, the misfit being taken up by elastic 'coherency strain'.

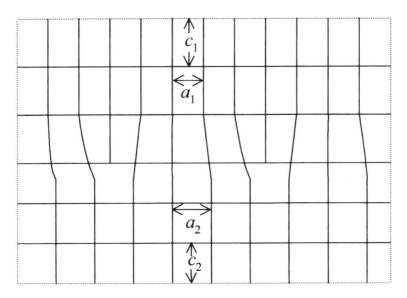

Figure 12.30 Interface between two orthorhombic crystals. The interface is normal to the plane of the figure

The quantity $(a_2 - a_1)/a_1$ may be defined as the *misfit*. When the misfit reaches 10%, the dislocations in Fig. 12.30 are separated by only ten atom spacings. The core regions of the dislocations are then on the point of overlapping and the dislocation array becomes useless for the purposes of calculating atom positions and the energy of the boundary by elasticity theory.

When the misfit is small, the energy of the boundary can be calculated. Its dependence on the misfit is very similar to the dependence of the energy of a low-angle grain boundary on its angle of tilt or twist. For a boundary between phases with identical elastic properties, Brookes [32] obtained the expression

$$E = \frac{\mu b \delta}{4\pi(1 - \nu)}(A_0 - \ln \delta) \qquad (12.34)$$

where δ is the misfit and A_0 has the same meaning as in Eqn (12.9). Figure 12.31 is a plot of a similar result obtained by Van der Merwe [33], who took explicit account of the dislocation core energy by using a Peierls model, i.e. by assuming that a sinusoidal force law operates between atoms on either side of the boundary (Section 7.5). For misfits greater than about 0.1 the energy stored in the bonds between these atoms exceeds the energy stored in the two elastic media.

The simple case of a misfit in one direction is less likely to be encountered in real materials than cases of a two-dimensional misfit, where more than one set of dislocations is required to take up the mismatch. For example, a misfit in the *b* dimensions of the orthorhombic crystals of Fig. 12.30 could be taken up by a set of edge dislocations orthogonal to that which matches the *a* dimensions. Square networks of this type have been seen at the faces of plates of UC_2 precipitated within a matrix of UC [34]. The structure of UC is the same as that of NaCl; the square pattern of U atoms in the $\{1\,0\,0\}$ planes almost matches that of the U atoms in the $(0\,0\,1)$ of UC_2, which has a b.c.t. structure. A square net of edge dislocations takes up the small mismatch at the faces of the $(0\,0\,1)$ plates of UC_2, which precipitate on the cube planes of UC.

Apart from its application to phase transformations, the theory of interphase boundaries is of interest in connection with the phenomenon of *epitaxial*

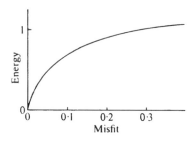

Figure 12.31 Energy of a boundary of the type shown in Fig. 12.30, between two large crystals having the same elastic constants. (After Van der Merwe [33])

growth — the growth of one crystal upon the surface of another in some definite orientation relationship. The earliest investigations were of growth from solution where, for example, one alkali halide will grow upon the surfaces of another, in the same orientation. In this case it was found that epitaxy was limited to crystals whose lattice parameters differed by less than about 15%. More frequently studied today is the epitaxial growth of a thin film of crystal by condensation from a vapour. When the film and substrate crystal structures differ only by having a slightly different lattice parameter in one direction within the interface, a dislocation array like that of Fig. 12.30 may take up the mismatch. The elastic energy of the array will be relatively unimportant in comparison with the misfit energy at the interface when the film is very thin, but a model such as that of Van der Merwe, which assumes a sinusoidal force law between the atoms at the interface, can still be used to study the energy of the array.

Although the results of early studies had suggested that epitaxy was restricted to crystals of the same type of bonding and having a small mismatch in the interface plane, many contrary examples were soon found. Table 12.6 lists some examples of epitaxial growth. It will be seen that f.c.c. metals grow epitaxially on cleaved mica and that they can be condensed epitaxially on to heated rock-salt with misfits of up to 38%. The uncertainty of predictions made solely on the basis of geometrical fit is demonstrated by the epitaxy of Ag on {0 0 1} surfaces of NaCl. With (0 0 1) Ag parallel to (0 0 1) NaCl, a misfit of atom sites of only 3% can be found by orienting [0 1 1] Ag parallel to [0 0 1] NaCl. However, the *observed* orientation relationship is [0 0 1] Ag parallel to [0 0 1] NaCl (Fig. 12.32). It is likely that a complete understanding of orientation relationships in epitaxy will depend on understanding how a deposit is nucleated on the substrate.

Table 12.6   Epitaxial deposition of f.c.c. metals. (After Pashley [35])

			Orientation relationship				
			Parallel planes		Parallel directions		Misfit
		Temperature					
Substrate	Metal	(°C)	Substrate	Metal	Substrate	Metal	(%)
NaCl	Ni	370	(0 0 1)	(0 0 1)	[1 0 0]	[1 0 0]	38
	Cu	300					36
	Ag	150					28
	Au	400					28
MoS$_2$	Ni	$\geq$120	(0 0 0 1)	(1 1 1)	[1 0 $\bar{1}$ 0]	[1 $\bar{1}$ 0]	6
	Cu	$\geq$50					3
	Ag	$\geq$20					9
	Au	$\geq$80					9
Mica	Ag	150	(0 0 1)	(1 1 1)	[1 0 0]	[1 $\bar{1}$ 0] or [2 $\bar{1}$ $\bar{1}$]	44
		250		(0 0 1)	[1 0 0]	[1 0 0]	21
	Au	450		(1 1 1)	[1 0 0]	[1 $\bar{1}$ 0] or [2 $\bar{1}$ $\bar{1}$]	44

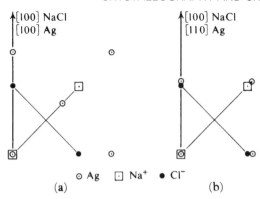

Figure 12.32 Epitaxy of Ag deposited on (0 0 1) of NaCl: (a) observed orientation relationship and (b) orientation relationship of small misfit

12.6 STRAINED LAYER EPITAXY OF SEMICONDUCTORS

Almost perfect expitaxy is obtained in some semiconductor crystals. The high purity of the materials, their similar crystal structures and the precision of modern growth techniques, such as molecular beam epitaxy (MBE), can lead to epitaxial layers grown on single-crystal substrates with no detectable defects at the interfaces. The interfaces are completely coherent in the sense defined on pages 393 and 394. The resulting layers can be in a state of biaxial strain. For example, for an InGaAs alloy layer grown on a (0 0 1) oriented GaAs substrate wafer, the biaxial strain in the plane of the layer changes the crystal structure from cubic to tetragonal because of the Poisson effect (Section 5.5), with the result that interesting electronic properties are realized in the layer. Similarly, InGaAs alloy layers grown on in the (1 1 1) surface for binary alloys such as GaAs undergo a trigonal distortion of the unit cell, which results in an interesting strain-induced piezoelectric effect.

Growth of perfect *pseudomorphic* strained layers is generally possible for layer misfits (or mismatches) below 2% and for layer thicknesses below a critical thickness. Above this critical thickness it is energetically favourable for some or all of the layer strain to be reduced or *relaxed* via the introduction of interfacial misfit dislocations as pointed out in Section 12.5. It is useful to be able to obtain the appropriate elasticity equations for layers grown in any orientation. Dunstan [36] provides the general method for achieving this. Figure 12.33 illustrates the special case of a (0 0 1) oriented strained layer. Consider that the substrate and the layer have cubic lattice parameters a_s and a_1 respectively when in their bulk forms. In order for the layer to grow coherently on the substrate it will be strained to adopt the in-plane lattice parameter of the substrate. This will result in a distortion of the lattice dimension perpendicular to the interface plane. The strained layer lattice parameters then become a_1, a_2 and a_3. For a fully strained layer

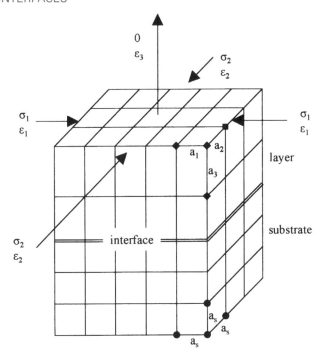

Figure 12.33 A strained epitaxial layer in the (0 0 1) orientation. The in-plane principal stresses and strains are taken to be compressive

$a_1 = a_2 = a_s$. For a relaxed layer a_1 and a_2 need not be equal and may take on values between a_1 and a_s. In general, whether fully strained or partially relaxed, the principle strains in the layer can be defined as follows:

$$\varepsilon_1 = \frac{a_1 - a_l}{a_l}, \qquad \varepsilon_2 = \frac{a_2 - a_l}{a_l}, \qquad \varepsilon_3 = \frac{a_3 - a_l}{a_l} \tag{12.35}$$

These strains are measurable, e.g. using X-ray diffraction techniques. In order to understand the relationships between these strains or to obtain the stress or strain energy in the layer it is necessary to use the equations developed in Chapter 5.

The stress–strain relationship for a layer in the (0 0 1) orientation using the two-suffix notation is

$$
\begin{pmatrix}
\sigma_1 \\
\sigma_2 \\
0 \\
0 \\
0 \\
0
\end{pmatrix}
=
\begin{pmatrix}
c_{11} & c_{12} & c_{12} & 0 & 0 & 0 \\
c_{12} & c_{11} & c_{12} & 0 & 0 & 0 \\
c_{12} & c_{12} & c_{11} & 0 & 0 & 0 \\
0 & 0 & 0 & c_{44} & 0 & 0 \\
0 & 0 & 0 & 0 & c_{44} & 0 \\
0 & 0 & 0 & 0 & 0 & c_{44}
\end{pmatrix}
\begin{pmatrix}
\varepsilon_1 \\
\varepsilon_2 \\
\varepsilon_3 \\
0 \\
0 \\
0
\end{pmatrix}
\tag{12.36}
$$

The free surface requires σ_3 to vanish and shear stresses are ruled out because of the translational symmetry in the layer. In pseudomorphic structures the in-plane strains ε_1 and ε_2 will normally be equal and have the value of the misfit, ε_0. The strain perpendicular to the interface is obtained from the third line of Eqn (12.36):

$$0 = c_{12}\varepsilon_1 + c_{12}\varepsilon_2 + c_{11}\varepsilon_3 \tag{12.37}$$

which rearranges to give

$$\varepsilon_3 = -\frac{2c_{12}}{c_{11}} \frac{\varepsilon_1 + \varepsilon_2}{2} \tag{12.38}$$

This equation is also used for the case when there is some relaxation of the strain via plastic relaxation. In the case of plastic relaxation the in-plane strains ε_1 and ε_2 will not necessarily be equal. The strain is also commonly expressed in the form

$$\varepsilon_3 = -\nu_2\varepsilon_{\text{ave}} \tag{12.39}$$

where

$$\varepsilon_{\text{ave}} = \frac{\varepsilon_1 + \varepsilon_2}{2}$$

This defines a two-dimensional Poisson's ratio.

The first two lines of Eqn (12.36) give the in-plane stress–strain relationship:

$$\sigma_1 = c_{11}\varepsilon_1 + c_{12}\varepsilon_2 + c_{12}\varepsilon_3$$
$$\sigma_2 = c_{12}\varepsilon_1 + c_{11}\varepsilon_2 + c_{12}\varepsilon_3 \tag{12.40}$$

Thus, adding and using Eqn (12.38) we have

$$\sigma_{\text{ave}} = (c_{11} + c_{12} - 2c_{12}^2/c_{11})\varepsilon_{\text{ave}}$$
$$= M\varepsilon_{\text{ave}} \tag{12.41}$$

which defines a two-dimensional Young's modules denoted by M. Table 12.7 gives values for some common semiconductors. For a pseudomorphic layer, σ_1, σ_2 and σ_{ave} are all equal and may be written as σ_0.

The elastic strain energy of the layer can be written in terms of the tensors as

$$E = \int_V \frac{1}{2}\sum \sigma_{ij}\varepsilon_{ij}\,dV \tag{12.42}$$

or per unit area for a pseudomorphic layer of thickness h as

$$E = 2 \times \frac{1}{2}h\sigma_0\varepsilon_0 = hM\varepsilon_0^2 \tag{12.43}$$

The elastic energy of the layer thus depends on the square of the misfit.

If dislocations were introduced into the interface between the layer and substrate the elastic strain energy may be reduced. The situation is a little

Table 12.7 Lattice parameters and stiffness constants for some common semi-conductors. (The stiffness constant values are taken from Dunstan [36])

Compound	Lattice parameter (Å)	C_{11} (GPa)	C_{12} (GPa)	C_{44} (GPa)	M_{001} (GPa)
Si	5.4310	166	64	79.5	181
Ge	5.6537	124	41	68	138
AlP	5.4580	160	75	40	165
AlAs	5.6610	120	57	59	123
AlSb	6.1353	88	43	41	89
GaP	5.4505	140	62	70	147
GaAs	5.6535	118	53.5	59	123
GaSb	6.0954	88	40	43	92
InP	5.8688	101	56	45.5	95
InAs	6.0585	83	45	39.5	79
InSb	6.4788	67	38	31	62
$In_x Ga_{1-x} As^a$	$0.4050x + 5.6535$	$-35x + 118$	$-8.5x 53.5$	$-19.5x + 59$	$-44x + 123$

[a] Where ternary alloys such as InGaAs exist as single-phase solutions throughout the entire composition range, Vegard's law can be used to interpolate for the lattice parameter and stiffness constant between the binary compound values.

different from that shown in Fig. 12.30 since the layer is of finite thickness h. The introduction of dislocations will lower the elastic energy because the coherency strain is relieved close to the dislocation. However, the dislocations possess self-energy and if they are spaced p apart and are close to the interface the additional energy per unit area of the interface, due to the dislocations, will be approximately

$$E_\perp = \frac{Gb^2}{4\pi(1-v)}\frac{1}{p}\ln\left(\frac{\beta h}{b}\right) \qquad (12.44)$$

where β is a constant (see Eqn 8.8). If the strain were completely relieved the value of p would be equal to a_s/ε_0 from Eqn (12.33).

The elastic strain energy of the layer increases as $h\varepsilon_0^2$ from Eqn (12.43) while the energy of the dislocation array will increase as ε_0 from Eqn (12.44). These equations therefore predict that for small values of h the pseudomorphic layer is stable with no dislocations present, but that above a certain critical thickness it should be energetically favourable to introduce dislocations to relieve the strain (see problem 12.14). The critical thickness is inversely proportional to ε_0 and Dunstan, Young and Dixon [37] have given a simple geometrical argument as to why this should be so. This is illustrated in Fig. 12.34. The strain field of a dislocation at a depth h from a surface must decay laterally to zero within a distance of approximately h. Consequently, a dislocation can only relieve the strain within the layer over a lateral distance mh, where m is a small number around 1 or 2. Outside this region the relaxation is zero and within it averages b/mh, where b is the Burgers vector. A dislocation will not be produced if the strain induced, b/mh, exceeds the strain in the layer.

(a)   (b)   (c)

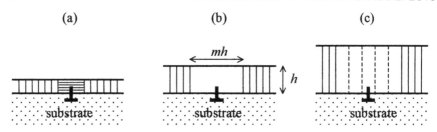

Figure 12.34 The relaxed region, with lateral dimension mh, around a misfit dislocation is shown for three layer thicknesses, h. Vertical hatching corresponds to strain as grown-in. Horizontal hatching indicates strain of the opposite sign. In a thin layer which is below the critical thickness, (a), relaxation induces an opposite strain in the region around the dislocation. When the layer is at critical thickness, (b), the strain is fully relieved around the dislocation. In the thickest layer, (c), the strain is only partly relieved

Thus the layer is prevented from relaxing at small thicknesses because the relaxation (b/mh) exceeds the strain ε_0. It may relax when $(b/mh) = \varepsilon_0$. Rearranging, we obtain a prediction for the critical thickness of

$$h_c = \frac{b}{m\varepsilon_0} \qquad (12.45)$$

If we take b as the atomic separation and $m = 1$, the equation becomes

$$h_c = \frac{1}{\varepsilon_0} \qquad (12.46)$$

where h_c is measured in units of monolayers. This equation is in agreement with experiment and is consistent with treatments based on the change of energy.

Layers may be grown which are considerably thicker than the critical thickness and remain strained for long periods at high temperature. One reason for this is the difficulty of producing dislocations with the correct Burgers vector to relieve the strain. This difficulty arises in face-centred cubic lattices strained biaxially on (1 0 0) as in Fig. 12.33 because the Burgers vectors of common glide dislocations and their slip planes are quite inconvenient for relieving this particular pattern of strain (see problem 12.15).

PROBLEMS

12.1 Sketch the structure or structures of surfaces of the following orientations in a c.p.h. metal: (a) $(1\,0\,\bar{1}\,2)$, (b) $(1\,1\,\bar{2}\,2)$.

12.2 Show that any dislocation line of Burgers vector **b** emerging through a surface of unit normal **n** produces a step whose height, measured normal to the surface, is $\mathbf{n} \cdot \mathbf{b}$. Show that the step cannot be removed by evaporation of the crystal except in one special case. Write down an expression for the height of the step in units of the spacing of the lattice planes parallel to the surface. Apply the expression to the case of a dislocation with a $\frac{1}{2}[1\,1\,0]$ Burgers vector in a crystal with an f.c.c. lattice when the dislocation emerges through the following surfaces: $(1\,1\,1)$, $(2\,1\,1)$, $(1\,0\,0)$, $(1\,1\,0)$.

12.3 According to Mackenzie, Moore and Nicholas, when the energy of a surface in an f.c.c. metal is estimated on the basis of the number of bonds between nearest neighbours which are broken, the energy of a surface of orientation $(h\,k\,l)$ is related to the energy of the $\{2\,1\,0\}$ surface by the equation

$$\gamma_{hkl} = \gamma_{210} \cos\theta$$

where θ is the angle between the surface $\{h\,k\,l\}$ and the nearest $\{2\,1\,0\}$ surface. Show that the Wulff plot for this model consists of portions of spheres whose diameters are the vectors representing $\{2\,1\,0\}$ energies on the plot and sketch the section of the plot for surfaces in the $[0\,0\,1]$ zone, giving the values of the energies of low-index surfaces in terms of γ_{210}.

12.4 A silver wire of radius 6.5×10^{-5} m stressed in tension at an elevated temperature was observed to contract slowly when the stress was 1.4×10^4 Pa and to extend at about the same rate when the stress was increased to 1.7×10^4 Pa. If the wire contained 60 grain boundaries running across the wire, per cm length of wire, estimate the surface free energy of the wire. Assume that the grain boundary free energy is one-third the surface free energy. What error is incurred if the effect of the grain boundaries is neglected, and is this error likely to be significant?

12.5 Derive a simple expression for the energy of a small-angle tilt boundary between two crystals in terms of the angle of tilt and the (isotropic) elastic properties of the material. Use your expression to investigate:

(a) whether two parallel boundaries of the same kind attract or repel one another;

(b) how the energy of a boundary of a given angle changes as the number of dislocations in it is reduced and their Burgers vectors are increased.

12.6 Consider the low-angle tilt boundary produced by a wall of edge dislocations corresponding to one of the predominant slip systems in each of the following materials: (a) NaCl, (b) Al, (c) Zn. Sketch the boundary structure on a plane normal to the edge dislocations, give the indices of the boundary plane and of the axis of tilt and give an expression for the angle of tilt in terms of the separation of the dislocations in the boundary.

12.7 In graphite, twinned regions are found in which the twin and matrix are related by a tilt of approximately $20.5°$ about an axis of the type $\langle 1\,\bar{1}\,0\,0\rangle$.

The axial ratio of graphite $c/a = 2.72$. Investigate the dislocation structure of the boundary and find the simplest structures employing (a) total dislocations and (b) partial dislocations.

12.8 Sketch and determine the structure of the coincidence lattice produced in a b.c.c. lattice by a rotation of $36.9°$ about a $\langle 1\,0\,0 \rangle$ axis.

12.9 A twin lamella in an f.c.c. metal intersects the surface of the metal at $90°$. After annealing, a ridge is observed along one twin boundary–surface junction and a groove along the other. If the twin boundary free energy is one-twentieth of the surface free energy, what difference do you expect to find between the angle at the base of the notch and the angle at the tip of the ridge? Carefully point out any assumptions or approximations in your calculation.

12.10 If a bubble is blown in a liquid, its radius of curvature r is related to the excess pressure p inside the bubble by the well-known equation

$$p = \frac{2\gamma}{r}$$

where γ is the surface free energy of the liquid. If this equation is applied without modification to a material whose surface energy is anisotropic (e.g. to a gas bubble in a crystal at an elevated temperature) it appears to predict that the orientations of highest surface free energy on the bubble will have the largest radius of curvature and therefore the largest extent. This conflicts with the requirement that the total free energy should be a minimum, i.e. it predicts a shape quite different from that of the Wulff body. Resolve this apparent paradox.

12.11 The surface of a crystal lies at $10°$ to a low-index plane. After annealing, it is observed that the surface has broken up into facets, one of which is the low-index surface and the other a surface of no special orientation. If the free energies of the original surface and of the non-special facet are both γ and the angle that the non-special facet makes with the low-index plane is $26°$, determine the free energy of the low-index surface and the effective free energy of the faceted surface (i.e. the free energy per unit area of the original surface). Assume that the free energy of the non-special facet does not vary with rotation about its present orientation.

12.12 The cleavage plane of mica contains potassium ions at the corners of a network of equilateral triangles (i.e. similar to a close packed array). If the K^+ ion separation is 5.2 Å, sketch the atomic fit obtained by epitaxially depositing silver with the close packed plane of the silver parallel to the cleavage plane of mica and the $\langle 1\,1\,2 \rangle$ of the silver parallel to a closest packed direction of the K^+ ions. Calculate the misfit in this direction. Show that, with this orientation relationship, the silver may grow epitaxially in either of two twin-related forms. If the silver deposits as isolated islands, having {1 1 1} surfaces, sketch the shape of islands of both forms of deposit.

12.13 Two f.c.c. crystals are rotated 0.5° about [1 1 $\bar{2}$]. Lattice parameter $a = 4$ Å. If the boundary is a symmetric tilt boundary made up of edge dislocations and the tilt angle = 0.5°
   (a) What is the boundary plane?
   (b) What is the length of dislocation line per unit area of boundary?
   (c) Calculate the angle at which the individual dislocations making up the boundary become indistinguishable.
   (d) Show that these dislocations are stable with respect to displacement of any one of them on its slip plane.

12.14 Use Eqns (12.43) and (12.44) to obtain an estimate of the critical thickness of a pseudomorphic layer below which it will not relax. Take M in equation (12.43) to be $2G(1 + \nu)/(1 - \nu)$, where G is the shear modulus. Hence show the relationship is of the same form as Eqn (12.45).

12.15 Take Fig. 12.33 to represent the lattice of a face-centred cubic crystal and consider how $a/2\langle 1\,1\,0\rangle$ dislocations moving on {1 1 1} can relieve the biaxial strain. What is the minimum number of slip systems that must operate in order to relieve the strain?

SUGGESTIONS FOR FURTHER READING

Dunstan, D. J., 'Strain and strain relaxation in semiconductors', *J. Mater. Sci.: Mater. in Electronics*, **8**, 337 (1997).
Mullins, W. W., 'Solid surface morphologies governed by capillarity', in *Metal Surfaces*, American Society in Metals, Metals Park, Ohio (1962).
Mykura, H., *Solid Surfaces and Interfaces*, Routledge and Kegan Paul, London, and Dover, New York (1966).
Nicholas, J. F., *An Atlas of Models of Crystal Surfaces*, Gordon and Breach (1965).
Nix, W. D., 'Mechanical properties of thin films', *Met. Trans.*, **A20**, 2217 (1989).
Pashley, D. W., 'The study of epitaxy in thin surface films', *Adv. Phys.*, **5**, 174 (1956).
Read, W. T., and Shockley, W., 'Dislocation models of crystal grain boundaries', *Phys. Rev.*, **78**, 275 (1950).
Shewmon, P. G., and Robertson, W. M., 'Variation of surface tension with orientation', in *Metal Surfaces*, American Society in Metals, Metals Park, Ohio (1962).
Smith, C. S., 'Grain shapes', in *Metal Interfaces*, American Society in Metals, Metals Park, Ohio (1952).
Sutton, A. P., and Balluffi, R. W., *Interfaces in Crystalline Materials*, Clarendon, Oxford (1995).

REFERENCES

1. J. K. Mackenzie, A. J. W. Moore and J. F. Nicholas, *J. Phys. Chem. Solids*, **23**, 185 (1962).

2. E. R. Funk, H. Udin and J.Wulff, *J. Metals*, **3**, 1206 (1951).
3. H. Udin, A. J. Shaler and J. Wulff, *Trans. Am. Inst. Mining Metall. and Petrol. Engrs*, **185**, 186 (1949). Corrected by H. Udin, *J. Metals*, **3**, 63 (1951).
4. J. J. Gilman, *J. Appl. Phys.*, **31**, 2208 (1960).
5. A. T. Price, H. A. Holl and A. P. Greenough, *Acta Met.*, **12**, 49 (1964).
6. A. R. C. Westwood and T. T. Hitch, *J. Appl. Phys.*, **39**, 3085 (1963).
7. A. R. C. Westwood and D. Goldheim, *J. Appl. Phys.*, **34**, 3335 (1963).
8. P. L. Gutshall and G. E. Gross, *J. Appl. Phys.*, **36**, 2459 (1965).
9. B. C. Allen, *Trans. Am. Inst. Mining Metall. and Petrol. Engrs*, **236**, 903 (1966).
10. S. V. Radcliffe, *J. Less Common Metals*, **3**, 360 (1961).
11. E. R. Hayward and A. P. Greenough, *J. Inst. Metals*, **88**, 217 (1959).
12. E. B. Greenhill and S. R. McDonald, *Nature, Lond.*, **171**, 37 (1953).
13. W. T. Read Jr, *Dislocations in Crystals*, McGraw-Hill Book Company (1953).
14. W. T. Read Jr and W. Shockley, 'Dislocation models of crystal grain boundaries,' *Phys. Rev.*, **78**, 275 (1950).
15. P. G. Shewmon, *Recrystallization Grain Growth and Textures*, American Society for Metals, Metals Park, ohio (1965), p. 165.
16. D. G. Brandon, B. Ralph, S. Rananathan and M. S. Wald, *Acta Met*, **12**, 813 (1964).
17. W. Bollmann, *Disc. Farad. Soc.*, **38**, 26 (1964).
18. A. T. Price, H. A. Holl and A. P. Greenough, *Acta Met.*, **12**, 49 (1964).
19. M. C. Inman and H. R. Tipler, *Met. Revs.*, **8**, 105 (1963).
20. D. P. Spitzer, *J. Phys. Chem.*, **66**, 31 (1962).
21. C. Herring, in *The Physics of Powder Metallurgy*, (ed. W. E. Kingston), McGraw-Hill, New York (1951).
22. H. Mykura, 'The variation of the surface tension of nickel with crystallographic orientation', *Acta Met.*, **9**, 570 (1961).
23. G. Rhead and M. McLean, *Acta Met.*, **12**, 401 (1964).
24. M. McLean and H. Mykura, *Phil. Mag.*, **14**, 1191 (1966).
25. W. M. Robertson and P. G. Shewmon, *Trans. Am. Inst. Mining Metall. and Petrol. Engrs*, **224**, 804 (1962).
26. W. W. Mullins, Chapter 2 of *Metal Surfaces, Structure and Kinetics*, American Society for Metals, Metals Park, Ohio (1962), ch. 2.
27. C. Herring, 'Some theorems on the free energies of crystal surfaces', *Phys. Rev.*, **82**, 87 (1951).
28. Lord Kelvin, *Mathematical and Physical Papers*, Vol. V, Cambridge (1911), p. 297.
29. C. S. Smith, 'grain shapes', in *Metal Interfaces*, american society in Metals, Metals Park, Ohio (1952).
30. N. A. Gjostein, H. A. Doman, H. I. Aaronson and E. Eichen, Ford Scientific Laboratory Report.
31. C. S. Smith, *Met. Revs.*, **9**, 1 (1964).
32. H. Brookes, 'Theory of internal boundaries', in *Metal Interfaces*, American Society for Metals, Metals Park, Ohio (1952).
33. J. H. Van der Merwe, *J. Appl. Phys.*, **34**, 123 (1963).
34. J. L. W. Whitton, 'Transmission electron microscopy of uranium monocarbide', *J. Nucl. Mater.*, **12**, 115 (1964).
35. D. W. Pashley, *Adv. Phys.*, **5**, 174 (1956).
36. D. J. Dunstan, 'Strain and strain relaxation in semiconductors', *J. Mater. Sci.: Mater. in Electronics*, **8**, 337–375 (1997).
37. D. J. Dunstan, S. Young and R. H. Dixon, 'Geometrical theory of critical thickness and relaxation in strained-layer growth', *J. Appl. Phys.*, **70** (6), 3038–3045 (1991).

Appendix 1

Crystallographic Calculations

A1.1 COORDINATE GEOMETRY

In Fig. A1.1 the plane has intercepts on the axes X, Y, Z of A, B, C respectively. The axes are not necessarily orthogonal. x, y, z are the coordinates of a point on the plane and p, the perpendicular from the plane to the origin, makes angles α, β, γ with the coordinate axes. From the figure

$$p = x \cos \alpha + y \cos \beta + z \cos \gamma \qquad \text{(A1.1)}$$

or

$$1 = \frac{x}{p/\cos\alpha} + \frac{y}{p/\cos\beta} + \frac{z}{p/\cos\gamma} \qquad \text{(A1.2)}$$

Now

$$A \cos\alpha = B \cos\beta = C \cos\gamma = p \qquad \text{(A1.3)}$$

Therefore the equation to the plane is

$$\frac{x}{A} + \frac{y}{B} + \frac{z}{C} = 1 \qquad \text{(A1.4)}$$

and that of the parallel plane through the origin,

$$\frac{x}{A} + \frac{y}{B} + \frac{z}{C} = 0 \qquad \text{(A1.5)}$$

A crystal plane of Miller indices $(h\,k\,l)$ has axial intercepts proportional to a/h, b/k and c/l and so it follows directly from the relations (A1.3) that

$$\frac{a}{h} \cos\alpha = \frac{b}{k} \cos\beta = \frac{c}{l} \cos\gamma \qquad \text{(A1.6)}$$

where α, β, γ are the angles between the normal to the plane of indices $(h\,k\,l)$ and the crystal axes. Equation (A1.6) is true in all crystal systems and is called *the equation to the normal*. In crystals with *orthogonal axes* we always have, of course,

$$\cos^2\alpha + \cos^2\beta + \cos^2\gamma = 1$$

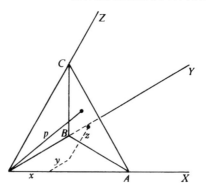

Figure A1.1

A1.2 SOLUTIONS OF SPHERICAL TRIANGLES

A spherical triangle is a triangle on the surface of a sphere of which the sides are great circles (Fig. A1.2). The sides of such a triangle are measured in terms of the *angle subtended* at the centre of the sphere. Any spherical triangle can be solved if any three of the sides or the angles at the corners are known. The sum of the three angles of a spherical triangle cannot be less than 180°.

In any spherical triangle the following relations hold between the angles A, B, C and the opposite sides a, b, c (Fig. A1.2):

$$\frac{\sin A}{\sin a} = \frac{\sin B}{\sin b} = \frac{\sin C}{\sin c} \tag{A1.7}$$

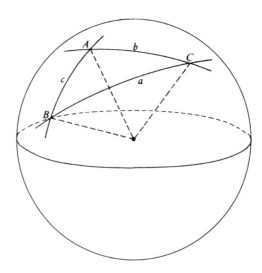

Figure A1.2

$$\cos a = \cos b \cos c + \sin b \sin c \cos A \qquad (A1.8)$$

$$\cos A = -\cos B \cos C + \sin B \sin C \cos a \qquad (A1.9)$$

and of course there are corresponding relations for $\cos b$, $\cos c$, $\cos B$ and $\cos C$.

If one of the sides or one of the angles of such a triangle is a right angle, the relations are greatly simplified and the triangle can be completely solved provided any two other quantities are known. If C is a right angle the equation for $\cos c$, similar to Eqn (A1.8), becomes

$$\cos c = \cos a \cos b \qquad (A1.10)$$

and that for $\cos C$ analogous to Eqn (A1.9) gives

$$\cos c = \cot A \cot B = \tan(90° - A)\tan(90° - B) \qquad (A1.11)$$

If c is a right angle we have correspondingly

$$\cos C = -\cot a \cot b = -\tan(90° - a)\tan(90° - b) \qquad (A1.12)$$

and

$$\cos C = -\cos A \cos B \qquad (A1.13)$$

A spherical triangle with either a side or an angle equal to a right angle is called a Napierian triangle. For any Napierian triangle there are two useful mnemonics, due to Napier, which recall the solutions of equations such as (A1.10) to (A1.13). These are illustrated in Figs. A1.3a and b. In Figs. A1.3a and b the top diagram shows the five parts of the triangle numbered in order, starting from the right angle. In each of the lower diagrams the same numbers (representing the values in degrees of the sides and angles of the triangle) are put into a diagram containing five compartments. In any Napierian triangle that is soluble, two quantities must be known besides the right angle. The two known values and any required unknown can always be grouped together so that either all three are in adjacent compartments of the figure — as in Fig. A1.4a, which illustrates one possible *adjacent* arrangement — or else the three quantities (one unknown and two knowns) are arranged in what could be called the *opposite* arrangement, an example of which is shown in Fig. A1.4b.

All of the formulae for the solution of a right-angled or a right-sided spherical triangle are then contained in the two statements:

$$\text{sIne of a mIddle part} = \begin{cases} \text{Product of tAngents of AdjAcent parts} \\ \text{or} \\ \text{Product of cOsines of OppOsite parts} \end{cases}$$

Middle part means the middle quantity in the adjacent case and the one opposite the other two in the opposite case. The middle part is ringed in each of the examples in Fig. A1.4.

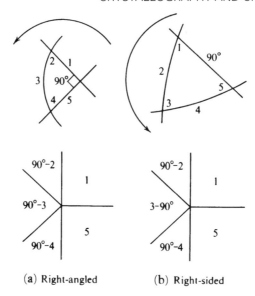

(a) Right-angled   (b) Right-sided

Figure A1.3

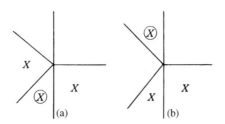

Figure A1.4

As an example of the application of Napier's rules consider the triangle shown in Fig. 1.17b which is right-angled at B and suppose that we knew only the sides $b = 54.73°$ and $c = 45°$. To find a the diagram would appear as in Fig. A1.5 and so

$$\sin(90° - 54.73°) = \cos a \cos 45°$$

or

$$\cos a = \frac{\sin 35.27°}{\cos 45°}$$

whence

$$a = 35.27°$$

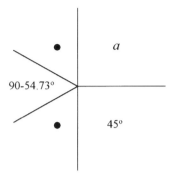

Figure A1.5

A1.3 THE SINE RATIO

Another useful aid in crystal calculations, which we prove in Section A2.2, is the sine ratio — also called the anharmonic ratio for planes in the same zone.

If four lattice planes of a crystal which we could designate P_1, P_2, P_3 and P_4 *all lie in the same zone* and no two of them are parallel, then, if we call the angle between P_1 and P_2, θ_{12}, that between P_1 and P_3, θ_{13}, the ratio

$$\frac{\sin\theta_{12}/\sin\theta_{13}}{\sin\theta_{42}/\sin\theta_{43}}$$

is easily calculated from the indices of the faces. This sine ratio is of some use in crystal calculations, either to find the angle between a certain face of known indices and other faces in a zone, or else to find the indices of a face from the angle it makes with other faces in the same zone. Let the indices of P_1 be $(h_1\,k_1\,l_1)$ and those of $P_2(h_2\,k_2\,l_2)$, etc. If we find the values of the indices of the zone $[U\,V\,W]$ in which P_1 and P_2 lie from the rule in Section 1.3, Eqn (1.9), we shall have

$$U = k_1 l_2 - l_1 k_2$$
$$V = l_1 h_2 - h_1 l_2$$
$$W = h_1 k_2 - k_1 h_2$$

Since these zone indices are derived from the indices of P_1 and P_2 we could call them U_{12}, V_{12} and W_{12}, and similarly the indices of the zone axis derived from P_2 and P_3 could be called U_{23}, V_{23}, W_{23} and so on. Adopting this notation the sine ratio is given by

$$\frac{\sin\theta_{12}/\sin\theta_{13}}{\sin\theta_{42}/\sin\theta_{43}} = \frac{U_{12}/U_{13}}{U_{42}/U_{43}} = \frac{V_{12}/V_{13}}{V_{42}/V_{43}} = \frac{W_{12}/W_{13}}{W_{42}/W_{43}} \qquad (A1.14)$$

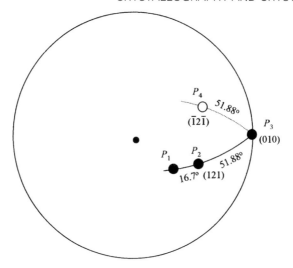

Figure A1.6

An example of the use of Eqn (A1.14) is as follows. Figure A1.6 shows a partial stereogram of a crystal in which four faces in the same zone have been located and the angles between them measured. These are marked upon the stereogram. It is required to find the indices of P_1, knowing the indices of the other three faces. If we call the indices of $P_1(h\,k\,l)$ then we have

$$\frac{\sin\theta_{12}/\sin\theta_{13}}{\sin\theta_{42}/\sin\theta_{43}} = \frac{\sin 16.70°/\sin 68.58°}{\sin 103.77°/\sin 51.88°} = \frac{0.2874/0.9311}{0.9713/0.7870} = 0.25 = \frac{1}{4}$$

If P_1 is $(h\,k\,l)$ and P_2 $(1\,2\,1)$ then forming the zone axis indices from the procedure on p. 13 we have

$$U_{12} = k - 2l, \qquad V_{12} = l - h, \qquad W_{12} = 2h - k$$

and similarly

$$U_{13} = \bar{l}, \qquad V_{13} = 0, \qquad W_{13} = h$$
$$U_{42} = 4, \qquad V_{42} = 0, \qquad W_{42} = \bar{4}$$
$$U_{43} = 1, \qquad V_{43} = 0, \qquad W_{43} = \bar{1}$$

Substituting for all of these quantities in Eqn (A1.14) we obtain

$$\frac{1}{4} = \frac{(k - 2l)/\bar{l}}{4/1} = \frac{(l - h)/0}{0/0} = \frac{(2h - k)/h}{\bar{4}/\bar{1}}$$

From these equations we find $l = h = k$. Therefore the indices of P_1 are $(1\,1\,1)$.

Note that Eqn (A1.14) can be used in any crystal system; it arises from the nature of the crystal lattice and because of this it is easily proved from the properties of the reciprocal lattice (see Section A2.2).

A1.4 MATRICES

A matrix is an array of quantities that can be helpful in writing down a set of equations in a compact form and in manipulating them. For example, the equations giving the coordinates (x'_1, x'_2, x'_3) of a point in a homogeneously strained body, which was at (x_1, x_2, x_3) before the strain was applied, are (Eqns 5.15)

$$x'_1 = (1 + e_{11})x_1 + e_{12}x_2 + e_{13}x_3$$

$$x'_2 = e_{21}x_1 + (1 + e_{22})x_2 + e_{23}x_3$$

$$x'_3 = e_{31}x_1 + e_{32}x_2 + (1 + e_{33})x_3$$

We define the matrix $\mathbf{D}$ as the following array of coefficients

$$\mathbf{D} = \begin{pmatrix} 1 + e_{11} & e_{12} & e_{13} \\ e_{21} & 1 + e_{22} & e_{23} \\ e_{31} & e_{32} & 1 + e_{33} \end{pmatrix} \tag{A1.15}$$

Equation (5.15) can now be written in the form

$$\begin{pmatrix} x'_1 \\ x'_2 \\ x'_3 \end{pmatrix} = \mathbf{D} \begin{pmatrix} x_1 \\ x_2 \\ x_3 \end{pmatrix} \tag{A1.16}$$

provided that the following meaning is attached to the multiplication of the column (x_1, x_2, x_3) by the matrix $\mathbf{D}$. The first step is to multiply each term in the first row of the matrix by the corresponding term in the column (x_1, x_2, x_3) and to add the products. This is easy to remember as 'first of row times first of column plus second of row times second of column plus third of row times third of column'. The sum is set equal to the first term in the resulting column, x'_1, which gives the first of Eqns (5.15). The second and third of Eqns (5.15) are obtained in a similar way from the second and third rows of the matrix. Equations (5.15) can now be written still more compactly as

$$\mathbf{x}' = \mathbf{D}\mathbf{x} \tag{A1.17}$$

where the position vectors $\mathbf{x}$ and $\mathbf{x}'$ are understood to be represented by columns for the purpose of the multiplication. The reason for writing the components of a vector in a column will become clear when we consider what meaning should be attached to the multiplication of one matrix by another. To take a specific example, suppose that two homogeneous strains are applied to a body, one after the other. If the first strain carries the point **x** to **x'**, then

$$\mathbf{x'} = \mathbf{D}_1\mathbf{x} \qquad (A1.18)$$

where $\mathbf{D}_1$ is the matrix representing the first strain. The second strain carries the point **x'** to the point **x''**, where

$$\mathbf{x''} = \mathbf{D}_2\mathbf{x'} \qquad (A1.19)$$

where $\mathbf{D}_2$ is the matrix representing the second strain. From Eqn (A1.18),

$$\mathbf{x''} = \mathbf{D}_2\mathbf{D}_1\mathbf{x} \qquad (A1.20)$$

When Eqns (A1.18) and (A1.19) are written out in full, it will be seen that Eqns (A1.20), describing the net effect of the two strains, can be written as

$$\mathbf{x''} = \mathbf{Rx} \qquad (A1.21)$$

where $\mathbf{R}(=\mathbf{D}_2\mathbf{D}_1)$ is a matrix whose terms can be obtained as follows. The term in the ith row and jth column of $\mathbf{R}$ is obtained by multiplying each term in the ith row of the left-hand matrix $\mathbf{D}_2$ by the corresponding term in the jth column of the right-hand matrix $\mathbf{D}_1$ and adding the products. For instance, if

$$\mathbf{D}_1 = \begin{pmatrix} a_{11} & a_{12} & a_{13} \\ a_{21} & a_{22} & a_{23} \\ a_{31} & a_{32} & a_{33} \end{pmatrix}, \qquad \mathbf{D}_2 = \begin{pmatrix} b_{11} & b_{12} & b_{13} \\ b_{21} & b_{22} & b_{23} \\ b_{31} & b_{32} & b_{33} \end{pmatrix}$$

and

$$\mathbf{R} = \mathbf{D}_2\mathbf{D}_1 = \begin{pmatrix} r_{11} & r_{12} & r_{13} \\ r_{21} & r_{22} & r_{23} \\ r_{31} & r_{32} & r_{33} \end{pmatrix}$$

then

$$r_{11} = b_{11}a_{11} + b_{12}a_{21} + b_{13}a_{31}$$

or, in general,

$$r_{ij} = b_{i1}a_{1j} + b_{i2}a_{2j} + b_{i3}a_{3j} \qquad (A1.22)$$

Compounding the rows of the left-hand matrix with the columns of the right-hand matrix in this way gives a general definition of the multiplication of two matrices. The operation can be performed with any two arrays of numbers *provided* that the number of columns of the left-hand array equals the number of rows of the right-hand array. The multiplication of a vector by a matrix, as in Eqn (A1.17), can be regarded as a special case of this rule where the vector is represented by a matrix with three rows and one column (i.e. a 3×1 matrix). Generally, therefore, the product of an $(m \times n)$ matrix with one of order $(n \times p)$ is a matrix of order $(m \times p)$.

It should be noted that in the definition of multiplication, the left-hand and right-hand matrices are treated differently so that in general

$$\mathbf{D}_1\mathbf{D}_2 \neq \mathbf{D}_2\mathbf{D}_1$$

In our example, the distortion produced by two successive *large* strains does in fact generally depend on the order in which they are applied.

Returning to the equation discussed at the beginning of this section, Eqns (5.15), we see that it can be written in the form

$$\mathbf{x}' = \mathbf{Ix} + \mathbf{Sx} \tag{A1.23}$$

where

$$\mathbf{I} = \begin{pmatrix} 1 & 0 & 0 \\ 0 & 1 & 0 \\ 0 & 0 & 1 \end{pmatrix} \quad \text{and} \quad \mathbf{S} = \begin{pmatrix} e_{11} & e_{12} & e_{13} \\ e_{21} & e_{22} & e_{23} \\ e_{31} & e_{32} & e_{33} \end{pmatrix}$$

Writing Eqn (A1.23) as

$$\mathbf{x}' = (\mathbf{I} + \mathbf{S})\mathbf{x}$$

and comparing with the original form

$$\mathbf{x}' = \mathbf{Dx}$$

we see that $\mathbf{D} = \mathbf{I} + \mathbf{S}$. In this case, and in general, one matrix is added to another merely by adding together corresponding terms in the two arrays. It is said that the two matrices are conformable for addition, when each has the same number of rows and each has the same number of columns. The matrix $\mathbf{I}$, defined above, has the special property that

$$\mathbf{IA} = \mathbf{A} = \mathbf{AI} \tag{A1.24}$$

where $\mathbf{A}$ is any 3×3 matrix. It is therefore called the unit matrix. A unit $n \times n$ matrix is defined in exactly the same way. Given a square $(n \times n)$ matrix $\mathbf{B}$, if we can find another $n \times n$ matrix that produces the unit matrix when it multiplies $\mathbf{B}$, we say that we have found the inverse or reciprocal of $\mathbf{B}$, written $\mathbf{B}^{-1}$. Thus

$$\mathbf{B}^{-1}\mathbf{B} = \mathbf{I} = \mathbf{BB}^{-1} \tag{A1.25}$$

The inverse of a matrix representing a homogeneous strain, for example, is the matrix representing the 'opposite' strain which will return the body to its undistorted state. As a second example, consider Hooke's law (Eqn 5.36). The six equations giving the stress in terms of the strain can be written in matrix form as

$$\begin{pmatrix} \sigma_1 \\ \sigma_2 \\ \sigma_3 \\ \sigma_4 \\ \sigma_5 \\ \sigma_6 \end{pmatrix} = \mathbf{C} \begin{pmatrix} \varepsilon_1 \\ \varepsilon_2 \\ \varepsilon_3 \\ \varepsilon_4 \\ \varepsilon_5 \\ \varepsilon_6 \end{pmatrix} \tag{A1.26}$$

where $\mathbf{C}$ is a 6×6 matrix of stiffness constants. Equation (A1.26) can be written more compactly as

$$\sigma = \mathbf{C}\varepsilon \tag{A1.27}$$

with the understanding that σ and ε are 6×1 matrices. Hooke's law can also be written in a form giving the strain in terms of the stress

$$\varepsilon = \mathbf{S}\sigma \tag{A1.28}$$

where $\mathbf{S}$ is a 6×6 matrix of compliances. The matrix $\mathbf{S}$ is in fact the inverse of the matrix $\mathbf{C}$, since by multiplying both sides of Eqn (A1.27) by $\mathbf{C}^{-1}$ we have

$$\mathbf{C}^{-1}\sigma = \mathbf{I}\varepsilon$$

or

$$\mathbf{C}^{-1}\sigma = \varepsilon$$

Therefore

$$\mathbf{C}^{-1} = \mathbf{S} \tag{A1.29}$$

Evidently, solving the six equations represented by Eqn (A1.26) simultaneously to obtain ε in terms of σ is equivalent to finding the inverse of the matrix $\mathbf{C}$.

In order for a matrix to possess an inverse it must be both square and non-singular. In order to find the inverse of a matrix $\mathbf{A}$ we have to find the modulus of the matrix, $|\mathbf{A}|$, which is the value obtained by treating $\mathbf{A}$ as a determinant and we have to write down the adjoint matrix to $\mathbf{A}$, adj $\mathbf{A}$. These steps follow from the theory of determinants and a very condensed proof can be found in the books by Nye and by Sands given at the end of Chapter 5. If the matrix has elements a_{ij}, so

$$\mathbf{A} = \begin{bmatrix} a_{11} & a_{12} & a_{13} & — & — & — & a_{1n} \\ a_{21} & a_{22} & a_{23} & — & — & — & a_{2n} \\ a_{31} & a_{32} & a_{33} & — & — & — & a_{3n} \\ — & — & — & — & — & — & — \\ — & — & — & — & — & — & — \\ — & — & — & — & — & — & — \\ a_{n1} & a_{n2} & a_{n3} & — & — & — & a_{nn} \end{bmatrix} \tag{A1.30}$$

then the cofactor of the element a_{ik} in $\mathbf{A}$ is defined as $(-1)^{i+k}$ times the value of the determinant formed by deleting the row and column in which a_{ik} occurs.

The matrix of cofactors, A_{ik}, so formed, which is clearly also a square matrix of the same order as $\mathbf{A}$, is

$$\begin{bmatrix} A_{11} & A_{12} & A_{13} & - & - & - & A_{1n} \\ A_{21} & A_{22} & A_{23} & - & - & - & A_{2n} \\ A_{31} & A_{32} & A_{33} & - & - & - & A_{3n} \\ - & - & - & - & - & - & - \\ - & - & - & - & - & - & - \\ - & - & - & - & - & - & - \\ A_{n1} & A_{n2} & A_{n3} & - & - & - & A_{nn} \end{bmatrix}$$

The adjoint matrix, adj $\mathbf{A}$, is then defined as the transposed matrix (i.e. the matrix obtained by interchanging rows and columns) of the cofactors of $\mathbf{A}$. Thus

$$\text{adj } A = \begin{bmatrix} A_{11} & A_{21} & A_{31} & - & - & - & A_{n1} \\ A_{12} & A_{22} & A_{32} & - & - & - & A_{n2} \\ A_{13} & A_{23} & A_{33} & - & - & - & A_{n3} \\ - & - & - & - & - & - & - \\ - & - & - & - & - & - & - \\ A_{1n} & A_{2n} & A_{3n} & - & - & - & A_{nn} \end{bmatrix} \tag{A1.31}$$

Having formed the adjoint matrix, adj $\mathbf{A}$, and having evaluated $|\mathbf{A}|$, the inverse of $\mathbf{A}$, i.e. $\mathbf{A}^{-1}$, is given by

$$\mathbf{A}^{-1} = \frac{\text{adj } \mathbf{A}}{|\mathbf{A}|} \tag{A1.32}$$

A set of n simultaneous equations in n unknowns can generally be solved, and so the inverse of a square matrix can generally be found. However, if it happens that one of the equations is merely a linear combination of some of the others, then there are only $n - 1$ essentially different equations and a solution is not possible. It follows that a matrix has no inverse if one of its rows is a linear combination of others, and it can be shown that the same is true if there is a linear relationship amongst its columns. Such a matrix is called a *singular* matrix. An illustration of these ideas occurs in Chapter 6. There it is shown that the strain produced by five simultaneously operating slip systems can be expressed in terms of the amounts of slip on the five systems $\alpha_1, \ldots, \alpha_5$ as follows:

$$\begin{pmatrix} \varepsilon_{11} - \varepsilon_{33} \\ \varepsilon_{22} - \varepsilon_{33} \\ \varepsilon_{12} \\ \varepsilon_{23} \\ \varepsilon_{31} \end{pmatrix} = \mathbf{A} \begin{pmatrix} \alpha_1 \\ \alpha_2 \\ \alpha_3 \\ \alpha_4 \\ \alpha_5 \end{pmatrix} \tag{A1.33}$$

where the terms in the matrix $\mathbf{A}$ are geometrical factors dependent on the orientations of the slip systems. More compactly,

$$\varepsilon = \mathbf{A}\alpha \tag{A1.34}$$

If the matrix **A** has an inverse we can write

$$\boldsymbol{\alpha} = \mathbf{A}^{-1}\boldsymbol{\varepsilon} \tag{A1.35}$$

Equation (A1.35) would enable us to calculate the amounts of slip needed to produce any strain. If the matrix **A** is singular, Eqn (A1.35) cannot be written: some strains cannot be produced. This corresponds to the five slip systems not being independent of one another.

A1.5 PROOF OF PROPERTIES OF THE STEREOGRAPHIC PROJECTION

Suppose R is the radius of the sphere of projection in Fig. A1.7. S is the point of projection and N diametrically opposite S. P is the pole of any point on the sphere. Let the angle NOP equal ϕ. The angle NPS is a right angle, since NS is a diameter of the sphere. The distance SP is therefore equal to $2R\cos\phi/2$, since $\widehat{OSP} = \frac{1}{2}\widehat{NOP}$. P' is the stereographic projection of P. The triangle SOP' has a right angle at O. Therefore the length SP' is equal to $R\sec\phi/2$. The product of the lengths SP, SP' is

$$SP \times SP' = 2R^2 \tag{A1.36}$$

This product is independent of the angle ϕ and hence is constant for all poles P on the surface of the sphere of projection. Any pole P and its stereographic projection P' are said to be the *inverse of one another* because their distances from a fixed point (S) are related by Eqn (A1.36).

The geometry of *inversion* was invented by L. J. Magnus in 1831. The formal definition of the geometry of inversion is as follows. Given a fixed sphere of

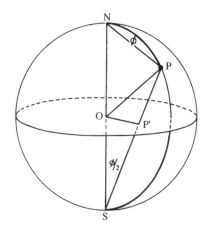

Figure A1.7

radius k and centre O we define the inverse of any point P (distinct from O) to be the point P' *on the ray* OP whose distance from O satisfies the equation

$$OP \times OP' = k^2 \qquad (A1.37)$$

P is clearly the inverse of P' itself and the point O is singular. Comparing this definition with our definition of the stereographic projection, we see that all poles P on the sphere of projection and their stereographic projections P' are such that P and P' are the inverse of one another in a sphere of centre S in Fig. A1.7 and of radius $\sqrt{2}R$. The sphere of inversion is *not* drawn in Fig. A1.7. The importance of the stereographic projection in crystallography arises because of the following features of the geometry of inversion. All spheres invert into spheres (or planes) and therefore all circles, being intersections of spheres, invert into circles or straight lines. The transformation inverts *angles into equal angles*. The stereographic projection is therefore said to be angle true. These properties are proved below. The mathematician defines the stereographic projection as that inversion which puts points of a plane into a one-to-one correspondence with the points on the surface of a sphere and we have chosen a particular example in which the centre of the sphere of inversion, S in Fig. A1.7, and the radius of the sphere of inversion are so chosen that all points on the surface of the *sphere of projection* invert into points that lie in a diametral plane of the sphere of projection. This plane is the plane of projection as we use it in this book.

A proof that any sphere inverts into a sphere is as follows. Let P be any point on a sphere σ of centre C. It is required to show that the inverse of σ is another sphere. Let O in Fig. A1.8 be the centre of the sphere of inversion and k its radius (the sphere of inversion is not shown in Fig. A1.8). Join PO and let the second intersection of PO with the sphere σ be Q. The product OP $\times$ OQ is independent of the position of P on the sphere.[†] Let the product OP $\times$ OQ equal p. Join OC and find the point D on OC such that OD/OC $= k^2/p$. Draw the sphere σ' of centre D and radius $(k^2/p) \times$ CP, i.e. a sphere of radius k^2/p times the radius of σ. Draw the radius of this sphere, DP', in the plane of OCD and P, parallel to CQ. Join OP'. Since

$$\frac{DP'}{QC} = \frac{k^2}{p}\frac{CP}{QC} = \frac{k^2}{p} = \frac{OD}{OC}$$

triangles OQC, OP' D are similar and hence OP and OP' coincide. It follows immediately that P' is the inverse of P, since

$$\frac{OP'}{OQ} = \frac{OD}{OC} = \frac{k^2}{p}$$

[†] This property follows from Euclid III.37 for a circle: if from a point outside a circle a secant and a tangent be drawn, the rectangle contained by the whole secant and the part outside the circle is equal to the square on the tangent.

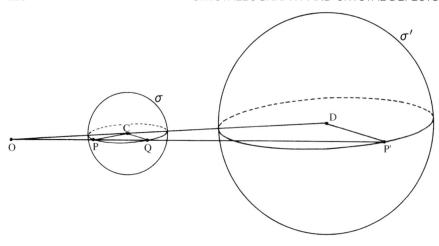

Figure A1.8

and

$$OQ = \frac{p}{OP}$$

Therefore

$$OP \times OP' = k^2$$

Since P was any point on σ we have proved that the sphere σ inverts to the sphere σ' and, incidentally, since any circle can be regarded as the intersection of two spheres, we have also shown that all circles invert to circles. It should be noted that, if the sphere σ passes through O, σ' is a plane (a sphere of infinite radius) and that a circle through O inverts into a line. It is also worth calling attention to the fact that D is not necessarily the inverse of C.

That the stereographic projection is angle true, that is to say, that the angle of intersection of two curves on the sphere of projection is the same as the angle between the two projected curves, is also easily shown from the geometry of inversion. To prove that the angle at which two curves cut is equal to the angle at which their inverse curves cut at the corresponding point of intersection, we proceed as follows. O is the centre of inversion (Fig. A1.9). PP' are the points of intersection of the two pairs of curves. Through O draw any line OQRR' Q' cutting the two given curves at Q, R and their inverses at Q', R'.

Since $OQ \cdot OQ' = OP \cdot OP' = OR \cdot OR'$, PQQ'P' and PRR'P' are cyclic quadrilaterals. Therefore

$$O\widehat{P}Q = O\widehat{Q}'P' \quad \text{and} \quad O\widehat{P}R = O\widehat{R}'P'$$

and so

$$Q\widehat{P}R = Q'\widehat{P}'R'$$

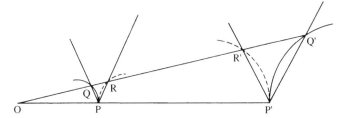

Figure A1.9

By taking the line OQ sufficiently close to OP the difference between these two angles and the angles between the two pairs of curves may be made as small as we please, thus proving the theorem.

Appendix 2

Vector Algebra and the Reciprocal Lattice

A2.1 VECTOR ALGEBRA

A vector is a quantity possessing both magnitude and direction. Both must be given to specify the vector. To add two vectors $\mathbf{A}$ and $\mathbf{B}$ we may represent them both in magnitude and direction by lines. Each is then represented by a certain displacement (Fig. A2.1) and their sum is the resultant displacement. Thus $\mathbf{C}$ in Fig. A2.1 is equivalent to the displacements $\mathbf{A}$ and $\mathbf{B}$ applied successively. We may therefore write

$$\mathbf{C} = \mathbf{A} + \mathbf{B} \qquad (A2.1)$$

and

$$\mathbf{B} = \mathbf{C} - \mathbf{A} \qquad (A2.2)$$

Thus reversal of sign of a vector is equivalent to a reversal of its direction. Note that in applying an equation like (A2.2) any change in origin is disregarded. A vector is represented by heavy type, thus $\mathbf{A}$. Its magnitude alone, without the idea of direction, is represented by ordinary type A or by $|\mathbf{A}|$.

A unit vector is a vector of unit magnitude in a given direction. Therefore, if $\mathbf{i}$ is a vector of unit magnitude in the direction of $\mathbf{A}$,

$$\mathbf{A} = \mathbf{i}|\mathbf{A}| = \mathbf{i}A \qquad (A2.3)$$

If $\mathbf{i}, \mathbf{j}, \mathbf{k}$ are *unit* vectors in the directions of three coordinate axes, x, y, z, not necessarily orthogonal, and if A_x, A_y, A_z are the components of the vector $\mathbf{A}$ referred to the three axes,

$$\mathbf{A} = A_x\mathbf{i} + A_y\mathbf{j} + A_z\mathbf{k} \qquad (A2.4)$$

(see Fig. A2.2).

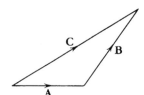

Figure A2.1

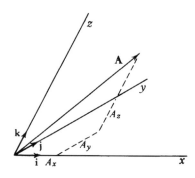

Figure A2.2

THE SCALAR PRODUCT

The scalar product of two vectors **A** and **B** is a scalar equal in magnitude to the product of the magnitudes of **A** and **B** and the cosine of the angle between the directions of **A** and **B**. It is denoted **A** · **B** (read **A** dot **B**). Thus

$$\mathbf{A} \cdot \mathbf{B} = AB\cos\theta \qquad (A2.5)$$

where θ is the angle between the vectors. Since $\cos(2\pi - \theta) = \cos\theta$ it does not matter whether θ is measured from **A** to **B** or from **B** to **A**. It follows from Eqn (A2.5) that the scalar product is commutative, i.e. **A** · **B** = **B** · **A**. If θ is the angle between the directions of **A** and **B**,

$$\cos\theta = \frac{\mathbf{A} \cdot \mathbf{B}}{|\mathbf{A}| \cdot |\mathbf{B}|} \qquad (A2.6)$$

The scalar product **A** · **B** is equal to the projection of **A** on the direction of **B** multiplied by the magnitude of **B**. If **n** is a unit vector, **A** · **n** is the projection of **A** on the direction of **n**. If **A** is a unit vector its projections on the axes are equal to its direction cosines. The scalar product of a vector with itself is always equal to the square of its magnitude. For *orthogonal axes* this gives

$$|\mathbf{A}|^2 = A_x^2 + A_y^2 + A_z^2 \qquad (A2.7)$$

Let A_x, A_y, A_z and B_x, B_y, B_z be the components of the vectors **A** and **B** referred to a set of axes. Then from Eqn (A2.4),

$$\mathbf{A} \cdot \mathbf{B} = (A_x\mathbf{i} + A_y\mathbf{j} + A_z\mathbf{k}) \cdot (B_x\mathbf{i} + B_y\mathbf{j} + B_z\mathbf{k}) \qquad (A2.8)$$

Scalar multiplication of vectors is distributive, for multiplication over addition.

If **i, j, k** are mutually perpendicular unit vectors, i.e. the *axes are orthogonal*, then

$$\mathbf{i} \cdot \mathbf{i} = \mathbf{j} \cdot \mathbf{j} = \mathbf{k} \cdot \mathbf{k} = 1 \qquad (A2.9)$$

$$\mathbf{i} \cdot \mathbf{j} = \mathbf{j} \cdot \mathbf{k} = \mathbf{k} \cdot \mathbf{i} = 0 \qquad (A2.10)$$

and so Eqn (A2.8) becomes

$$\mathbf{A} \cdot \mathbf{B} = A_x B_x + A_y B_y + A_z B_z \qquad (A2.11)$$

If α_1, β_1, γ_1 and α_2, β_2, γ_2 are the angles which the directions of **A** and **B** make with the axes respectively then from Eqns (A2.6), (A2.7) and (A2.10) the cosine of the angle between **A** and **B** is given by

$$\cos\theta = \cos\alpha_1 \cos\alpha_2 + \cos\beta_1 \cos\beta_2 + \cos\gamma_1 \cos\gamma_2 \qquad (A2.12)$$

for *orthogonal* axes.

THE VECTOR PRODUCT

The vector product of two vectors **A** and **B** is itself a vector equal to $(|\mathbf{A}||\mathbf{B}| \sin\theta)\mathbf{l}$, where **l** is a unit vector perpendicular to the plane containing both **A** and **B** and in such a direction that a right-handed screw driven in the direction of **l** would carry **A** into **B** through the angle θ, where θ is the angle less than 180° between **A** and **B** (Fig. A2.3). The symbol $\mathbf{A} \times \mathbf{B}$ (read **A** cross **B**) is the vector product of **A** and **B**, sometimes written as **A, B** or $\mathbf{A} \wedge \mathbf{B}$. It is clear from the definition of vector product that the magnitude of $\mathbf{A} \times \mathbf{B}$ is the area of the parallelogram of which **A** and **B** are adjacent sides, and that

$$\mathbf{A} \times \mathbf{B} = -\mathbf{B} \times \mathbf{A} \qquad (A2.13)$$

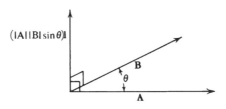

Figure A2.3

If two vectors are parallel, their vector product vanishes ($\theta = 0$). If they are normal to one another the magnitude of the vector product equals the product of their magnitudes. Thus, if the axes are rectangular,

$$\mathbf{i} \times \mathbf{i} = \mathbf{j} \times \mathbf{j} = \mathbf{k} \times \mathbf{k} = 0$$

$$\mathbf{i} \times \mathbf{j} = \mathbf{k}, \qquad \mathbf{j} \times \mathbf{k} = \mathbf{i}, \qquad \mathbf{k} \times \mathbf{i} = \mathbf{j} \qquad \text{(A2.14)}$$

$$\mathbf{j} \times \mathbf{i} = -\mathbf{k}, \qquad \mathbf{k} \times \mathbf{j} = -\mathbf{i}, \qquad \mathbf{i} \times \mathbf{k} = -\mathbf{j}.$$

Using these relations, we find that the components of the vector product of $\mathbf{A}$ and $\mathbf{B}$ are

$$\mathbf{A} \times \mathbf{B} = (A_x\mathbf{i} + A_y\mathbf{j} + A_z\mathbf{k}) \times (B_x\mathbf{i} + B_y\mathbf{j} + B_z\mathbf{k})$$

$$= \mathbf{i}(A_yB_z - A_zB_y) + \mathbf{j}(A_zB_x - A_xB_z) + \mathbf{k}(A_xB_y - A_yB_x) \quad \text{(A2.15)}$$

The reversal of sign in the coefficient multiplying $\mathbf{j}$ should be noted. The quantities multiplying $\mathbf{i}, \mathbf{j}$ and $\mathbf{k}$ are the components of the vector product of $\mathbf{A}$ and $\mathbf{B}$. The relationship (A2.15) may be memorized easily by writing it in the form of a determinant:

$$\mathbf{A} \times \mathbf{B} = \begin{vmatrix} \mathbf{i} & \mathbf{j} & \mathbf{k} \\ A_x & A_y & A_z \\ B_x & B_y & B_z \end{vmatrix} \qquad \text{(A2.16)}$$

The expression $\mathbf{A} \cdot [\mathbf{B} \times \mathbf{C}]$ is a scalar and is called the scalar triple product. If the components of $[\mathbf{B} \times \mathbf{C}]$ are written out using Eqn (A2.15) and then Eqn (A2.8) is employed, we have $\mathbf{A} \cdot [\mathbf{B} \times \mathbf{C}] = A_x(B_yC_z - B_zC_y) + A_y(B_zC_x - B_xC_z) + A_z(B_xC_y - B_yC_x)$, so

$$\mathbf{A} \cdot [\mathbf{B} \times \mathbf{C}] \begin{vmatrix} A_x & A_y & A_z \\ B_x & B_y & B_z \\ C_x & C_y & C_z \end{vmatrix} \qquad \text{(A2.17)}$$

$\mathbf{A} \cdot [\mathbf{B} \times \mathbf{C}]$ represents the volume of a parallelepiped of which $\mathbf{A}, \mathbf{B}, \mathbf{C}$ are three concurrent edges. It should be noted that change in the cyclic order of the vectors in the scalar triple product results in a change of sign of the product. Thus

$$\mathbf{A} \cdot [\mathbf{B} \times \mathbf{C}] = \mathbf{C} \cdot [\mathbf{A} \times \mathbf{B}] = \mathbf{B} \cdot [\mathbf{C} \times \mathbf{A}]$$

$$= -\mathbf{A} \cdot [\mathbf{C} \times \mathbf{B}] = -\mathbf{C} \cdot [\mathbf{B} \times \mathbf{A}] = -\mathbf{B} \cdot [\mathbf{A} \times \mathbf{C}]$$

The expression $\mathbf{A} \times [\mathbf{B} \times \mathbf{C}]$ is a vector called the vector triple product. It is a vector in the plane containing $\mathbf{B}$ and $\mathbf{C}$. By writing out the expressions for the components of $[\mathbf{B} \times \mathbf{C}]$ it is found that

$$\mathbf{A} \times [\mathbf{B} \times \mathbf{C}] = \mathbf{B}(\mathbf{A} \cdot \mathbf{C}) - \mathbf{C}(\mathbf{A} \cdot \mathbf{B}) \qquad \text{(A2.18)}$$

It should be noted that $\mathbf{A} \times [\mathbf{B} \times \mathbf{C}]$ is not equal to $[\mathbf{A} \times \mathbf{B}] \times \mathbf{C}$, so the order of multiplication must be shown by the brackets.

A2.2 THE RECIPROCAL LATTICE

In calculating such things as the angle between two lattice planes of given indices, or in transforming the indices of a given plane or direction from those corresponding to one unit cell to those appropriate to another, it is often extremely useful to use a device called the reciprocal lattice. This was first developed for carrying out lattice sums and then became of great use in diffraction problems and in the theory of the behaviour of electrons in crystals. We will introduce it in a formal way and then illustrate its utility by deriving some formulae of general use.

Suppose we have a crystal lattice and select a primitive unit cell and let the translation vectors be **a, b** and **c**. Then we define three vectors reciprocal to **a, b** and **c**, and call these **a***, **b*** and **c***. The defining relations are

$$\mathbf{a}^* = \frac{\mathbf{b} \times \mathbf{c}}{\mathbf{a} \cdot [\mathbf{b} \times \mathbf{c}]}, \quad \mathbf{b}^* = \frac{\mathbf{c} \times \mathbf{a}}{\mathbf{b} \cdot [\mathbf{c} \times \mathbf{a}]} \quad \text{and} \quad \mathbf{c}^* = \frac{\mathbf{a} \times \mathbf{b}}{\mathbf{c} \cdot [\mathbf{a} \times \mathbf{b}]} \tag{A2.19}$$

From these definitions we note that **a*** is normal to the plane containing **b** and **c** and its magnitude is equal to the reciprocal of the spacing of the planes of Miller indices (1 0 0) in the real lattice (Fig. A2.4). **a***, **b*** and **c*** are taken to define the primitive unit cell of the lattice which is reciprocal to the real crystal lattice with cell edge vectors **a, b** and **c**. Once **a***, **b*** and **c*** have been found from the relations (A2.19), then this reciprocal lattice can be constructed. The utility of the reciprocal lattice depends upon the following properties:

(a) A lattice vector of the reciprocal lattice such as

$$\mathbf{r}^* = h\mathbf{a}^* + k\mathbf{b}^* + l\mathbf{c}^*$$

is normal to the planes of Miller indices $(h\,k\,l)$ in the real lattice. Thus if h, k, l are given small integral values, the vectors $\mathbf{r}^*(h\,k\,l)$ so obtained represent the normals to low index planes.

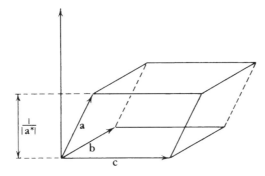

Figure A2.4

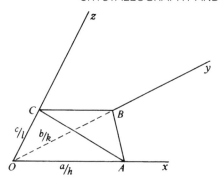

Figure A2.5

(b) The magnitude of $\mathbf{r}^*$, i.e. $|\mathbf{r}^*|$, is equal to the reciprocal of the spacing of the planes of the real lattice of Miller indices $(h\,k\,l)$.

To prove these two propositions we note from the definition of $\mathbf{a}^*$, $\mathbf{b}^*$ and $\mathbf{c}^*$ that $\mathbf{a}^* \cdot \mathbf{a} = 1$, $\mathbf{b}^* \cdot \mathbf{b} = 1$ and $\mathbf{c}^* \cdot \mathbf{c} = 1$, but that $\mathbf{a}^* \cdot \mathbf{b} = \mathbf{a}^* \cdot \mathbf{c} = \mathbf{b}^* \cdot \mathbf{a} = \mathbf{b}^* \cdot \mathbf{c} = \mathbf{c}^* \cdot \mathbf{a} = \mathbf{c}^* \cdot \mathbf{b} = 0$. The two properties of $\mathbf{r}^*(h\,k\,l)$ then follow quite simply. In Fig. A2.5, ABC is the plane of indices $(h\,k\,l)$ in the real crystal. The vector $(\mathbf{a}/h - \mathbf{b}/k)$ represents the line BA in this plane and also $(\mathbf{a}/h - \mathbf{c}/l)$ represents the line CA in the same plane. If $\mathbf{r}^*$ is normal to the plane ABC it must be normal to both AB and CA, which are non-parallel directions in ABC. The dot product of $\mathbf{r}^*$ with these two vectors must be equal to nothing. This is clearly so since, for example,

$$\mathbf{r}^* \cdot \left(\frac{\mathbf{a}}{h} - \frac{\mathbf{b}}{k}\right) = (h\mathbf{a}^* + k\mathbf{b}^* + l\mathbf{c}^*) \cdot \left(\frac{\mathbf{a}}{h} - \frac{\mathbf{b}}{k}\right) = \frac{h}{h}\mathbf{a} \cdot \mathbf{a}^* - \frac{k}{k}\mathbf{b} \cdot \mathbf{b}^* = 0$$

This proves proposition (a).

We prove proposition (b) by noting that the spacing of the lattice planes of indices $(h\,k\,l)$ equals the projection of $\mathbf{a}/h$ on to their normal. By proposition (a) a unit vector along the normal to this plane is $\mathbf{r}^*/r^*$. Then the spacing of the planes d_{hkl} is given by

$$d_{hkl} = \frac{\mathbf{a}}{h} \cdot \frac{\mathbf{r}^*}{|\mathbf{r}^*|} = \frac{h\mathbf{a} \cdot \mathbf{a}^*}{h|\mathbf{r}^*|} = \frac{1}{|\mathbf{r}^*|}$$

We can now use these properties of lattice vectors of the reciprocal lattice to prove some useful results, some of which have been derived by more lengthy methods in the text.

The plane $(h\,k\,l)$ is in the zone $u\,v\,w$ if and only if

$$hu + kv + lw = 0 \qquad\qquad (A2.20)$$

This is so because if $(h\,k\,l)$ is in the zone we must have $\mathbf{r}^*(h\,k\,l)$ normal to $[u\,v\,w]$, i.e.

$$(h\mathbf{a}^* + k\mathbf{b}^* + l\mathbf{c}^*) \cdot (u\mathbf{a} + v\mathbf{b} + w\mathbf{c}) = 0$$

or

$$hu\mathbf{a}^* \cdot \mathbf{a} + kv\mathbf{b}^* \cdot \mathbf{b} + lw\mathbf{c}^* \cdot \mathbf{c} = 0$$

i.e.

$$hu + kv + lw = 0$$

To find the indices of the zone axis common to the two planes of indices $(h_1\,k_1\,l_1)$, $(h_2\,k_2\,l_2)$, if $[u\,v\,w]$ defines the required axis we must have

$$uh_1 + vk_1 + wl_1 = 0 = uh_2 + vk_2 + wl_2$$

and therefore

$$u{:}v{:}w = (k_1 l_2 - k_2 l_1){:}(l_1 h_2 - l_2 h_1){:}(h_1 k_2 - h_2 k_1) \qquad \text{(A2.21)}$$

The condition that three planes $(h_1\,k_1\,l_1)$, $(h_2\,k_2\,l_2)$, $(h_3\,k_3\,l_3)$ shall be in the same zone is that the corresponding reciprocal lattice vectors shall be coplanar. If the corresponding reciprocal lattice vectors are $\mathbf{r}_1^*$, $\mathbf{r}_2^*$, $\mathbf{r}_3^*$ then this implies they define a parallelepiped of zero volume, i.e. $\mathbf{r}_1^* \cdot [\mathbf{r}_2^* \times \mathbf{r}_3^*]$ equals zero. From Eqn (A2.17) this implies

$$\begin{bmatrix} h_1 & k_1 & l_1 \\ h_2 & k_2 & l_2 \\ h_3 & k_3 & l_3 \end{bmatrix} = 0 \qquad \text{(A2.22)}$$

The condition that a given lattice point, say $\mathbf{r}(=u\mathbf{a} + v\mathbf{b} + w\mathbf{c})$ from the origin, should lie in a given lattice plane $(h\,k\,l)$ is useful to know in drawing arrangements of lattice points and/or atoms in lattice planes. If $\mathbf{r}$ lies in the pth plane from the origin of indices $(h\,k\,l)$ then the projection of $\mathbf{r}$ on to the normal to the lattice planes $(h\,k\,l)$ must equal p times the interplanar spacing, i.e.

$$\mathbf{r} \cdot \frac{\mathbf{r}^*(h\,k\,l)}{|\mathbf{r}^*(h\,k\,l)|} = \frac{p}{|\mathbf{r}^*(h\,k\,l)|}$$

Therefore, multiplying out,

$$hu + kv + lw = p \qquad \text{(A2.23)}$$

The angle between two sets of planes, with Miller indices $(h_1\,k_1\,l_1)$ and $(h_2\,k_2\,l_2)$, must be equal to that between the two reciprocal vectors

$$\mathbf{r}_1^* = h_1\mathbf{a}^* + k_1\mathbf{b}^* + l_1\mathbf{c}^*$$

and

$$\mathbf{r}_2^* = h_2\mathbf{a}^* + k_2\mathbf{b}^* + l_2\mathbf{c}^*$$

We have then

$$\mathbf{r}_1^* \cdot \mathbf{r}_2^* = |\mathbf{r}_1^*||\mathbf{r}_2^*| \cos \phi$$

where ϕ is the required angle. Therefore

$$\cos \phi = d_1 d_2 (h_1 \mathbf{a}^* + k_1 \mathbf{b}^* + l_1 \mathbf{c}^*) \cdot (h_2 \mathbf{a}^* + k_2 \mathbf{b}^* + l_2 \mathbf{c}^*) \qquad (A2.24)$$

where d_1 and d_2 are the corresponding spacings of the planes.

Spacings of lattice planes can be derived from the relation

$$\frac{1}{d_{hkl}} = \mathbf{r}^*(hkl) = h\mathbf{a}^* + k\mathbf{b}^* + l\mathbf{c}^* \qquad (A2.25)$$

Expanded versions of Eqn (A2.24) and of Eqn (A2.25) are given for the various crystal systems in Appendix 3.

The sine ratio, relation (A1.14), is easily proved using the reciprocal lattice. If $P_1(h_1 k_1 l_1)$, $P_2(h_2 k_2 l_2)$, $P_3(h_3 k_3 l_3)$, $P_4(h_4 k_4 l_4)$ are four planes all in the same zone and if $\mathbf{r}_1^*$, $\mathbf{r}_2^*$, $\mathbf{r}_3^*$, $\mathbf{r}_4^*$ are the reciprocal lattice vectors normal to $(h_1 k_1 l_1)$, $(h_2 k_2 l_2)$, etc., such that

$$\mathbf{r}_1^* = h_1 \mathbf{a}^* + k_1 \mathbf{b}^* + l_1 \mathbf{c}^*, \qquad \text{etc.}$$

then

$$\begin{aligned} [\mathbf{r}_1^* \times \mathbf{r}_2^*] &= (h_1 \mathbf{a}^* + k_1 \mathbf{b}^* + l_1 \mathbf{c}^*) \times (h_2 \mathbf{a}^* + k_2 \mathbf{b}^* + l_2 \mathbf{c}^*) \\ &= (k_1 l_2 - k_2 l_1)[\mathbf{b}^* \times \mathbf{c}^*] + (l_1 h_2 - l_2 h_1)[\mathbf{c}^* \times \mathbf{a}^*] \\ &\quad + (h_1 k_2 - h_2 k_1)[\mathbf{a}^* \times \mathbf{b}^*] \\ &= |\mathbf{r}_1^*||\mathbf{r}_2^*| \sin \theta_{12} \mathbf{z} \end{aligned} \qquad (A2.26)$$

where $\mathbf{z}$ is a unit vector parallel to the zone axis of the zone containing P_1, P_2, P_3, and

$$\begin{aligned} |\mathbf{r}_1^*||\mathbf{r}_2^*| \sin \theta_{12} \mathbf{z} \cdot \mathbf{a}^* &= (k_1 l_2 - k_2 l_1)[\mathbf{b}^* \times \mathbf{c}^*] \cdot \mathbf{a}^* \\ &= U_{12} V_r \end{aligned}$$

where V_r is the volume of the unit cell of the reciprocal lattice. This follows because

$$[\mathbf{c}^* \times \mathbf{a}^*] \cdot \mathbf{a}^* = [\mathbf{a}^* \times \mathbf{b}^*] \cdot \mathbf{a}^* = 0.$$

Similarly, from the vector product $[\mathbf{r}_1^* \times \mathbf{r}_3^*]$ we find that

$$|\mathbf{r}_1^*||\mathbf{r}_3^*| \sin \theta_{13} \mathbf{z} \cdot \mathbf{a}^* = U_{13} V_r$$

If similar expressions involving θ_{42} and θ_{43} are found the validity of Eqn (A1.14) follows immediately.

Appendix 3

Planar Spacings and Interplanar Angles

A3.1 PLANAR SPACINGS

By definition, two successive planes of indices $(h\,k\,l)$ make intercepts on the crystal axes of na/h, nb/k, nc/l and $(n+1)a/h$, $(n+1)b/k$, $(n+1)c/l$ respectively, where n is an integer. The perpendicular distance between successive planes, or planar spacing d_{hkl}, has been shown to be given by the inverse of the magnitude of the corresponding reciprocal lattice vector (Eqn A2.25), that is

$$d_{hkl} = \frac{1}{|h\mathbf{a}^* + k\mathbf{b}^* + l\mathbf{c}^*|} \tag{A3.1}$$

Now

$$
\begin{aligned}
|h\mathbf{a}^* + k\mathbf{b}^* + l\mathbf{c}^*|^2 &= (h\mathbf{a}^* + k\mathbf{b}^* + l\mathbf{c}^*) \cdot (h\mathbf{a}^* + k\mathbf{b}^* + l\mathbf{c}^*) \\
&= h^2\mathbf{a}^{*2} + k^2\mathbf{b}^{*2} + l^2\mathbf{c}^{*2} + 2kl\mathbf{b}^* \cdot \mathbf{c}^* + 2lh\mathbf{c}^* \cdot \mathbf{a}^* + 2hk\mathbf{a}^* \cdot \mathbf{b}^* \\
&= h^2a^{*2} + k^2b^{*2} + l^2c^{*2} + 2klb^*c^* \cos\alpha^* + 2lhc^*a^* \cos\beta^* \\
&\quad + 2hka^*b^* \cos\gamma^*
\end{aligned}
\tag{A3.2}
$$

where α^* is the angle between the reciprocal lattice vectors $\mathbf{b}^*$ and $\mathbf{c}^*$, β^* the angle between $\mathbf{c}^*$ and $\mathbf{a}^*$ and γ^* the angle between $\mathbf{a}^*$ and $\mathbf{b}^*$. The interplanar spacing can be evaluated in terms of the parameters of the direct lattice by substituting for $\mathbf{a}^*$, $\mathbf{b}^*$ and $\mathbf{c}^*$ from Eqns (A2.19) and for α^*, β^* and γ^* from the following equations:

$$
\begin{aligned}
\cos\alpha^* &= \frac{\cos\beta\cos\gamma - \cos\alpha}{\sin\beta\sin\gamma} \\[4pt]
\cos\beta^* &= \frac{\cos\alpha\cos\gamma - \cos\beta}{\sin\alpha\sin\gamma} \\[4pt]
\cos\gamma^* &= \frac{\cos\alpha\cos\beta - \cos\gamma}{\sin\alpha\sin\beta}
\end{aligned}
\tag{A3.3}
$$

Equations (A3.3) will be recognized as a particular application of Eqn (A1.8). In crystal systems other than the triclinic, some simplifications occur, as follows.

Monoclinic

Since

$$\alpha^* = \alpha = 90°$$

$$\gamma^* = \gamma = 90°$$

$$d_{hkl}^2 = \frac{1}{h^2 a^{*2} + k^2 b^{*2} + l^2 c^{*2} + 2lh c^* a^* \cos \beta^*}$$

where

$$\beta^* = 180° - \beta, \qquad a^* = \frac{1}{a \sin \beta}, \qquad b^* = \frac{1}{b}, \qquad c^* = \frac{1}{c \sin \beta}$$

Orthorhombic

Since

$$\alpha^* = \beta^* = \gamma^* = 90°$$

$$d_{hkl}^2 = \frac{1}{h^2 a^{*2} + k^2 b^{*2} + l^2 c^{*2}}$$

where

$$a^* = \frac{1}{a}, \qquad b^* = \frac{1}{b}, \qquad c^* = \frac{1}{c}$$

Trigonal

Since

$$\alpha^* = \beta^* = \gamma^*$$

$$a^* = b^* = c^*$$

$$d_{hkl}^2 = \frac{1}{[h^2 + k^2 + l^2 + 2(kl + lh + hk) \cos \alpha^*] a^{*2}}$$

where

$$\cos(\alpha^*/2) = \frac{1}{2 \cos(\alpha/2)}, \qquad a^* = \frac{1}{a \sin \alpha \sin \alpha^*}$$

Tetragonal

Since

$$\alpha^* = \beta^* = \gamma^* = 90°$$

$$a^* = b^*$$

$$d_{hkl}^2 = \frac{1}{(h^2 + k^2) a^{*2} + l^2 c^{*2}}$$

where

$$a^* = \frac{1}{a}, \qquad c^* = \frac{1}{c}$$

Hexagonal

Since

$$\alpha^* = \beta^* = 90°, \qquad \gamma^* = 60°$$

$$a^* = b^*$$

$$d^2_{hkl} = \frac{1}{(h^2 + k^2 + hk)a^{*2} + l^2 c^{*2}}$$

where

$$a^* = \frac{2}{a\sqrt{3}}, \qquad c^* = \frac{1}{c}$$

Cubic

Since

$$\alpha^* = \beta^* = \gamma^* = 90°$$

$$a^* = b^* = c^*$$

$$d^2_{hkl} = \frac{1}{(h^2 + k^2 + l^2)a^*}$$

where

$$a^* = \frac{1}{a}$$

It should be appreciated that the value of d_{hkl}, given by Eqn (A3.1), is the perpendicular distance between planes making intercepts of na/h, nb/k, nc/l and $(n + 1)a/h$, $(n + 1)b/k$, $(n + 1)c/l$ on the crystal axes, whether or not these planes coincide with sheets of lattice points. If h, k and l are integers having no common factor, then they are the Miller indices of a lattice plane. The spacing of successive lattice planes of Miller indices $(h\,k\,l)$ in the stack of these planes which builds up the complete lattice is given by Eqn (A3.1), provided that the Bravais lattice is primitive, i.e. provided that the vectors $\mathbf{a}$, $\mathbf{b}$ and $\mathbf{c}$ define a cell that contains a single lattice point. If the unit cell of the Bravais lattice is body-centred, face-centred or base-centred, Eqn (A3.1) may give twice the spacing of successive lattice planes, depending on the values of h, k and l. If the unit cell is body-centred, Eqn (A3.1) gives twice the lattice-plane spacing whenever $(h + k + l)$ is an odd number. For example, the spacing of lattice planes of Miller indices $(1\,0\,0)$ in a b.c.c. lattice is easily seen to be one-half the value of d_{100} given by Eqn (A3.1). The rule for obtaining the true lattice-plane spacing from

Table A3.1 Cubic interplanar angles. (From *International Tables for X-Ray Crystallography* [1])

	100	110	111	210	211	221	310	311
100	2 of 90°	2 of 45° 90°	3 of 54.73°	26.57° 63.43° 90°	35.27° 2 of 65.90°	2 of 48.18° 70.53°	18.43° 71.57° 90°	25.23° 2 of 72.45°
110	4 of 45° 2 of 90°	4 of 60° 90°	3 of 35.27° 3 of 90°	18.43° 2 of 50.77° 3 of 71.57°	2 of 30° 54.73° 2 of 73.22° 90°	19.47° 2 of 45° 2 of 76.37° 90°	26.57° 2 of 47.87° 63.43° 2 of 77.08°	2 of 31.48° 3 of 64.77° 90°
111	4 of 54.73°	2 of 35.27° 2 of 90°	3 of 70.53°	2 of 39.23° 2 of 75.03°	19.47° 2 of 61.87° 90°	15.80° 54.73° 2 of 78.90°	2 of 43.08° 2 of 68.58°	29.50° 2 of 58.52° 79.97°
210	4 of 26.57° 4 of 63.43° 4 of 90°	2 of 18.43° 4 of 50.77° 6 of 71.57°	6 of 39.23° 6 of 75.03°	3 of 36.87° 53.13° 4 of 66.42° 2 of 78.47° 90°	2 of 24.10° 2 of 43.08° 4 of 56.78° 2 of 79.48° 2 of 90°	2 of 26.57° 2 of 41.82° 2 of 53.40° 2 of 63.43° 2 of 72.65° 2 of 90°	8.13° 2 of 31.95° 2 of 45° 2 of 64.90° 2 of 73.57° 3 of 81.87°	2 of 19.28° 4 of 47.60° 2 of 66.13° 4 of 82.25°
211	4 of 35.27° 8 of 65.90°	4 of 30° 2 of 54.73° 4 of 73.22° 2 of 90°	3 of 19.47° 6 of 61.87° 3 of 90°	2 of 24.1° 2 of 43.08° 4 of 56.47° 2 of 79.48° 2 of 90°	2 of 33.55° 2 of 48.18° 2 of 60° 70.53° 4 of 80.40°	2 of 17.72° 35.27° 2 of 47.12° 2 of 65.90° 3 of 74.20° 2 of 82.18°	2 of 25.35° 4 of 49.80° 2 of 58.92° 2 of 75.03° 2 of 82.58°	10.03° 4 of 42.40° 3 of 60.50° 2 of 75.75° 2 of 90°

221	8 of 48.18° 4 of 70.53°	2 of 19.47° 4 of 45° 4 of 76.37° 2 of 90°	3 of 15.80° 3 of 54.73° 6 of 78.90°	2 of 26.57° 2 of 41.82° 2 of 53.40° 2 of 63.43° 2 of 72.65° 2 of 90°	2 of 17.72° 3527° 2 of 47.12° 2 of 65.90° 3 of 74.20° 2 of 82.18°	2 of 27.27° 38.95° 2 of 63.62° 2 of 83.62° 2 of 90°	2 of 32.52° 2 of 42.45° 4 of 58.20° 2 of 65.07° 2 of 83.57°	2 of 25.23° 3 of 45.28° 2 of 59.83° 4 of 72.45° 84.23°
310	4 of 18.43° 4 of 71.57° 4 of 90°	2 of 26.57° 4 of 47.87° 2 of 63.43° 4 of 77.08°	6 of 43.08° 6 of 68.58°	8.13° 2 of 31.95° 2 of 45° 2 of 64.90° 2 of 73.57° 3 of 81.87°	2 of 25.35° 4 of 49.80° 2 of 58.92° 2 of 75.03° 2 of 82.58°	2 of 32.52° 2 of 42.45° 4 of 58.20° 2 of 65.07° 2 of 83.57°	2 of 25.85° 36.87° 53.13° 4 of 72.55° 2 of 84.27° 90°	2 of 17.55° 2 of 40.28° 2 of 55.10° 2 of 67.58° 2 of 79° 2 of 90°
311	4 of 25.23° 8 of 72.45°	4 of 31.48° 6 of 64.77° 2 of 90°	3 of 29.50° 6 of 58.52° 3 of 79.97°	2 of 19.28° 4 of 47.60° 2 of 66.13° 4 of 82.25°	10.03° 4 of 42.40° 3 of 60.50° 2 of 75.75° 2 of 90°	2 of 25.23° 3 of 45.28° 2 of 59.83° 4 of 72.45° 84.23°	2 of 17.55° 2 of 40.28° 2 of 55.10° 2 of 67.58° 2 of 79° 2 of 90°	2 of 35.10° 3 of 50.48° 2 of 62.97° 4 of 84.78°

Eqn (A3.1) in a body-centred (I) lattice is therefore to double the Miller indices whenever ($h + k + l$) is odd. In a face-centred (F) lattice, the rule is to double the Miller indices if either h or k or l is even since the spacing of all lattice planes except those with h, k and l all odd numbers is one-half d_{hkl}. Zero counts as an even number. For a cell which is centred on, say, the (0 0 1) face, the rule is to double the Miller indices when $h + k$ is an odd number.

A3.2 INTERPLANAR ANGLES

It has been shown that the angle between two lattice planes is equal to the angle between the corresponding reciprocal lattice vectors (see Eqn A2.24). This leads to the following general formula for the angle ϕ between the planes ($h\,k\,l$) and ($h'\,k'\,l'$):

$$\cos\varphi = d_{hkl}d_{h'k'l'}[hh'a^{*2} + kk'b^{*2} + ll'c^{*2} + (kl' + lk')b^*c^*\cos\alpha^*$$
$$+ (hl' + lh')a^*c^*\cos\beta^* + (hk' + kh')a^*b^*\cos\gamma^*] \tag{A3.4}$$

where d_{hkl} is given by Eqns (A3.1) and (A3.2). In crystal systems other than the triclinic some simplification occurs. Only the hexagonal and cubic formulae will be written out here.

Hexagonal

$$\cos\phi = d_{hkl}d_{h'k'l'}\{[hh' + kk' + \tfrac{1}{2}(hk' + kh')]a^{*2} + ll'c^{*2}\}$$

where

$$a^* = \frac{2}{a\sqrt{3}}, \qquad c^* = \frac{1}{c}$$

Cubic

$$\cos\phi = \frac{hh' + kk' + ll'}{\sqrt{h^2 + k^2 + l^2}\sqrt{h'^2 + k'^2 + l'^2}}$$

The values of some interplanar angles in cubic crystals are listed in Table A3.1. Values of interplanar angles in the hexagonal metals Mg, Zn and Cd are listed by Salkovitz [2] and in Re by Feng [3]. Values for tetragonal crystals of various c/a ratios are listed by Cherin [4], of Sn and In by Chandrasekhar and Veal [5] and of Ga by Wilson [6].

REFERENCES

1. *International Tables for X-Ray Crystallography*, Vol. II, Kynoch Press (1952).
2. E. I. Salkovity, *Trans. Am. Inst. Mining Metall. and Petrol. Engrs*, **191**, 64 (1951).

3. C. Feng, *J. Less Common Metals*, **4**, 103 (1962).
4. P. Cherin, *Trans. am. Inst. Mining Metall. and Petrol. Engrs*, **239**, 927 (1967).
5. B. S. Chandrasekhar and B. W. Veal, *Trans. Am. Inst. Mining Metall. and Petrol. Engrs*, **221**, 202 (1961).
6. C. G. Wilson, *Trans. am. Inst. Mining. Metall. and Petrol. Engrs*, **224**, 1293 (1962).

Appendix 4

Transformation of Indices Following a Change of Unit Cell

A4.1 CHANGE OF INDICES OF PLANES

When studying crystals different choices for the unit cell of a given crystal may be convenient for different considerations and so it is often necessary to know how the Miller indices of planes and the indices of directions alter when the choice of unit cell is altered. In this appendix we state how to obtain the transformation formulae and give a number of examples of their use. At the end of the appendix we prove the validity of the formulae for the general case using the reciprocal lattice. The formulae may be applied without having read the proof.

Let $(h\,k\,l)$ be the indices of a set of lattice planes referred to the unit cell of translation vectors $\mathbf{a}$, $\mathbf{b}$, $\mathbf{c}$, which we will call the 'old' cell. We wish to find the indices of the same set of planes referred to the 'new' cell, which has translation vectors $\mathbf{A}$, $\mathbf{B}$, $\mathbf{C}$.

We write $\mathbf{A}$, $\mathbf{B}$ and $\mathbf{C}$ in terms of $\mathbf{a}$, $\mathbf{b}$ and $\mathbf{c}$ so that

$$\mathbf{A} = s_{11}\mathbf{a} + s_{12}\mathbf{b} + s_{13}\mathbf{c}$$
$$\mathbf{B} = s_{21}\mathbf{a} + s_{22}\mathbf{b} + s_{23}\mathbf{c} \qquad \text{(A4.1)}$$
$$\mathbf{C} = s_{31}\mathbf{a} + s_{32}\mathbf{b} + s_{33}\mathbf{c}$$

Then $(H\,K\,L)$ are given in terms of $(h\,k\,l)$ as follows:

$$H = s_{11}h + s_{12}k + s_{13}l$$
$$K = s_{21}h + s_{22}k + s_{23}l \qquad \text{(A4.2)}$$
$$L = s_{31}h + s_{32}k + s_{33}l$$

This can be written out in matrix form

$$\begin{pmatrix} H \\ K \\ L \end{pmatrix} = \begin{pmatrix} s_{11} & s_{12} & s_{13} \\ s_{21} & s_{22} & s_{23} \\ s_{31} & s_{32} & s_{33} \end{pmatrix} \begin{pmatrix} h \\ k \\ l \end{pmatrix} \tag{A4.3}$$

The rule for multiplying out the right-hand side is 'first of row times first of column plus second of row times second of column plus third of row times third of column'.

If we wish to write $(h\,k\,l)$ in terms of $(H\,K\,L)$, i.e. to write the old indices in terms of the new, we must first express the old cell vectors $\mathbf{a}$, $\mathbf{b}$, $\mathbf{c}$ in terms of the new ones as

$$\mathbf{a} = t_{11}\mathbf{A} + t_{12}\mathbf{B} + t_{13}\mathbf{C}$$
$$\mathbf{b} = t_{21}\mathbf{A} + t_{22}\mathbf{B} + t_{23}\mathbf{C} \tag{A4.4}$$
$$\mathbf{c} = t_{31}\mathbf{A} + t_{32}\mathbf{B} + t_{33}\mathbf{C}$$

and then $(h\,k\,l)$ are given in terms of the new indices by the equations

$$h = t_{11}H + t_{12}K + t_{13}L$$
$$k = t_{21}H + t_{22}K + t_{23}L \tag{A4.5}$$
$$l = t_{31}H + t_{32}K + t_{33}L$$

or in matrix form

$$\begin{pmatrix} h \\ k \\ l \end{pmatrix} = \begin{pmatrix} t_{11} & t_{12} & t_{13} \\ t_{21} & t_{22} & t_{23} \\ t_{31} & t_{32} & t_{33} \end{pmatrix} \begin{pmatrix} H \\ K \\ L \end{pmatrix} \tag{A4.6}$$

An example may make the procedure for the use of Eqns (A4.2) and (A4.5) clear. In dealing with hexagonal crystals it is sometimes convenient to use the orthorhombic cell shown in Fig. A4.1, which is sometimes called an orthohexagonal cell. The conventional hexagonal cell has cell edge vectors $\mathbf{a}$, $\mathbf{b}$, $\mathbf{c}$ as shown and the *new* orthorhombic cell has edges $\mathbf{A}$, $\mathbf{B}$, $\mathbf{C}$, where it is clear from the figure that

$$\mathbf{A} = 2 \cdot \mathbf{a} + 1 \cdot \mathbf{b} + 0 \cdot \mathbf{c}$$
$$\mathbf{B} = 0 \cdot \mathbf{a} + 1 \cdot \mathbf{b} + 0 \cdot \mathbf{c} \tag{A4.7}$$
$$\mathbf{C} = 0 \cdot \mathbf{a} + 0 \cdot \mathbf{b} + 1 \cdot \mathbf{c}$$

If $(h\,k\,l)$ are the (Miller) indices of a plane referred to the hexagonal cell and $(H\,K\,L)$ the indices of the same plane referred to the orthorhombic cell then, comparing Eqns (A4.2) and (A4.7),

$$H = 2h + k$$
$$K = k$$
$$L = l$$

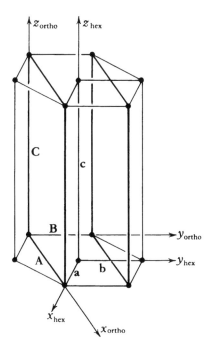

Figure A4.1

The transformation matrix is

$$\begin{pmatrix} 2 & 1 & 0 \\ 0 & 1 & 0 \\ 0 & 0 & 1 \end{pmatrix}$$

If we wish to write $(h\,k\,l)$ in terms of $(H\,K\,L)$ we must first write **a, b, c** in terms of **A, B, C** obtaining (see Fig. A4.1)

$$\mathbf{a} = \tfrac{1}{2} \cdot \mathbf{A} - \tfrac{1}{2} \cdot \mathbf{B} + 0 \cdot \mathbf{C}$$
$$\mathbf{b} = 0 \cdot \mathbf{A} + 1 \cdot \mathbf{B} + 0 \cdot \mathbf{C} \qquad\qquad \text{(A4.8)}$$
$$\mathbf{c} = 0 \cdot \mathbf{A} + 0 \cdot \mathbf{B} + 1 \cdot \mathbf{C}$$

and so

$$h = \tfrac{1}{2}(H - K)$$
$$k = K$$
$$l = L$$

The matrix for writing $(h\,k\,l)$ in terms of $(H\,K\,L)$ is

$$\begin{pmatrix} \frac{1}{2} & -\frac{1}{2} & 0 \\ 0 & 1 & 0 \\ 0 & 0 & 1 \end{pmatrix}$$

A4.2 CHANGE OF INDICES OF DIRECTIONS

The transformation of *directions* proceeds as follows. Let $[u\,v\,w]$ be the indices of a given direction in terms of the old unit cell of cell edge vectors $\mathbf{a}$, $\mathbf{b}$, $\mathbf{c}$ and $[U\,V\,W]$ those of the same direction in terms of the cell of cell edge vectors $\mathbf{A}$, $\mathbf{B}$, $\mathbf{C}$. Then if $\mathbf{A}$, $\mathbf{B}$, $\mathbf{C}$ are each written in terms of $\mathbf{a}$, $\mathbf{b}$, $\mathbf{c}$ according to Eqns (A4.1) and $\mathbf{a}$, $\mathbf{b}$, $\mathbf{c}$ in terms of $\mathbf{A}$, $\mathbf{B}$, $\mathbf{C}$ according to Eqns (A4.4), it follows that

$$\begin{pmatrix} U \\ V \\ W \end{pmatrix} = \begin{pmatrix} t_{11} & t_{21} & t_{31} \\ t_{12} & t_{22} & t_{32} \\ t_{13} & t_{23} & t_{33} \end{pmatrix} \begin{pmatrix} u \\ v \\ w \end{pmatrix} \tag{A4.9}$$

and that

$$\begin{pmatrix} u \\ v \\ w \end{pmatrix} = \begin{pmatrix} s_{11} & s_{21} & s_{31} \\ s_{12} & s_{22} & s_{32} \\ s_{13} & s_{23} & s_{33} \end{pmatrix} \begin{pmatrix} U \\ V \\ W \end{pmatrix} \tag{A4.10}$$

As well as being used to transform vectors, i.e. directions (or zone axis symbols), Eqns (A4.9) and (A4.10) are the necessary relations for transforming the coordinates of positions in the two unit cells.

The positions of the s's and t's in the formulae (A4.2) and (A4.5) and their positions in (A4.9) and in (A4.10) should be very carefully noted. Not only are the matrices for changing 'new' to 'old' interchanged when one considers directions instead of planes but, in addition, the rows and columns are interchanged.

If V_1 is the volume of the cell $\mathbf{a}$, $\mathbf{b}$, $\mathbf{c}$ and V_2 that of $\mathbf{A}$, $\mathbf{B}$, $\mathbf{C}$, then the determinants of the arrays of s's and t's relate the volumes of the two cells:

$$V_1{:}V_2 = 1{:} \begin{vmatrix} s_{11} & s_{12} & s_{13} \\ s_{21} & s_{22} & s_{23} \\ s_{31} & s_{32} & s_{33} \end{vmatrix} = \begin{vmatrix} t_{11} & t_{12} & t_{13} \\ t_{21} & t_{22} & t_{23} \\ t_{31} & t_{32} & t_{33} \end{vmatrix}{:}1 \tag{A4.11}$$

A4.3 INTERCHANGE OF RHOMBOHEDRAL AND HEXAGONAL INDICES

As a further example of the transformation of indices we will consider the relationship between rhombohedral and hexagonal cells in the trigonal crystal system. A trigonal crystal may possess either a rhombohedral primitive unit cell (Fig. 1.19k) or else a cell of the shape of Fig. 1.19j. A hexagonal crystal must

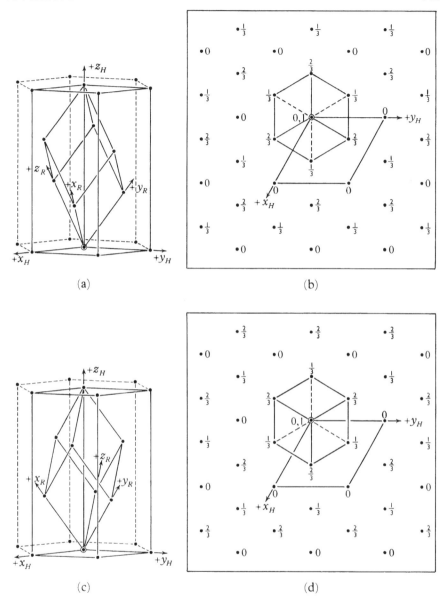

Figure A4.2 Unit cells in the rhombohedral lattice. (a) Obverse setting of rhombohedron and corresponding hexagonal non-primitive unit cell. (b) Plan of obverse setting (- - - - - lower edges, ———— upper edges of rhombohedron). (c) Reverse setting of rhombohedron and corresponding hexagonal non-primitive unit cell. (d) Plan of reverse setting (- - - - - lower edges, ———— upper edges of rhombohedron)

possess a primitive unit cell of the type shown in Fig. 1.19j. When hexagonal axes are used with a crystal that has a true rhombohedral Bravais lattice, then if the smallest unit cells are chosen, the smallest hexagonal cell is triply-primitive. We will consider this case. The rhombohedral axis can be oriented in two different ways relative to the corresponding hexagonal axes. These are shown in Figs. A4.2a,b,c and d.

Consider the *obverse* setting with the rhombohedral cell edge vectors $\mathbf{a}_R$, $\mathbf{b}_R$, $\mathbf{c}_R$ and let the hexagonal cell edge vectors be $\mathbf{a}_H$, $\mathbf{b}_H$, $\mathbf{c}_H$. Let the indices of a plane referred to the hexagonal cell be $(H\,K\,L)$ and those of the same plane referred to the rhombohedral cell be $(h\,k\,l)$. Writing the hexagonal cell vectors in terms of the rhombohedral we have

$$\mathbf{a}_H = 1 \cdot \mathbf{a}_R + (-1) \cdot \mathbf{b}_R + 0 \cdot \mathbf{c}_R$$

$$\mathbf{b}_H = 0 \cdot \mathbf{a}_R + 1 \cdot \mathbf{b}_R + (-1) \cdot \mathbf{c}_R \qquad (A4.12)$$

$$\mathbf{c}_H = 1 \cdot \mathbf{a}_R + 1 \cdot \mathbf{b}_R + 1 \cdot \mathbf{c}_R.$$

Hence the transformation matrix equation for the indices of planes is from Eqns (A4.2) or (A4.3):

$$\begin{pmatrix} H \\ K \\ L \end{pmatrix} = \begin{pmatrix} 1 & \bar{1} & 0 \\ 0 & 1 & \bar{1} \\ 1 & 1 & 1 \end{pmatrix} \begin{pmatrix} h \\ k \\ l \end{pmatrix} \qquad (A4.13)$$

Hence, for example, $(1\,0\,\bar{1})_R = (1\,1\,0)_H$ and $(2\,0\,\bar{1})_R = (2\,1\,1)_H$. The reverse transformation is obtained by writing $\mathbf{a}_R$, $\mathbf{b}_R$, $\mathbf{c}_R$ in terms of $\mathbf{a}_H$, $\mathbf{b}_H$, $\mathbf{c}_H$, so obtaining

$$\mathbf{a}_R = \tfrac{2}{3} \cdot \mathbf{a}_H + \tfrac{1}{3} \cdot \mathbf{b}_H + \tfrac{1}{3} \cdot \mathbf{c}_H$$

$$\mathbf{b}_R = -\tfrac{1}{3} \cdot \mathbf{a}_H + \tfrac{1}{3} \cdot \mathbf{b}_H + \tfrac{1}{3} \cdot \mathbf{c}_H \qquad (A4.14)$$

$$\mathbf{c}_R = -\tfrac{1}{3} \cdot \mathbf{a}_H + -\tfrac{2}{3} \cdot \mathbf{b}_H + \tfrac{1}{3} \cdot \mathbf{c}_H$$

The transformation matrix is therefore

$$\begin{pmatrix} h \\ k \\ l \end{pmatrix} = \begin{pmatrix} \tfrac{2}{3} & \tfrac{1}{3} & \tfrac{1}{3} \\ -\tfrac{1}{3} & \tfrac{1}{3} & \tfrac{1}{3} \\ -\tfrac{1}{3} & -\tfrac{2}{3} & \tfrac{1}{3} \end{pmatrix} \begin{pmatrix} H \\ K \\ L \end{pmatrix} \qquad (A4.15)$$

For example, substituting $H = 1$, $K = 1$, $L = 1$ we obtain $h = \tfrac{4}{3}$, $k = \tfrac{1}{3}$, $l = -\tfrac{2}{3}$ or $(1\,1\,1)_H = (4\,1\,\bar{2})_R$.

The matrices for transforming directions $[u\,v\,w]$, referred to the rhombohedral lattice, to $[U\,V\,W]$, referred to the hexagonal lattice, can be found from (A4.15) by interchanging rows and columns. Thus

$$\begin{pmatrix} U \\ V \\ W \end{pmatrix} = \begin{pmatrix} \frac{2}{3} & -\frac{1}{3} & -\frac{1}{3} \\ \frac{1}{3} & \frac{1}{3} & -\frac{2}{3} \\ \frac{1}{3} & \frac{1}{3} & \frac{1}{3} \end{pmatrix} \begin{pmatrix} u \\ v \\ w \end{pmatrix} \tag{A4.16}$$

For example, $[1\,1\,1]_R = [0\,0\,1]_H$ and $[1\,1\,0]_R$ yields $U = \frac{1}{3}$, $V = \frac{2}{3}$, $W = \frac{2}{3}$ or $[1\,2\,2]_H$.

The reverse transformation giving $[u\,v\,w]$ from $[U\,V\,W]$ is accomplished from the equation

$$\begin{pmatrix} u \\ v \\ w \end{pmatrix} = \begin{pmatrix} 1 & 0 & 1 \\ \bar{1} & 1 & 1 \\ 0 & \bar{1} & 1 \end{pmatrix} \begin{pmatrix} U \\ V \\ W \end{pmatrix} \tag{A4.17}$$

The matrices for accomplishing the transformation of indices of planes and of directions between the hexagonal cell and the *reverse* rhombohedral cell can be deduced from Figs A4.2c and d.

A4.4 PROOF OF TRANSFORMATION EQUATIONS

We use the notation of the previous sections of this appendix. The cell with translation vectors $\mathbf{a}$, $\mathbf{b}$, $\mathbf{c}$ has a reciprocal cell with translation vectors $\mathbf{a}^*$, $\mathbf{b}^*$, $\mathbf{c}^*$, where $\mathbf{a}^*$, $\mathbf{b}^*$, $\mathbf{c}^*$ are related to $\mathbf{a}$, $\mathbf{b}$, $\mathbf{c}$ by equations such as (A2.19). The cell with translation vectors $\mathbf{A}$, $\mathbf{B}$, $\mathbf{C}$ has a corresponding reciprocal cell with vectors $\mathbf{A}^*$, $\mathbf{B}^*$, $\mathbf{C}^*$.

A given lattice point of the reciprocal lattice represents a plane in real space and so must be the same distance from the origin regardless of the choice of unit cell. Therefore,

$$h\mathbf{a}^* + k\mathbf{b}^* + l\mathbf{c}^* = H\mathbf{A}^* + K\mathbf{B}^* + L\mathbf{C}^* \tag{A4.18}$$

If we form the dot product of both sides of this equation first with $\mathbf{A}$, then with $\mathbf{B}$ and then with $\mathbf{C}$, we obtain

$$\mathbf{A} \cdot (h\mathbf{a}^* + k\mathbf{b}^* + l\mathbf{c}^*) = H$$
$$\mathbf{B} \cdot (h\mathbf{a}^* + k\mathbf{b}^* + l\mathbf{c}^*) = K \tag{A4.19}$$
$$\mathbf{C} \cdot (h\mathbf{a}^* + k\mathbf{b}^* + l\mathbf{c}^*) = L$$

If we now substitute for $\mathbf{A}$ in the first of Eqns (A4.19) by using (A4.1) and multiply out the dot product, we have

$$H = s_{11}h + s_{12}k + s_{13}l$$

Expressions for K and L are obtained by substituting for $\mathbf{B}$ and $\mathbf{C}$, thus obtaining Eqns (A4.2). To obtain Eqns (A4.5) we form the dot product of both sides of (A4.18) first with **a**, then with **b** and **c** and proceed similarly, substituting for $\mathbf{A}^*$, $\mathbf{B}^*$ and $\mathbf{C}^*$.

To prove Eqns (A4.9) and (A4.10) we note that directions must be the same when referred to either of the unit cells and therefore

$$u\mathbf{a} + v\mathbf{b} + w\mathbf{c} = U\mathbf{A} + V\mathbf{B} + W\mathbf{C} \qquad (A4.20)$$

To obtain the expression for u in Eqns (A4.10) we form the dot product of both sides of (A4.20) with $\mathbf{a}^*$ and have

$$u = (U\mathbf{A} + V\mathbf{B} + W\mathbf{C}) \cdot \mathbf{a}^*$$

$$= [U(s_{11}\mathbf{a} + s_{12}\mathbf{b} + s_{13}\mathbf{c}) + V(s_{21}\mathbf{a} + s_{22}\mathbf{b} + s_{23}\mathbf{c})$$

$$+ W(s_{31}\mathbf{a} + s_{32}\mathbf{b} + s_{33}\mathbf{c})] \cdot \frac{\mathbf{b} \times \mathbf{c}}{\mathbf{a} \cdot [\mathbf{b} \times \mathbf{c}]}$$

Multiplying out the right-hand side we obtain

$$u = Us_{11} + Vs_{21} + Ws_{31}$$

The continuation of this procedure to obtain the remaining equations of (A4.10) and of (A4.9) is obvious.

Appendix 5

Crystal Structure Data

A5.1 CRYSTAL STRUCTURES OF THE ELEMENTS, INTERATOMIC DISTANCES AND IONIC RADII AT ROOM TEMPERATURE

Atomic number	Element	Structure	Shortest distance between atoms (Å)	Ion	Ionic radius[a] (Å)
1	H	—	—	H^-	1.54
3	Li	b.c.c.	3.04	Li^+	0.60
4	Be	c.p.h.	2.24	Be^{2+}	0.31
5	B	Trigonal	1.71	—	—
6	C (graphite)	Hexagonal	1.42	—	—
	C (diamond)	Cubic	1.54	—	—
8	O	—	—	O^{2-}	1.40
9	F	—	—	F^-	1.36
11	Na	b.c.c.	3.71	Na^+	0.95
12	Mg	c.p.h.	3.21	Mg^{2+}	0.65
13	Al	f.c.c.	2.86	Al^{3+}	0.50
14	Si	Cubic (diamond)	2.35	Si^{4+}	0.41
15	P (black)	Orthorhombic	2.17	—	—
16	S	Orthorhombic	2.12	S^{2-}	1.84
17	Cl	—	—	Cl^-	1.81
19	K	b.c.c.	4.63	K^+	1.33
20	Ca	f.c.c.	3.94	Ca^{2+}	0.99
21	Sc	f.c.c.	3.21	Sc^{3+}	0.81
22	Ti	c.p.h.	2.95	Ti^{3+}	0.76
				Ti^{4+}	0.68

Atomic number	Element	Structure	Shortest distance between atoms (Å)	Ion	Ionic radius[a] (Å)
23	V	b.c.c.	2.63	V^{2+}	0.88
				V^{3+}	0.74
				V^{4+}	0.60
24	Cr	b.c.c.	2.50	Cr^{2+}	0.84
				Cr^{3+}	0.63
				Cr^{4+}	0.56
25	Mn	Cubic	2.24	Mn^{2+}	0.80
				Mn^{3+}	0.66
				Mn^{4+}	0.54
26	Fe	b.c.c.	2.48	Fe^{2+}	0.80
				Fe^{3+}	0.64
27	Co	c.p.h.	2.51	Co^{2+}	0.72
				Co^{3+}	0.63
28	Ni	f.c.c.	2.49	Ni^{2+}	0.69
				Ni^{3+}	0.62
29	Cu	f.c.c.	2.56	Cu^{+}	0.96
30	Zn	c.p.h.	2.66	Zn^{2+}	0.74
31	Ga	Orthorhombic	2.44	Ga^{3+}	0.62
32	Ge	Cubic (Diamond)	2.45	Ge^{4+}	0.53
33	As	trigonal	2.90	—	—
34	Se	Hexagonal	2.32	Se^{2-}	1.98
35	Br	—	—	Br^{-}	1.95
37	Rb	b.c.c.	4.90	Rb^{+}	1.48
38	Sr	f.c.c.	4.30	Sr^{2+}	1.13
39	Y	c.p.h.	3.62	Y^{3+}	0.93
40	Zr	c.p.h.	3.20	Zr^{4+}	0.80
41	Nb(Cb)	b.c.c.	2.86	—	—
42	Mo	b.c.c.	2.72	—	—
43	Tc	c.p.h.	2.70	—	—
44	Ru	c.p.h.	2.65	Ru^{4+}	0.63
45	Rh	f.c.c.	2.69	Rh^{3+}	0.68
46	Pd	f.c.c.	2.75	Pd^{2+}	0.80
				Pd^{4+}	0.65
47	Ag	f.c.c.	2.89	Ag^{+}	1.26
48	Cd	c.p.h.	2.98	Cd^{2+}	0.97
49	In	Tetragonal	3.25	In^{3+}	0.81

Atomic number	Element	Structure	Shortest distance between atoms (Å)	Ion	Ionic radius^a (Å)
50	Sn (white)	Tetragonal	3.02	Sn^{4+}	0.71
51	Sb	Trigonal	2.90	—	—
52	Te	Hexagonal	2.87	Te^{2-}	2.21
53	I	Orthorhombic	2.71	I^-	2.16
55	Cs	b.c.c.	5.25	Cs^+	1.69
56	Ba	b.c.c.	4.35	Ba^{2+}	1.35
57	La	c.p.h.	3.75	La^{3+}	1.15
58	Ce	f.c.c.	3.65	Ce^{3+}	1.02
59	Pr	hexag.	3.64	Pr^{3+}	1.00
60	Nd	hexag.	3.66	Nd^{3+}	0.99
63	Eu	b.c.c.	3.96	Eu^{3+}	0.97
64	Gd	c.p.h.	3.56	Gd^{3+}	0.97
65	Tb	c.p.h.	3.53	Tb^{3+}	1.00
66	Dy	c.p.h.	3.50	Dy^{3+}	0.99
67	Ho	c.p.h.	3.52	Ho^{3+}	0.97
68	Er	c.p.h.	3.55	Er^{3+}	0.96
69	Tm	c.p.h.	3.48	Tm^{3+}	0.95
70	Yb	f.c.c.	3.87	Yb^{3+}	0.94
71	Lu	c.p.h.	3.46	Lu^{3+}	0.93
72	Hf	c.p.h.	3.16	Hf^{4+}	0.78
73	Ta	b.c.c.	2.86	—	—
74	W	b.c.c.	2.74	W^{4+}	0.66
75	Re	c.p.h.	2.74	Re^{4+}	0.72
76	Os	c.p.h.	2.68	Os^{4+}	0.65
77	Ir	f.c.c.	2.71	Ir^{4+}	0.64
78	Pt	f.c.c.	2.77	Pt^{4+}	0.65
79	Au	f.c.c.	2.88	Au^+	1.37
80	Hg	Trigonal	3.00	Hg^{2+}	1.10
81	Tl	c.p.h.	3.41	Tl^+	1.44
				Tl^{3+}	0.95
82	Pb	f.c.c.	3.49	Pb^{2+}	1.21
				Pb^{4+}	0.84
83	Bi	Trigonal	3.11	—	—
84	Po	Simple cubic	3.35	—	—
89	Ac	f.c.c.	3.75	Ac^{3+}	1.11
90	Th	f.c.c.	3.60	Th^{3+}	1.08
				Th^{4+}	0.95

Atomic number	Element	Structure	Shortest distance between atoms (Å)	Ion	Ionic radius[a] (Å)
91	Pa	b.c. tetragonal	3.21	Pa^{3+}	1.06
				Pa^{4+}	0.91
92	U	Orthorhombic	2.77	U^{3+}	1.04
				U^{4+}	0.89
93	Np	Orthorhombic	2.60	Np^{3+}	1.02
				Np^{4+}	0.88
94	Pu	Monoclinic	3.28	Pu^{3+}	1.01
				Pu^{4+}	0.86

[a]The radius of an ion depends on the number of neighbouring ions of opposite sign, or in other words, the coordination number. The values in the table apply to an ion with a coordination number of 6. In structures where the coordination number is 4 the radius should be decreased by about 7%, with coordination number 8 the radius should be increased by 3% and with coordination number 12 by 6%.

A5.2 CRYSTALS WITH THE SODIUM CHLORIDE STRUCTURE

The data in Sections A5.2 to A5.8 are taken from more extensive tables in *Crystal Structures* by R.G. Wyckoff, Interscience, New York, 1963.

Crystal	a_0 (Å)	Crystal	a_0 (Å)
Antimonides		*Borides*	
CeSb	6.40	ZrB	4.65
LaSb	6.47	*Bromides*	
ScSb	5.86	AgBr	5.77
SnSb	6.13	KBr	6.60
ThSb	6.32	LiBr	5.50
USb	6.19	NaBr	5.97
Arsenides		RbBr	6.85
CeAs	6.06	*Carbides*	
LaAs	6.13	HfC	4.46
ScAs	5.49	NbC	4.47
SnAs	5.68	TaC	4.45
ThAs	5.97	TiC	4.32
UAs	5.77	UC	4.96

Crystal	a_0 (Å)
VC	4.18
ZrC	4.68
Chlorides	
AgCl	5.55
KCl	6.29
LiCl	5.13
NaCl	5.64
RbCl	6.58
Cyanides	
KCN	6.53
NaCN	5.89
RbCN	6.82
Fluorides	
AgF	4.92
CsF	6.01
KF	5.35
LiF	4.03
NaF	4.62
RbF	5.64
Hydrides	
CsH	6.38
KH	5.70
LiH	4.08
NaH	4.88
RbH	6.04
Iodides	
KI	7.07
LiI	6.00
NH_4I	7.26
NaI	6.47
RbI	7.34
Nitrides	
CeN	5.01
CrN	4.14
LaN	5.30
NbN	4.70
ScN	4.44
TiN	4.23
UN	4.88

Crystal	a_0 (Å)
VN	4.13
ZrN	4.61
Oxides	
BaO	5.52
CaO	4.81
CdO	4.70
CoO	4.27
FeO	4.28–4.31
MgO	4.21
MnO	4.45
NbO	4.21
NiO	4.17
SrO	5.16
TaO	4.42–4.44
TiO	4.18
UO	4.92
ZrO	4.62
Phosphides	
CeP	5.90
LaP	6.01
ThP	5.82
UP	5.59
ZrP	5.27
Selenides	
BaSe	6.60
CaSe	5.91
CeSe	5.98
LaSe	6.06
MgSe	5.45
MnSe	5.45
PbSe	6.12
SnSe	6.02
SrSe	6.23
ThSe	5.87
USe	5.75
Sulphides	
BaS	6.39
CaS	5.69
CeS	5.78

Crystal	a_0 (Å)
LaS	5.84
MgS	5.20
MnS	4.45
PbS	5.94
SrS	6.02
ThS	5.68
US	5.48
ZrS	5.25
Tellurides	
BaTe	6.99

Crystal	a_0 (Å)
BiTe	6.47
CaTe	6.34
CeTe	6.35
LaTe	6.41
PbTe	6.45
SnTe	6.31
SrTe	6.47
UTe	6.16

A5.3 CRYSTALS WITH THE CAESIUM CHLORIDE STRUCTURE

Crystal	a_0 (Å)
AgCd	3.33
AgMg	3.28
CuBe	2.70
CuZn	2.94
CsBr	4.29
CsCl	4.12
CsCN	4.25
CsI	4.57
TlBr	3.97
TlCl	3.83
TlCN	3.82
TlI	4.20

A5.4 CRYSTALS WITH THE SPHALERITE STRUCTURE

Crystal	a_0 (Å)
AgI	6.47
BN	3.62
BeS	4.85
BeSe	5.07
BeTe	5.54
CdS	5.82

Crystal	a_0 (Å)
InAs	6.06
AiAs	5.66
GaSb	6.12
HgS	5.85
HgSe	6.08
HgTe	6.43

Crystal	a_0 (Å)	Crystal	a_0 (Å)
CdTe	6.48	InAs	6.04
CuBr	5.69	InP	5.87
CuCl	5.40	InSb	6.48
CuF	4.26	SiC	4.35
CuI	6.04	ZnS	5.41
GaAs	5.65	ZnSe	5.67
GaP	5.45	ZnTe	6.09

A5.5 CRYSTALS WITH THE WURTZITE STRUCTURE

Crystal	a_0 (Å)	c_0 (Å)
AgI	4.58	7.49
AlN	3.11	4.98
BeO	2.70	4.38
CdS	4.14	6.75
CdSe	4.30	7.02
NH_4F	4.39	7.02
SiC	3.08	5.05
TaN	3.05	4.94
ZnO	3.25	5.21
ZnS	3.81	6.23
ZnSe	3.98	6.53
ZnTe	4.27	6.99

A5.6 CRYSTALS WITH THE NICKEL ARSENIDE STRUCTURE

Crystal	a_0 (Å)	c_0 (Å)	Crystal	a_0 (Å)	c_0 (Å)
CoS	3.37	5.16	FeSb	4.06	5.13
CoSb	3.87	5.19	FeSe	3.64	5.96
CoSe	3.63	5.30	FeTe	3.80	5.65
CoTe	3.89	5.36	MnAs	3.71	5.69
CrS	3.45	5.75	MnBi	4.30	6.12
CrSb	4.11	5.44	MnSb	4.12	5.78
CrSe	3.68	6.02	MnTe	4.14	6.70
FeS	3.44	5.88	NiAs	3.60	5.01

Crystal	a_0 (Å)	c_0 (Å)	Crystal	a_0 (Å)	c_0 (Å)
NiSb	3.94	5.14	PtSb	4.13	5.47
NiSe	3.66	5.36	PtSn	4.10	5.43
NiSn	4.05	5.12	VP	3.18	6.22
NiTe	3.96	5.35	VS	3.36	5.81
PtB[a]	3.36	4.06	VSe	3.58	5.98
PtBi	4.31	5.49	VTe	3.94	6.13

[a] Anti-nickel arsenide structure.

A5.7 CRYSTALS WITH THE FLUORITE STRUCTURE

Crystal	a_0 (Å)	Crystal	a_0 (Å)
$BaCl_2$	7.34	Li_2S	5.71
BaF_2	6.20	Li_2Se	6.01
Be_2B	4.67	Li_2Te	6.50
Be_2C	4.33	Na_2O	5.55
CaF_2	5.46	Na_2S	6.53
CdF_2	5.39	Na_2Se	6.81
CeH_2	5.59	Na_2Te	7.31
CeO_2	5.41	NbH_2	4.56
$CoSi_2$	5.36	$NiSi_2$	5.39
HfO_2	5.12	Rb_2O	6.74
HgF_2	5.54	Rb_2S	7.65
K_2O	6.44	$SrCl_2$	6.98
K_2S	7.39	SrF_2	5.80
K_2Se	7.68	ThO_2	5.60
K_2Te	8.15	UO_2	5.47
Li_2O	4.62	ZrO_2	5.07

A5.8 CRYSTALS WITH THE RUTILE STRUCTURE

Crystal	a_0 (Å)	c_0 (Å)	Crystal	a_0 (Å)	c_0 (Å)
CoF_2	4.70	3.18	NbO_2	4.77	2.96
FeF_2	4.70	3.31	PbO_2	4.95	3.38
GeO_2	4.40	2.86	SnO_2	4.74	3.19
MgF_2	4.62	3.05	TaO_2	4.71	3.06
MnF_2	4.87	3.31	TiO_2	4.59	2.96
MnO_2	4.40	2.87	WO_2	4.86	2.77
MoO_2	4.86	2.79	ZnF_2	4.70	3.13

Data taken from *An Introduction to Polymer Science* by H.-G. Elias, VCH, Weinheim (1997), ch. 8, p. 281.

Polymer	Smallest constitutional unit (CRU)	Mod.[a]	N_u[b]	a (Å)	b (Å)	c (Å)	α	β	γ	Helix[d]	Crystal system
Poly(ethylene)	$-[(CH_2)]_n-$	I	2	7.42	4.95	2.54	90	90	90	1_1	Orthorhombic
		II	2	8.09	2.53[c]	4.79	90	90	107.9	1_1	Monoclinic
Poly(vinylchloride)	$-[CH_2CHCl]_n-$		4	10.4	5.30	5.10	90	90	90	2_1	Orthorhombic
Poly(tetrafluoroethylene) (PTFE)	$-[(CF_2)]_n-$	I	15	5.66	5.66	19.50	90	87	120	15_7	Hexagonal
		II	13	9.52	5.59	17.06	88	90	92	13_6	Triclinic
Poly(propylene)	$-[CH(CH_3)]_n-$		8	14.5	5.60	7.40	90	90	90	2_1	Orthorhombic
		I(α)	12	6.65	20.9	6.50	90	99.5	90	3_1	Monoclinic
		II(β)	18	19.08	11.01	6.49	90	90	90	3_1	Orthorhombic
		III(γ)	12	6.38	6.38	6.33	89	100	90	3_1	Triclinic
Poly(isobutylene)	$-[CH_2C(CH_3)_2]_n-$		16	6.88	11.91	18.60	90	90	90	8_3	Orthorhombic
Poly(1-butylene)	$-[CH_2CH(C_2H_5)]_n-$	I	18	17.7	17.70	6.51	90	90	90	3_1	Orthorhombic
		II	44	14.85	14.85	20.60	90	90	90	11_3	Tetragonal
		III		12.38	8.92	7.45	90	90	90	4_1	Orthorhombic
Poly(styrene)	$-[CH_2CH(C_6H_5)]_n-$	I	18	21.9	21.9	6.65	90	90	120	3_1	Trigonal
Poly(oxymethylene)	$-[CH_2O]_n-$	I	9	4.46	4.46	17.30	90	90	90	9_5	Trigonal
Poly(oxyethylene)	$-[CH_2CH_2O]_n-$	I	28	8.03	12.09	19.48	90	125.4	90	3_1	Monoclinic
Poly(oxypropylene)	$-[CH(CH_3)CH_2O]_n-$		4	10.52	4.68	7.10	90	90	90	2_1	Orthorhombic
Poly(glycine)	$-[NH\ CH_2CO]_n-$	I	2	4.77	4.77	7.0	90	90	66	2_1	Monoclinic
		II	3	4.8	4.8	9.3	90	90	120	3_1	Hexagonal
Poly(ε-caprolactam)	$-[NHCO(CH_2)_5]_n-$	α	8	9.60	17.18[c]	8.05	90	68.6	90	2_1	Monoclinic
		β	1	4.8	4.8	8.6	90	90	120	1_1	Hexagonal
Poly(hexamethylene adipamide) (Nylon 66)	$-[NH(CH_2)_6-NHCO(CH_2)_4CO]_n-$	I	1	4.9	5.4	17.3	48	77	63	1_1	Triclinic

[a] The different possible conformations of polymer chains are named modifications (Mod.) I, II, III and α, β, γ. [b] N_u = number of monomer units per unit cell; a, b, c = lattice constants; α, β, γ = angles of unit cell. [c] Fibre axis (when not c). [d] For a helical polymer chain the helix is described by two numbers B_N, where B = integral number of conformational repeat units per N turns and N = number of turns needed to return to original position.

Answers to Problems

CHAPTER 1

1.2 (a) See Fig. 3.11, (b) two, (c) and (d) four 0 at $\sqrt{2a^2(\frac{1}{2} - u)^2 + c^2/4} \approx$ 1.94 Å and two 0 at $\sqrt{2}ua \approx 2$ Å.

1.3 (a) I, (b) P, (c) I.

1.5 (a) $[1\bar{1}0]$, $(1\,1\,1)$, $(1\,1\,\bar{1})$, $(\bar{1}\,\bar{1}\,\bar{1})$, $(\bar{1}\,\bar{1}\,1)$, (c) $a/\sqrt{2}$, (d) 70.53° or 109.47° $(=180° - 70.53°)$.

1.6 (a) 0°, (b) 40.4°.

1.7 (a) $[\bar{2}\,1\,1]$, (b) 35.27°.

1.8 (a) $c/a = 0.985$, (b) 31.42°.

1.9 See Table 1.2.

1.11 See diagrams in Fig. 2.42.

1.12 See diagrams in Figs. 1.28 and A4.2.

1.14 See diagrams in Fig. 2.42.

2.2 (d) Approximately 10 000 miles.

2.3 (b) $(0\,0\,1)\hat{}(0\,1\,1) = 62.25°(1\,0\,0)\hat{}(1\,1\,0) = 39.05°$.

2.3 (c) All special except $\{1\,1\,1\}$ and $\{1\,1\,3\}$.

2.5 (b) $\bar{3}m$, 32; $\bar{3}$.

2.5 (c) $\bar{3}$, $\bar{3}m$.

2.6 (b) $4/m$, $4/mmm$.

2.6 (c) $4/mmm$, 42, $\bar{4}2m$, $4/m$, $\bar{4}$.

2.8 (122).

2.9 $(3\,0\,2)$, 72.08°, $[\bar{2}\,\bar{2}\,3]$, 90°.

2.10 (b) (i) $m3$, 432, $m3m$, (ii) 23, $\bar{4}3m$, (iii) $m3$, 432, $m3m$.

2.11 Trigonal, orthorhombic, monoclinic, tetragonal.

2.12 $32.6°$, $(1\bar{2}11)$.

2.13 $(21\bar{3}4)$.

2.14 $a{:}b{:}c = 0.6900{:}1{:}0.4124$; $\beta = 99.3°$.

2.15 0.8543.

2.16 (a) $2mm$, 222, m, 2; (b) 222, 2.

2.17 (b) One choice is
$$\begin{pmatrix} 0 & -\frac{1}{2} & \frac{1}{2} \\ \frac{1}{2} & 0 & -\frac{1}{2} \\ 1 & 1 & 1 \end{pmatrix}.$$

(c) Hexagonal cell has a volume of $\frac{3}{4}$ that of the cubic cell.

(d) $(1\bar{1}08)$, $(01\bar{1}2)$, $(11\bar{2}0)$.

CHAPTER 3

3.2 (a) $\bar{4}3m$, (b) $m3m$, (c) $\bar{4}3m$, (d) $3m$, (e) $\bar{3}m$, (f) $\bar{6}m2$.

3.3 (i) $a/\sqrt{3}$, (ii) $a/(2\sqrt{3})$.

3.6 Simple cubic.

3.7 (ii) $\beta A\beta$ $\alpha B\alpha$ $\beta A\beta$.

(iii) $(\frac{1}{3}, \frac{1}{6}, \frac{1}{4})$.

(iv) Mo at $\pm(\frac{1}{3}, \frac{1}{6}, \frac{1}{4})$.

S at $\pm(\frac{1}{3}, \frac{1}{6}, -\frac{1}{4} + z)$ and $\pm(\frac{1}{3}, \frac{1}{6}, -\frac{1}{4} - z)$.

3.8 (a) One, (b) 2.94 Å, 3.26 Å, 2.83 Å, 2.55 Å.

(c) 5.53 g cm^{-3}.

3.9 Interstitial.

3.10 4.6×10^{21} iron vacancies per cm^3.

CHAPTER 4

4.1 D_{12} is the flux of atoms in direction 1 when a unit (negative) concentration gradient is imposed along axis 2.

4.2 No, because they do not transform, when the axes are changed, according to Eqn (4.18).

4.3
$$\begin{pmatrix} 12 & 5 & 0 \\ 5 & 7 & 0 \\ 0 & 0 & 3 \end{pmatrix} + \begin{pmatrix} 0 & 1 & 0 \\ -1 & 0 & 0 \\ 0 & 0 & 0 \end{pmatrix}$$

4.5
$$\begin{pmatrix} S_{11} & S_{12} & 0 \\ -S_{12} & S_{11} & 0 \\ 0 & 0 & S_{33} \end{pmatrix}$$

4.7 (a)

$$
\begin{array}{cc}
 & \begin{array}{ccc} Ox_1 & Ox_2 & Ox_3 \end{array} \\
\begin{array}{c} Ox'_1 \\ Ox'_2 \\ Ox'_3 \end{array} &
\begin{pmatrix}
\dfrac{1}{2} & -\dfrac{\sqrt{3}}{2} & 0 \\[2mm]
\dfrac{\sqrt{3}}{2} & \dfrac{1}{2} & 0 \\[2mm]
0 & 0 & 1
\end{pmatrix}
\end{array}
$$

(b)
$$
\begin{pmatrix}
25 & 0 & 0 \\
0 & 16 & 0 \\
0 & 0 & 9
\end{pmatrix} \times 10^8 \ \Omega^{-1} \ \text{m}^{-1}.
$$

(d) At $30°$ to x'_1, $22.75 \times 10^8 \Omega^{-1} \ \text{m}^{-1}$.
   At $60°$ to x'_1, $18.25 \times 10^8 \Omega^{-1} \ \text{m}^{-1}$.
(f) At $48°$ to x'_1.

CHAPTER 5

5.2
$$
\begin{pmatrix}
e & 0 & 0 \\
0 & -e & 0 \\
0 & 0 & 0
\end{pmatrix}, \qquad
\begin{pmatrix}
0 & -e & 0 \\
-e & 0 & 0 \\
0 & 0 & 0
\end{pmatrix}
$$

5.3
$$
\begin{pmatrix}
e & 0 & 0 \\
0 & -ve & 0 \\
0 & 0 & -ve
\end{pmatrix}, \qquad
\begin{pmatrix}
e/2(1-v) & -e/2(1+v) & 0 \\
-e/2(1+v) & e/2(1+v) & 0 \\
0 & 0 & -ve
\end{pmatrix}
$$

5.6 $\sigma/2$.

5.16 $\dfrac{1}{S_{11}}, \quad \dfrac{S_{11}+S_{12}}{S_{11}^2+S_{11}S_{12}-2S_{12}^2}, \quad \dfrac{S_{11}}{S_{11}^2+S_{11}S_{12}-2S_{12}^2}, \quad \dfrac{1}{S_{11}+2S_{12}}$

CHAPTER 6

6.1 (a) No, (b) yes.
6.3 24, (a) $\langle 1\,0\,0\rangle$, (b) $\langle 1\,1\,1\rangle$, (c) $\langle 1\,0\,1\rangle$.
6.4 1.46.
6.5 3:3.

6.6 (a)
$$
\begin{pmatrix}
0 & 0 & 0 \\
0 & \alpha & 0 \\
0 & 0 & -\alpha
\end{pmatrix}, \quad \text{(b)}
\begin{pmatrix}
0 & \alpha'/2 & -\alpha'/2 \\
\alpha'/2 & 0 & 0 \\
-\alpha'/2 & 0 & 0
\end{pmatrix}
$$

(c) 2, (d) one.
6.7 [1 0 0] and [0 1 0] and [0 0 1] in a crystal belonging to one of the orthogonal systems (cubic, tetragonal, orthogonal).
6.8 (a) $(\bar{1}\,1\,1)[1\,0\,1]$, (b) $(\bar{1}\,1\,1)[1\,0\,1]$ and $(1\,\bar{1}\,1)[0\,1\,1]$, (c) $(\bar{1}\,1\,1)[1\,0\,1]$ and $(\bar{1}\,1\,1)[1\,1\,0]$.

6.9 (a) $(1\,0\,1)[\bar{1}\,0\,1]$ and $(\bar{1}\,0\,1)[1\,0\,1]$, (b) $(0\,1\,1)[0\,\bar{1}\,1]$ and $(0\,\bar{1}\,1)[0\,1\,1]$, (c) $(1\,0\,1)[\bar{1}\,0\,1]$ and $(\bar{1}\,0\,1)[1\,0\,1]$.

6.10 (a) $\langle 1\,1\,1 \rangle$, (b) no, (c) $\langle 0\,0\,0\,1 \rangle$ and $\langle u\,v\,t\,0 \rangle$.

6.12 (a) $\begin{pmatrix} \sigma & \tau & 0 \\ \tau & 0 & 0 \\ 0 & 0 & 0 \end{pmatrix}$, (b) $\sigma/2$, (c) $(-\tau + \sigma/2)$.

6.14 (a) $[\bar{1}\,2\,\bar{1}\,0]$, (b) 8.3×10^5 Pa.

6.15 (a) $(1\,1\,1)[\bar{1}\,0\,1]$, (b) $(1\,1\,1)[\bar{1}\,0\,1]$ and $(\bar{1}\,\bar{1}\,1)[0\,1\,1]$, (c) $[\bar{1}\,1\,2]$, (d) 13.1 cm.

CHAPTER 7

7.3 2.82×10^{-12} cm^3.

7.4 (a) $2 \times 10^{-7} \sigma$Pa, (b) zero.

7.5 (a) $(1\,\bar{1}\,0)$, (b) $(0\,0\,1)$.

7.6 Couple $= \dfrac{\mu b}{2}(R^2 - r_0^2)$.

7.7 $\varepsilon_{11} = \varepsilon_{22} = \dfrac{-b}{2\pi x_2}\left(\dfrac{\mu}{\lambda + 2\mu}\right)$,

dilatation $= \pm 0.0153$.

7.10 Width $= \dfrac{a}{2A^{1/2}}$ ($a =$ lattice parameter).

7.11 $\dfrac{1}{l}\dfrac{dl}{dt} = \tfrac{1}{2}\rho b \bar{v}$.

7.13 Force $= \dfrac{\mu b_1 b_2}{2}$.

7.15 (a) 4, (b) 3, (c) 3, (d) 4.

CHAPTER 8

8.1 $L/10$.

8.2 σbl, μb^2.

8.4 Jointed extrinsic fault.

8.5 Two.

8.6 (a) 1.6×10^{11} cm^{-1}, (b) 1.6×10^{10} cm^{-1}.

8.7 20.5 Å, 2.5 Å, 7.6 Å.

8.8 Simple cubic: $\langle 1\,1\,0 \rangle$, $\langle 1\,1\,1 \rangle$; b.c.c. none; f.c.c. $\langle 1\,0\,0 \rangle$; hexagonal $\tfrac{1}{3}\langle 1\,1\,\bar{2}\,3 \rangle$.

8.9 $3\mu b^2/4\pi\gamma$.

8.14 Identical $\tfrac{1}{2}\langle 1\,1\,1 \rangle$ dislocations will form pairs.

CHAPTER 9

9.1 245.

9.2 1.25 eV, 2.1 kcal/K.

9.3 Concentration in impure crystal is 300 times that in pure crystal.

9.4 $x = 0.94$.

9.5 1.09 eV, true $E_f = 1.11$ eV.

9.6 24.

9.7 1440, 4.

9.8 2.

9.10 Fractional increase $= 1.3 \times 10^{-4}$.

9.11 (a) $4mm$, (b) $\bar{3}m$.

  (a) Tensile stress along $\langle 1\,1\,1 \rangle$, (b) along $\langle 1\,0\,0 \rangle$.

9.12 (a) 48, (b) 16, number of orientations equals ratio of multiplicities.

9.13 (a) 8, (b) group $[1\,1\,1]$, $[1\,\bar{1}\,1]$, $[1\,1\,\bar{1}]$, $[1\,\bar{1}\,\bar{1}]$ from $[\bar{1}\,1\,1]$, $[\bar{1}\,1\,\bar{1}]$, $[\bar{1}\,\bar{1}\,1]$, $[\bar{1}\,\bar{1}\,1]$,
  (c) 4 lying in $(1\,\bar{1}\,0)$ from remaining 4.

CHAPTER 10

10.1 70.53°.

10.4 $(1\,1\,0)$, $(1\,\bar{1}\,2)$ and $(\bar{1}\,1\,2)$, $\begin{pmatrix} \bar{1} & 1 & 1 \\ 1 & \bar{1} & 1 \\ 2 & 2 & 0 \end{pmatrix}$, $(0\,1\,1)$, referring to axes reflected
  in K_1.

10.6 $[1\,1\,0]$, $[1\,\bar{1}\,1]$ and $[\bar{1}\,1\,1]$, $\begin{pmatrix} \bar{1} & 1 & 2 \\ 1 & \bar{1} & 2 \\ 1 & 1 & 0 \end{pmatrix}$, $[0\,1\,0]$, referring to axes reflected
  in K_1.

10.8 93.73°.

10.9 Compression for $c/a > \sqrt{3}$, elongation for $c/a < \sqrt{3}$.

10.10 Yes.

10.12 4°.

10.13 10%.

CHAPTER 11

11.1 4, 2 (twins). Component of **b** normal to $(1\,1\,1)$ must equal $2a/\sqrt{3}$.

11.3 12 $\langle 1\,1\,0 \rangle$ vectors. Yes $\begin{pmatrix} 1 & 0 & 0 \\ -\frac{1}{2} & \frac{1}{2} & -\frac{1}{2} \\ 0 & 1 & 1 \end{pmatrix}$.

11.5 1.57.

11.6 0.18°, 0.003.

11.7 2, 4, 6.

11.10 $\begin{pmatrix} 1 & 0 & 0 \\ 0 & \frac{1}{2} & -\frac{1}{2} \\ 0 & \frac{1}{2} & \frac{1}{2} \end{pmatrix}$

CHAPTER 12

12.2 1, 3, 1 and 2 lattice plane spacings.

12.3 $\gamma_{(0\,0\,1)} = 0.894\gamma_{(2\,1\,0)}$
$\gamma_{(1\,1\,1)} = 0.775\gamma_{(2\,1\,0)}$
$\gamma_{(1\,1\,0)} = 0.949\gamma_{(2\,1\,0)}$.

12.4 1.116 J m^{-2}, 15%.

12.5 (a) Attract
(b) Energy increases as $b^2 \ln b$.

12.6 Boundary plane (a) {1 1 0}, (b) {1 1 0}, (c) {1 1 $\bar{2}$ 0}.

Axis (a) $\langle 0\,0\,1 \rangle$, (b) $\langle 2\,1\,1 \rangle$, (c) $\langle \bar{1}\,1\,0\,0 \rangle$.

Angle of tilt (a) $2\sin^{-1}\dfrac{a}{2\sqrt{2}d}$, (b) $2\sin^{-1}\dfrac{a}{2\sqrt{2}d}$, (c) $2\sin^{-1}\dfrac{a}{2d}$.

12.7 See E. J. Freise and A. Kelly, *Proc. Roy. Soc.*, **A264**, 269 (1961).

12.9 Difference in angle in radians is of the order of the ratio of the twin boundary to surface free energy.

12.11 Effective surface free energy $= 0.96\gamma$.

12.13 (a) {1 1 0}, (b) $3.1 \times 10^5 \text{ cm}^{-1}$, (c) about 8°.

12.15 Two.

Subject Index